Heat Capacity and Thermal Expansion at Low Temperatures

THE INTERNATIONAL CRYOGENICS MONOGRAPH SERIES

Current volumes in this series

APPLIED SUPERCONDUCTIVITY, METALLURGY, AND PHYSICS OF TITANIUM ALLOYS • *E. W. Collings*
Volume 1: Fundamentals
Volume 2: Applications

CRYOCOOLERS • *G. Walker*
Part 1: Fundamentals
Part 2: Applications

CRYOGENIC PROCESS ENGINEERING • *Klaus D. Timmerhaus and Thomas M. Flynn*

CRYOGENIC REGENERATIVE HEAT EXCHANGERS • *Robert A. Ackermann*

HEAT CAPACITY AND THERMAL EXPANSION AT LOW TEMPERATURES • *T. H. K. Barron and G. K. White*

HELIUM CRYOGENICS • *Steven W. Van Sciver*

MODERN GAS-BASED TEMPERATURE AND PRESSURE MEASUREMENTS • *Franco Pavese and Gianfranco Molinar*

POLYMER PROPERTIES AT ROOM AND CRYOGENIC TEMPERATURES • *Günther Hartwig*

SAFETY IN THE HANDLING OF CRYOGENIC FLUIDS • *Frederick J. Edeskuty and Walter F. Stewart*

STABILIZATION OF SUPERCONDUCTING MAGNETIC SYSTEMS • *V. A. Al'tov, V. B. Zenkevich, M. G. Kremlev, and V. V. Sychev*

THERMODYNAMIC PROPERTIES OF CRYOGENIC FLUIDS • *Richard T Jacobsen, Steven G. Penoncello, and Eric W. Lemmon*

Heat Capacity and Thermal Expansion at Low Temperatures

T. H. K. Barron

University of Bristol
Bristol, United Kingdom

and

G. K. White

CSIRO
Lindfield, Australia

Kluwer Academic / Plenum Publishers
New York, Boston, Dordrecht, London, Moscow

ISBN 0-306-46198-6

233 Spring Street, New York, N.Y. 10013

10 9 8 7 6 5 4 3 2 1

A C.I.P. record for this book is available from the Library of Congress

Printed in the United States of America

Preface

The birth of this monograph is partly due to the persistent efforts of the General Editor, Dr. Klaus Timmerhaus, to persuade the authors that they encapsulate their forty or fifty years of struggle with the thermal properties of materials into a book before they either expired or became totally senile. We recognize his wisdom in wanting a monograph which includes the closely linked properties of heat capacity and thermal expansion, to which we have added a little 'cement' in the form of elastic moduli. There seems to be a dearth of practitioners in these areas, particularly among physics postgraduate students, sometimes temporarily alleviated when a new generation of exciting materials are found, be they heavy fermion compounds, high-temperature superconductors, or fullerenes. And yet the needs of the space industry, telecommunications, energy conservation, astronomy, medical imaging, etc., place demands for more data and understanding of these properties for all classes of materials — metals, polymers, glasses, ceramics, and mixtures thereof.

There have been many useful books, including *Specific Heats at Low Temperatures* by E. S. Raja Gopal (1966) in this Plenum Cryogenic Monograph Series, but few if any that covered these related topics in one book in a fashion designed to help the cryogenic engineer and cryophysicist.

We hope that the introductory chapter will widen the horizons of many without a solid state background but with a general interest in physics and materials. The next two chapters deal with basic theory (including the often neglected thermodynamics of anisotropic materials), and with experimental techniques; the experimental physicist and engineer should be helped also by the tables of data in the Appendix C, with their attached references. The remaining chapters cover specific properties of various classes of material.

Finally we hope that this monograph will help meet the information needs in cryogenics that were envisioned by the Founding Editor and mentor to one of us, the late Dr. Kurt Mendelssohn, F.R.S.

Acknowledgments

We thank our many colleagues who have helped in various ways. They include in the CSIRO National Measurement Laboratory Dr. S. Collocott (especially for writing the section on heat capacity measurements), Mr. C. Andrikidis, Dr. P. Ciddor, Dr. J. G. Collins, Mr. Stuart Morris (Drawing Office), and Mr. Curtis Simmonds (Library); and in the University of Bristol Drs. N. L. Allan, R. Evans, B. L. Gyorffy, A. G. Orpen, R. M. Richardson, M. Springford, M. B. Taylor, C. J. Wormald and Mr. C. E. Sims. Elsewhere we have had help and advice from Drs. C. A. Swenson (Ames), G. D. Barrera, J. A. O. Bruno (Buenos Aires), N. E. Phillips, R. E. Fisher, P. Richards (Berkeley), I. Jackson (Canberra), M. L. Klein (Pennsylvania), J. A. Rayne (Pittsburgh), E. R. Cowley and G. K. Horton (Rutgers). We have also drawn on previous unpublished work produced in collaboration with Dr. R. W. Munn (UMIST) and with the late Dr. J. A. Morrison (McMaster). We thank Mrs. Gillian Barron for collating the bibliography and unifying its format.

Contents

Chapter 1

Introduction

1.1. THERMODYNAMIC PROPERTIES AT LOW TEMPERATURES

This book is concerned with the properties of materials at low temperatures, their measurement and the basic physics underlying them. These topics are complementary. Research in physics involves the use and often the construction of equipment, and even theorists need to appreciate what kinds of measurement are practicable and what is their precision and reliability. Conversely, the cryogenic engineer benefits from a fundamental understanding of the physical effects he is exploiting and of the materials he is using.

Heat capacity, thermal expansion and elasticity are all thermodynamic properties. The principles of thermodynamics apply universally: in general both the experimental techniques used at low temperatures and the underlying theory apply also at ambient and higher temperatures, and so to the technology of everyday life. Consider, for instance, a domestic electric storage heater: a thermally insulated core is heated electrically during the night at low cost, and the heat is released during the following day to bring the surrounding room to a comfortable temperature. This simple example illustrates the equivalence of heat and electrical energy, the use of an adiabatic enclosure with facility for controlled breakdown of insulation (permitting the exchange of heat between the core and the air), and the definition and measurement of temperature. All these are also essential concepts for cryogenics. Furthermore, the efficiency of the heater is critically dependent on the relative heat capacities of the core and the air in the room.

The understanding of thermodynamics in terms of atomic and molecular behavior is similarly universal: the general principles (statistical mechanics) apply at all temperatures. Despite this, working at low temperatures does tend to have special characteristics: in particular, heat capacities are often low, so that absorption of unwanted energy due to inadequate insulation or to external vibrations can wreak havoc with temperature control; also, changes in crystal dimensions and elastic properties may be small, requiring high precision for their measurement.

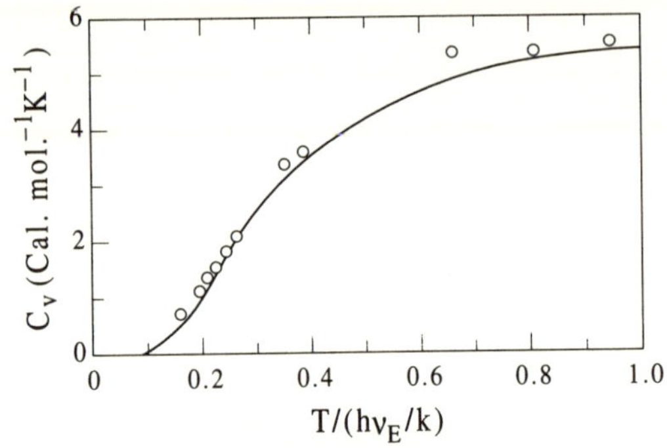

Fig. 1.1. Heat capacity of a mole of harmonic oscillators of frequency ν_E as a function of temperature. o, experimental points plotted by Einstein for diamond, with $h\nu_E/k = 1326$ K. ($3R = 24.94$ J·mol^{-1}·K^{-1} = 5.96 cal·mol^{-1}·K^{-1}).

Such characteristics are in accord with the Third Law of thermodynamics, which governs thermodynamic behavior as temperature approaches the absolute zero. The Third Law and its consequences are in turn due to the quantum nature of matter, and in particular to the consequent discrete energy levels of physical systems. As long ago as 1907, in a discussion of the abnormally low heat capacity of diamond, Einstein [Ein07] pointed out that at sufficiently low temperatures ($kT \ll h\nu$, the spacing between energy levels), none of the higher energy levels of a harmonic oscillator is excited. The system is then in its quantum ground state; its energy no longer changes with temperature, and its entropy and heat capacity have fallen to zero (Fig. 1.1). The same is true of bulk physical systems (in which the thermal expansion also falls to zero), except of course that the approach of the heat capacity to zero will not be the same as that for a harmonic oscillator. To demonstrate this, and to give a taste of the variety of low temperature behavior exhibited by different substances, let us now look briefly at some examples of increasing complexity. Fuller discussion of these examples will be given in later chapters.

KCl. Figure 1.2 shows the temperature dependence of the heat capacity C_P and volumetric thermal expansion coefficient β of the cubic crystal potassium chloride. At room temperature C_P has flattened off to a value of about 50 J·mol^{-1}·K^{-1}, in agreement with the empirical law of Dulong and Petit [Dul819] that the heat capacity of many solids is about 25 J·g-at^{-1}·K^{-1}. The way in which C_P decreases to zero (as T^3) is more gradual than the exponential decrease seen in Fig. 1.1. This is because there are many different vibrations of a crystal structure, giving a vibrational spectrum varying from the low frequencies of sound waves to higher frequencies typically in the infra red region 10^{11}–10^{13} Hz. To a first approximation the heat capacity is the sum of harmonic contributions from all these vibrations, each contribution having the type of temperature dependence shown in Fig. 1.1. As the temperature is lowered, the contributions of the highest frequency vibrations are the first to decrease, and then

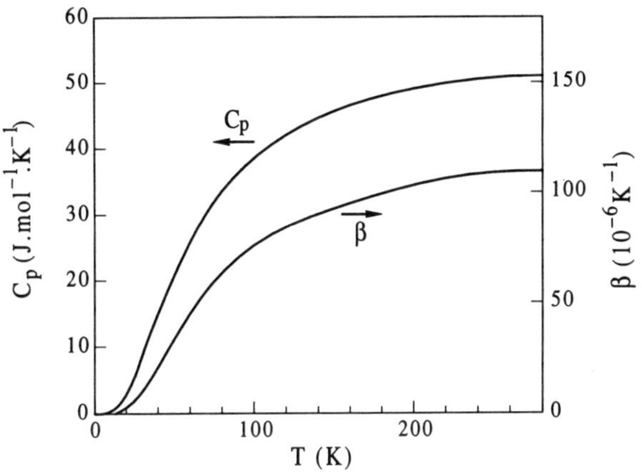

Fig. 1.2. C_P [Ber57] and β [Whi73] for KCl below room temperature.

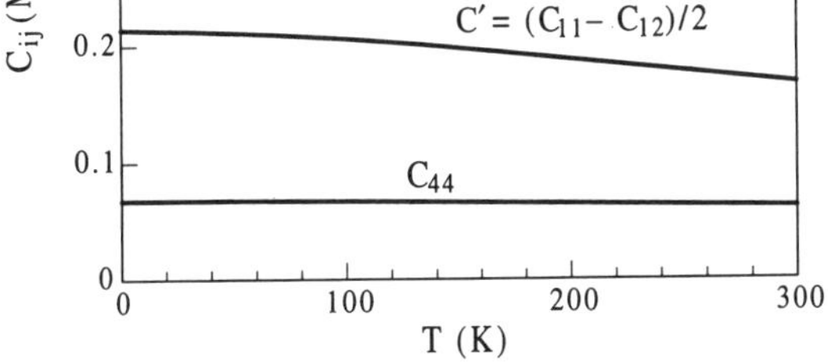

Fig. 1.3. Elastic stiffnesses for KCl below room temperature [Nor58].

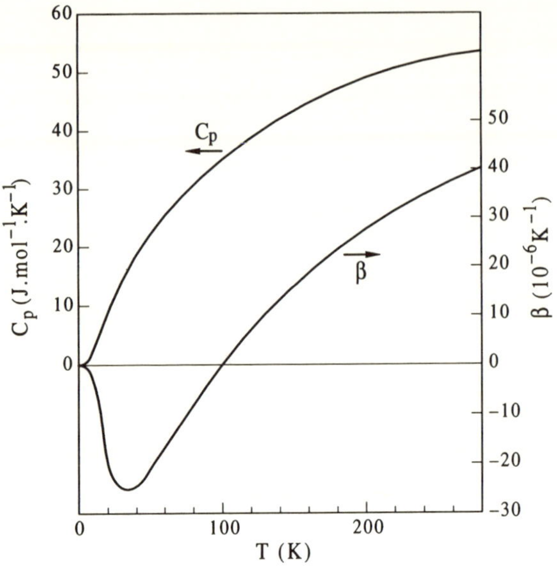

Fig. 1.4. C_P and β for cubic CuCl below room temperature [Bar77a].

successively those of lower frequency, until at very low temperatures we see only the contributions of the acoustic vibrations, giving a characteristic T^3 dependence (see Section 2.6). Unlike C_P and β, the elastic stiffnesses $c_{\lambda\mu}$ tend to non-zero limits as $T \to 0$ (Fig. 1.3). At very low temperatures the departure from these limits is usually small and hard to measure accurately.

CuCl. In Fig. 1.2 we have seen that the heat capacity and thermal expansion of KCl have qualitatively similar temperature dependence. Our next example, cuprous chloride, shows that this need not be so (Fig. 1.4): once again both C_P and β approach zero as T^3, but whereas C_P remains positive at all temperatures (as required for thermodynamic stability), β becomes negative at low temperatures. It is clear from this example that although all vibrations contribute similarly to the heat capacity, their effect on the thermal expansion can be very different. Since it is only the lower frequency vibrations that are excited at low temperatures, we can deduce that for CuCl such vibrations on balance contract the crystal lattice. There is nothing anomalous about this. Negative expansion is quite common, especially at low temperatures (see, e.g., Section 5.5.1), and low expansion materials can be produced by balancing the factors that make for positive and negative expansion (Section 5.7). We note also that there is no law for thermal expansion analogous to that of Dulong and Petit for heat capacity: the room temperature values of β for KCl and CuCl are quite different.

It should be noted that both KCl and CuCl are cubic crystals, with isotropic thermal expansion, whereas non-cubic crystals have anisotropic thermal expansion (e.g., Fig. 1.9).

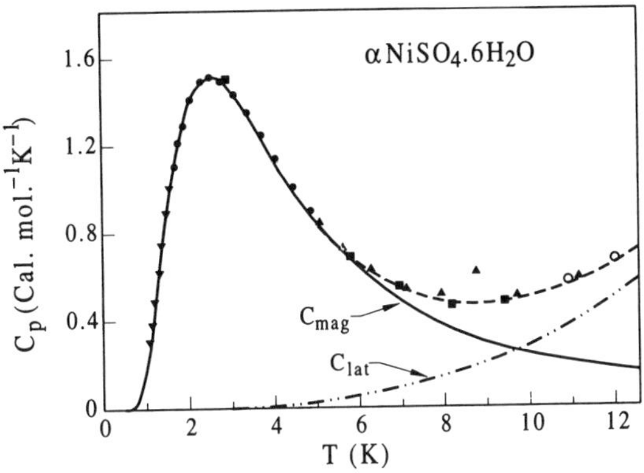

Fig. 1.5. Measured C_P for α-$NiSO_4.6H_2O$ below 12 K. - - -, total; —, Schottky component; dot–dash, vibrational component [Sto64]. (1 cal = 4.184 J).

α-NiSO₄·6H₂O. The strong peak in C_P superimposed on the vibrational T^3 dependence at low temperatures shown in Fig. 1.5 is a simple example of a non-vibrational contribution to the thermal properties. The contribution builds up exponentially as T increases, and then falls off as T^{-2} at high temperatures. Such behavior is typical of an assembly of so-called *Schottky* systems, which are essentially non-interacting localized systems (e.g., ions, atomic nuclei, etc.) which can exist in only a small number of energy states (see Section 2.5.3). In α-nickel sulphate the energy states arise from the three-fold degeneracy of the magnetic Ni^{++} ion, which in this non-cubic crystal is split into three closely spaced energy levels. At very low temperatures all the ions have the lowest energy; as T increases some ions become excited to the higher levels, but the resulting heat capacity dies away at higher temperatures as all three energies become equally likely.

In this example the Ni^{++} ions are well separated from each other by the water of crystallization; the interaction between neighboring spins is small compared to the splitting of the degeneracy by the crystal field, satisfying the criterion for Schottky systems. Much sharper peaks in C_P are seen when the degeneracy is lifted by interactions between the systems, as for example in some forms of ferromagnetism (Section 5.11). For both types of system effects will be seen also in the thermal expansion and (usually less marked) in the elasticity.

Cu. In simple metals the conduction electrons contribute small terms to the heat capacity and thermal expansion that are proportional to the temperature. At room temperature these terms are swamped by the vibrational terms, but at low temperatures (typically $T \sim 1$ K) almost all the vibrations cease to contribute, leaving the electronic contribution dominant. The electronic and vibrational terms can be conveniently displayed by plotting C_P/T and α/T or β/T against T^2. The low temperature heat capacity of copper is thus shown in Fig. 1.6: the intercept at $T = 0$ gives the coefficient

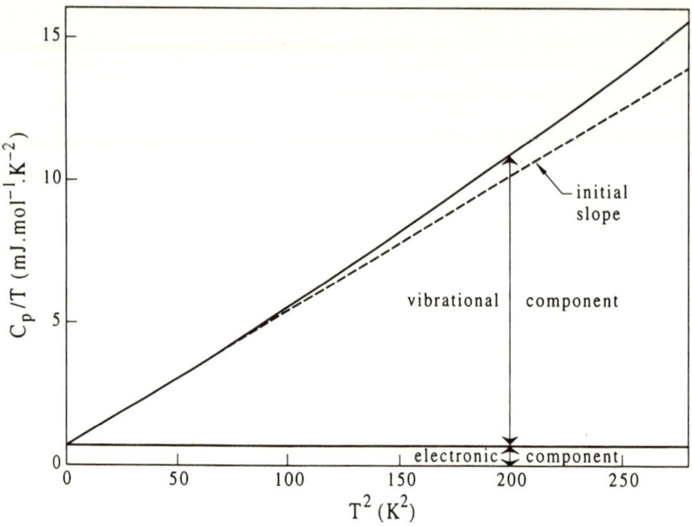

Fig. 1.6. C_P/T plotted against T^2 for Cu at low temperatures [Hol72].

of the electronic term $\Gamma_e T$, and the initial slope the coefficient of the vibrational term aT^3. Note that it is only below 4 K that the electronic term begins to dominate. For the thermal expansion experimental precision is insufficient to permit extension to temperatures where the electronic term is dominant (see Section 3.3).

Pd. The *d*-electrons in transition metals such as palladium enhance the electronic density of states, giving electronic contributions to the heat capacity and thermal expansion much larger than those in copper (Fig. 1.7). In particular, the electronic thermal expansion is large enough at low temperatures to be determined quite accurately.

Ga. A similar plot (of C_P/T against T^2) for the superconductor gallium shows more complex behavior (Fig. 1.8). Instead of showing the T-dependence of a normal metal, C_P rises exponentially at low temperatures to a peak at the superconducting transition temperature T_c; it then falls discontinuously to the normal (non-superconducting) value. The rise at low temperatures is rather similar to that seen in the Schottky peak of Fig. 1.5, but the discontinuous drop at T_c is in marked contrast to the long high temperature tail seen there; above T_c all trace of superconductivity has disappeared. Thermal expansion coefficients of superconductors also have a discontinuity at T_c, but (unlike C_P) they may either increase or decrease at the transition.

Figure 1.8 also shows that the normal T-dependence is observed at lower temperatures if the superconductivity is suppressed by applying a magnetic field H. For gallium, however, a further non-vibrational contribution then appears at very low

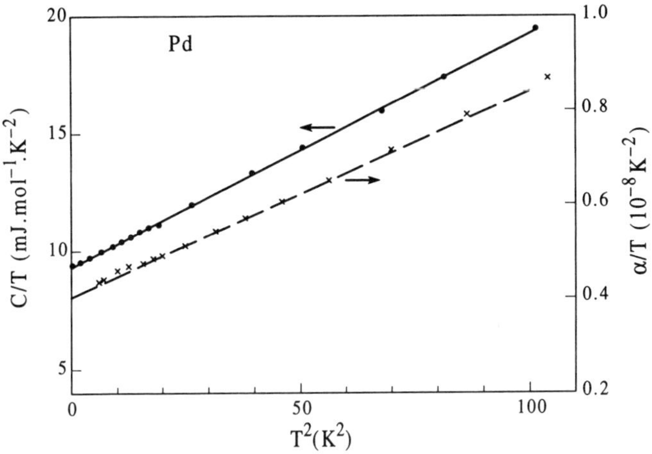

Fig. 1.7. C_P/T and α/T plotted against T^2 for Pd at low temperatures. Analysis gives $\Theta_0 = 274$ K, $\Gamma_e = 9.4$ mJ·mol^{-1}·K^{-2}, $\gamma_e = 2.2$, $\gamma_0 = 2.25$. Data for C_P from [Vea64], and for α from [Whi70].

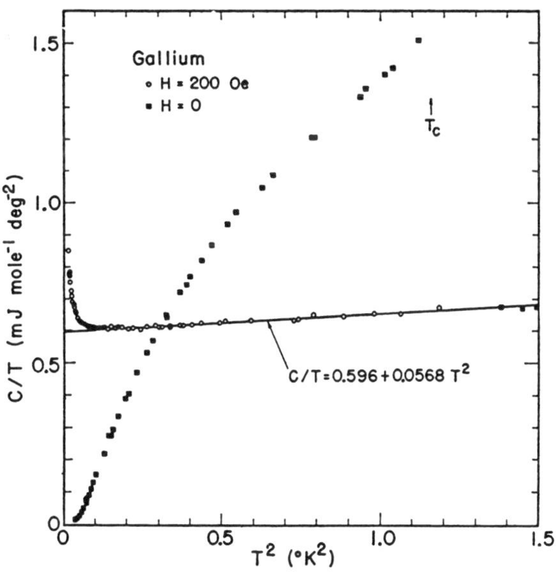

Fig. 1.8. C_P/T plotted against T^2 for Ga at low temperatures: circles, normal state; filled squares, superconducting state. The magnetic field of 20 mT (200 Oe) suppresses superconductivity [Phi64].

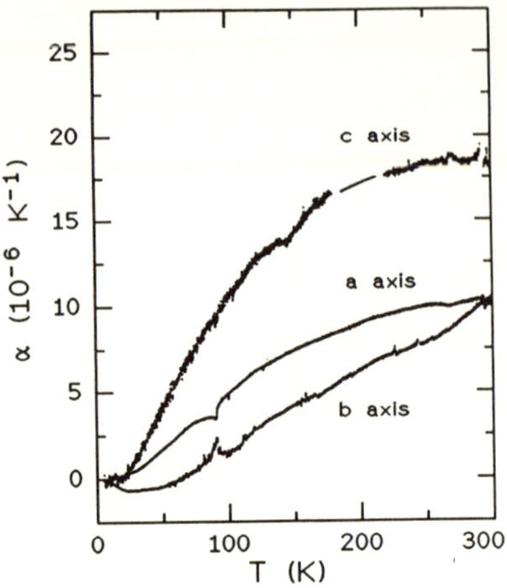

Fig. 1.9. Linear expansion coefficients of $YBa_2Cu_3O_7$ [Mei91].

temperatures, which has been identified as the high temperature tail of a Schottky contribution arising from the lifting of the nuclear spin degeneracy by the crystal field (Section 2.5.3). That this contribution is not seen when $H = 0$ is a striking example of the importance of kinetics: in the superconducting state at low temperatures the interaction between the nuclear spin system and the lattice vibrations is so small that thermodynamic equilibrium is not established within the time of the experiment.

$YBa_2Cu_3O_7$. More complex forms of superconductivity than that originally seen in simple metals at very low temperatures have been found in various types of substance (Section 6.5), including the high-T_c ceramic oxides (Section 5.10). Like many of these oxides, $YBa_2Cu_3O_7$ has orthorhombic symmetry with independent coefficients of linear expansion along the three crystallographic directions (Fig. 1.9). At T_c ($\approx 92K$) there is a peak of about 2% in C_P, but the effect on the expansion coefficients is different in each direction: negative for α_a, positive for α_b and undetected for α_c.

Invar systems. Magnetic solids of various types provide some of the most complex and difficult systems to understand, and some of them are also of great technical importance. These include the alloys "Invar" and "Elinvar," which have respectively very small thermal expansion and very small change of elasticity over a wide range of temperature. A century ago Guillaume [Gui897] reported that properties of Fe/Ni alloys were critically dependent on concentration, and later measurements alloying iron with other metals have shown that a dominant factor in determining properties is their high sensitivity to the number of conduction electrons per atom (Fig. 1.10). Invar, an iron/nickel alloy with 35%Ni, can be seen in this figure to have a very low thermal expansion at room temperature.

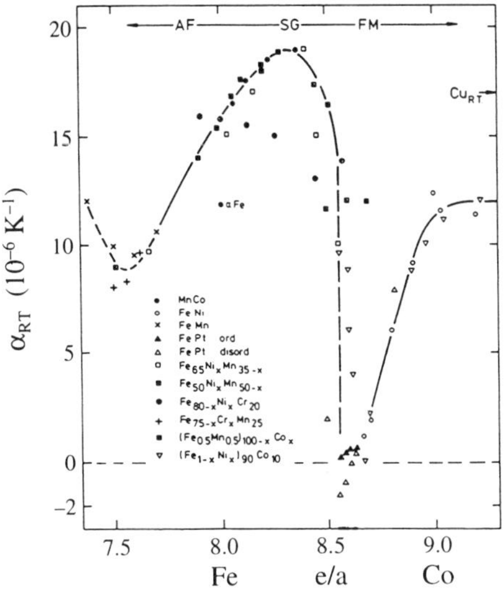

Fig. 1.10. Values of α at room temperature as function of electron/atom ratio for various Invar-like systems. AF–SG–FM denote antiferromagnetic, spin-glass and ferromagnetic regimes [Was90].

$CeAl_3$. "Heavy fermion" metallic compounds are another class of solid with properties difficult to understand. Their name arises from the very high effective masses of their conduction electrons (see Section 6.6), corresponding to an electronic heat capacity much greater than that of a normal metal, and large effects also in other properties. For $CeAl_3$ below 1 K, the heat capacity is about three orders of magnitude greater than that for copper. The expansion coefficient β is negative, and about five orders of magnitude greater than that for copper, becoming positive above about 1 K. Perhaps most strikingly, even the elasticity changes appreciably between 0 and 3 K (Fig. 1.11). This behavior should not be regarded as typical, however; as a class heavy fermion compounds display a very varied range of behavior, including superconductivity and different forms of magnetism.

Ice. This important solid is mentioned here to illustrate the problems that can occur when the relaxation time needed to reach thermodynamic equilibrium becomes comparable to the time taken to perform a measurement. Figure 1.12 shows the results of sensitive measurements of the heat capacity of three samples of normal (hexagonal) ice between 70 and 160 K. At high temperatures the water molecules are in thermodynamic equilibrium and randomly oriented. As the temperature is lowered short-range order begins to set in, but concurrently the time needed to reach equilibrium increases rapidly; relaxation times have been estimated to be about an hour at 108 K and a week at 89.4 K. In such circumstances, the apparent heat capacity depends both on the time allowed for the measurement and the previous history of the sample.

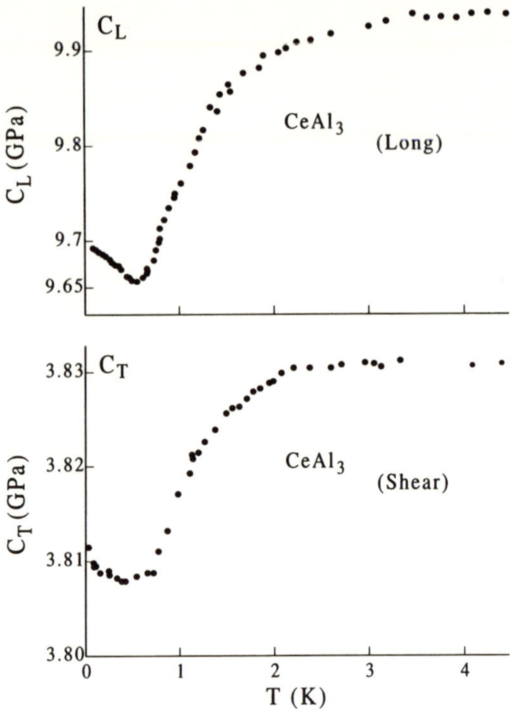

Fig. 1.11. Elastic moduli of longitudinal and transverse waves for polycrystalline $CeAl_3$ below 4 K [Nik80].

In this example, the small amount of orientational ordering (at most about 2%) has no detectable effect on the heat capacity and thermal expansion of the different metastable states obtained at lower temperatures. We should therefore stress that the properties of systems which have a range of possible metastable states are in general dependent on previous history.

Let us sum up what we have seen in these examples. At low temperatures heat capacity and thermal expansion change by many orders of magnitude, but elasticity tends to a finite limit. As the temperature decreases, vibrational contributions become less, and other contributions may become dominant; sometimes it is easy to identify separate contributions, but sometimes this is not possible. Effects of specific mechanisms are seen in all three properties, but are more marked in heat capacity and thermal expansion than in elasticity; heat capacities are always positive, but thermal expansion may be negative. Electronic effects, sometimes interacting with the vibrations, may give rise to a rich complexity of behavior, especially in non-cubic crystals; this can be critically dependent on composition. Finally, properties may depend on the rate of measurement if thermodynamic equilibrium is not achieved within the time-scale of the experiment.

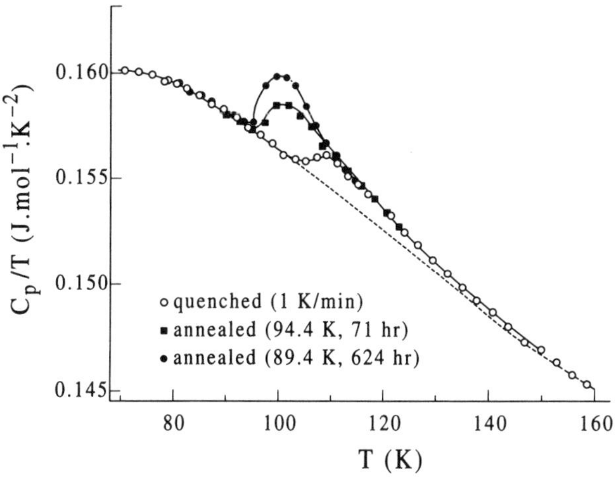

Fig. 1.12. Measurements of the heat capacity of differently annealed samples of ice. - - -, estimated behavior if there was no orientational ordering [Hai72].

1.2. IMPLICATIONS FOR DESIGN OF EQUIPMENT

The consequences of the Third Law of Thermodynamics and rapid decrease of C_P and β towards zero as $T \to 0$ have obvious implications for the process of measuring these two quantities. The traditional method of measuring the heat capacity C_P is to apply a measured heat pulse ΔQ and determine the temperature rise, $\Delta T = T_1 - T_2$, thus obtaining a value for C_P of $\Delta Q/\Delta T$ at the average temperature $(T_1 + T_2)/2$. Likewise to determine the thermal expansion coefficient we heat the specimen and measure the associated changes in length (per unit length) or volume (per unit volume) and the change in temperature. At normal temperatures (around the Debye theta and above — see Section 1.3.3) these properties often vary slowly with temperature (e.g., Fig. 1.2), enabling relatively large intervals ΔT to be used. For example, rather insensitive methods such as X-ray lattice spacings determined at 50 or 100 K intervals may suffice to give expansion coefficients to $\pm 1\%$.

At low temperatures C and β vary rapidly with T; and to obtain meaningful data, intervals of ΔT must then normally be much less than T itself, certainly $\leq 0.1T$. This requires sensitive thermometry, close temperature control (minimum heat leakage), accurate control and measurement of ΔQ (for C_P) and very sensitive 'dilatometry' (for β or α). Consider orders of magnitude when measuring C_P at $T = \Theta/100$, that is around 3 or 4 K for most solids. Ignoring 'anomalous' features like Schottky bumps or heavy fermion effects, the lattice heat capacity will be about $1944(T/\Theta)^3 \simeq 2 \times 10^{-3}$ J·g-at^{-1}·K^{-1} $\sim 10^{-4}$ J·cm^{-3}·K^{-1}. This should not present a measurement problem: an electric current generating a few μW in a resistive element attached to the sample (say ~ 1 cm^3 in size) for a few seconds will produce a measurable temperature rise of a few hundredths of a kelvin. For an accurate result we need to ensure that the heat pulse goes into the specimen (*i.e.*, no leakage) and that the thermometer records faithfully the temperature of the specimen. Germanium or carbon resistance thermometers with resolutions of few μK are suited to the purpose at these temperatures (see details in Ch. 3).

At lower temperatures, in the millikelvin range, lattice heat capacity is many orders of magnitude smaller, and unwanted heat sources in the form of external vibrations, eddy currents, microwaves (and even cosmic rays) affect the stability and accuracy of C_P measurements. Usually the sample is loosely coupled thermally to a cooling stage (*e.g.*, dilution refrigerator) via a heat link, and some type of transient method (see Ch. 3) is used. This may involve either (i) a heat pulse and measurement of the subsequent decay in temperature as heat leaks away to the cooling system or (ii) an ac heat input with a phase-sensitive detector. Such measurements of heat capacity sometimes concern thin film samples of only a milligram or so deposited on a sapphire substrate with a thin-film Ge or Si thermometer. Total heat capacity at such temperatures may then be as low as 10^{-10} or 10^{-11} J·K^{-1}, requiring highly sensitive measurement of voltage signals. Below these temperatures the lattice heat capacity continues to fall as T^3 until even the lowest frequency modes are no longer excited ($T \sim 10^{-6}$–10^{-7} K).

You may ask how and why we bother to measure at such extremes of temperature? We digress with an illustration due to the late Sir Francis Simon, an eminent thermodynamicist and low temperature physicist ... the so-called 'desert' picture. In a really featureless desert there are no points of interest and exploration would be very difficult and pointless. If on the other hand there is a feature of interest, say an oasis, this is worth exploring and often also provides the means to make exploration possible. In the present context the featureless desert corresponds thermodynamically to a material at low temperatures which has lost practically all its entropy, so that its state is hardly distinguishable from its state at absolute zero. However, at similar temperatures another material may still have appreciable sources of entropy, provided for example by nuclear spins or heavy fermion effects; these correspond to the oases. Their existence can make cooling and temperature measurement in this range both feasible and physically significant (see Ch. 3).

Turning to thermal expansion at low temperatures, the major problems arise from the limited resolving power of length measurement. Even with the best inductive or capacitative detectors, it is difficult to detect reliably length changes of less than 0.01 Å (10^{-12}m), that is, one hundredth of an atomic diameter. This difficulty is hardly surprising, since 0.01 Å is already much smaller than the scale of irregularities on a crystal surface, or even the amplitude of the zero-point vibrations of surface atoms. The result is that for a copper sample of 100 mm length at a temperature of $\Theta_D/100$ (where $\alpha \sim 10^{-9}$ K^{-1}) a temperature increase of 0.1 K will increase l by only 0.1 Å (10^{-11}m). If our limit of measurement is 0.01 Å the accuracy of measurement of α will be only about 10%. Clearly we have no hope of determining thermal expansions at temperatures below 1 K except for systems having Schottky bumps or other large non-vibrational effects.

1.3. USEFUL THEORETICAL CONCEPTS

Historically, interpretation of the thermodynamic behavior of materials developed progressively, as early theories were found inadequate to account fully for the wide range of observed behavior becoming available experimentally. But for a comprehensive treatment it is better to start with a general conceptual framework into which most aspects of both theory and experiment can then be fitted, including the early theories. Such a framework is provided by thermodynamics and its general interpretation by statistical mechanics, as described in Chapter 2. Specific models, nearly always approximate, can then be used to interpret the properties of individual materials or classes of material. For complete generality, this framework would need to be extended to take account of measurements on substances which are not in thermodynamic equilibrium.

However, we do not need to establish this entire framework before discussing some of the concepts and procedures that occur most frequently in the presentation and interpretation of experimental data. For example, several early theories are still in common use, such as the classical theory of dilute gases, the Debye theory of heat capacities of solids, and the simple Grüneisen equation of state. Their importance is due not only to their graphic simplicity, but also to their use as standards to which the behavior of real materials can be compared. In this section we discuss some of the key concepts arising from this early work.

1.3.1. Grüneisen Function and Grüneisen Parameters

The Grüneisen Function. Empirically, heat capacity, thermal expansion, and elasticity are qualitatively correlated. We have seen that the magnitudes of the heat capacity and thermal expansion vary similarly with temperature. In addition, substances that are elastically stiff tend to have low thermal expansion. Neither of these correlations is surprising: the greater the heat capacity, the more energy is absorbed per unit increase of temperature, and it is this energy that causes the thermal expansion; and resistance to thermal expansion will be greater in a stiff material. To make these considerations quantitative, and to understand what additional factor affects the thermal expansion, we need precise definitions of the properties involved. For simplicity, we here consider only fluids and solids maintained under hydrostatic pressure.

Elastic behavior is described in terms of stress (force per unit area) and strain (relative change of dimensions), and defined either by *compliances*, which give the response of a material under specified conditions to applied stress, or (reciprocally) in terms of *stiffnesses*, which describe its resistance to applied strain. In our case the pressure P is the only stress and the relative change in volume the only strain, and it suffices to use one compliance, the compressibility χ, or its reciprocal stiffness, the bulk modulus B. These may be defined under isothermal or adiabatic conditions:

$$\chi_T = -(\partial \ln V/\partial P)_T = 1/B_T, \qquad \chi_S = -(\partial \ln V/\partial P)_S = 1/B_S \tag{1.1}$$

Similarly, the heat capacity may be defined under conditions of constant volume or constant pressure:

$$C_V = (\partial U/\partial T)_V, \qquad C_P = (\partial H/\partial T)_P \tag{1.2}$$

where U is the internal energy and H is the enthalpy $U+PV$. The volumetric expansion coefficient is defined by

$$\beta = (\partial \ln V/\partial T)_P \tag{1.3}$$

Imagine now the thermal expansion at constant pressure to proceed in two stages:

1. The temperature is raised by dT while the volume is held constant at V; this requires an input of energy per unit volume

$$(1/V)dU = (C_V/V)dT$$

and causes the pressure to change by

$$dP = (C_V/V)[\partial P/\partial(U/V)]_V dT$$

2. The pressure is allowed to relax to its original value while the temperature is held constant at $T+dT$; the final change of relative volume is thus

$$\frac{dV}{V} = \chi_T dP = \left\{\frac{\chi_T C_V}{V}\left[\frac{\partial P}{\partial(U/V)}\right]_V\right\} dT \tag{1.4}$$

Equation (1.4) makes the rôles of heat capacity and compressibility explicit, and also makes it clear that by themselves they are insufficient to determine the expansivity. We need also to know the sign and magnitude of the thermal pressure caused by a given increase of energy density, expressed by a third thermodynamic quantity, called the *Grüneisen function* because it first appeared as a parameter in an early model of E. Grüneisen [Gru12]:

$$\gamma(T,V) = \left[\frac{\partial P}{\partial(U/V)}\right]_V \tag{1.5}$$

The thermal expansion coefficient can then be expressed as

$$\beta = \gamma \chi_T C_V/V \tag{1.6}$$

Equation (1.5) has been chosen here as the definition of the Grüneisen function because it brings out most clearly its rôle in determining thermal expansion. Other thermodynamically equivalent expressions are given in Section 2.2.3. In particular, experimental values of γ are usually obtained from the expression $\beta V/(\chi_S C_P)$, and for this reason γ is sometimes called the Grüneisen *ratio*.

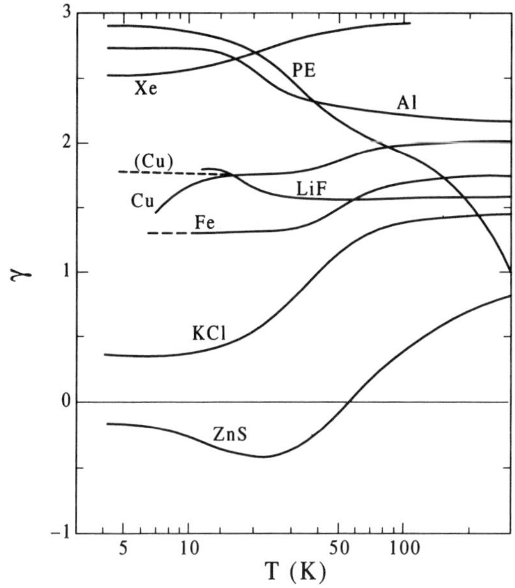

Fig. 1.13. Temperature variation of Grüneisen functions for selected solids. PE denotes a sample of polyethylene of 80% crystallinity. - - -, vibrational function $\gamma_{vib}(T)$.

The Grüneisen function is dimensionless, and unlike the expansivity it usually has the same order of magnitude over the entire experimental range. It thus provides a sensitive way of plotting experimental data: different materials have their characteristic signatures in the shapes of the $\gamma(T)$ plots (see Fig. 1.13), and trends within the same class of material (e.g., alkali halides) are clearly displayed (Fig. 5.5). For some materials γ varies little over wide ranges of temperature, with values typically between 1 and 3; the heat capacity and expansivity then vary similarly with temperature. But for other materials very different behavior is observed, especially at low temperatures: γ may be negative, causing the material to contract on heating; and it may vary strongly with temperature (e.g., Fig. 5.8), sometimes with very large positive or negative values. The behavior of polyethylene in Fig. 1.13 is due to its more complex structure: the higher frequency modes have little effect on the thermal expansion but contribute to the heat capacity at higher temperatures (Section 5.9).

Although it is a well defined thermodynamic function dependent on both temperature and volume, γ is sometimes called the "Grüneisen constant," a term that is particularly misleading in the cryogenic range of temperatures (Fig. 1.13). To call it the "Grüneisen parameter" is less objectionable, although in this book we reserve this term for other quantities related to the parameter γ in Grüneisen's original theory.

Grüneisen Parameters. Grüneisen's original approximation took the same frequency ν_{vib} for all the vibrations. The volume derivative of this frequency was described dimensionlessly by a parameter

$$\gamma = -d\ln \nu_{vib}/d\ln V \tag{1.7}$$

For his model the Grüneisen function defined in Eq. (1.5) has this value at all temperatures, and so the model predicted that the ratio $\beta V/(\chi_T C_V)$ should be approximately independent of temperature. Indeed, it can be shown that whenever all the energy states of a system scale to a single characteristic energy $E_c(V)$, the Grüneisen function is independent of temperature with the value $-d\ln E_c/d\ln V$. For example, in an ideal gas of single particles all the energy states scale as $V^{-\frac{2}{3}}$, giving $\gamma = \frac{2}{3}$ in both the classical and the quantum limits.

Generally, however, the energy states do not all scale in the same way, and there are separate Grüneisen parameters for different frequencies or characteristic energies: e.g.,

$$\gamma_j = -d\ln \nu_j/d\ln V, \qquad \gamma_c = -d\ln E_c/d\ln V \tag{1.8}$$

These can sometimes be determined by spectrocopic measurements under pressure.

1.3.2. Additive Contributions

When discussing experimental examples in Section 1.1 we referred to various contributions to the thermodynamic properties: vibrational, electronic and so forth. Ideally such contributions will be distinguishable and independent of each other only if the free energy can be expressed as the sum of distinct components:

$$F = F_{vib} + F_e + \dots \tag{1.9}$$

Although this is not exact for real materials, it is often true to a good approximation (Section 2.3), and is a feature of nearly all theoretical models; when necessary, interaction between the different components of a model is considered as a further refinement. It follows from Eq. (1.9) that all derivatives of F with respect to temperature and strain are similarly additive; among these are pressure P, entropy S, isothermal elastic moduli (e.g., B_T) and heat capacity C_V. Thus

$$C_V = C_1 + C_2 + C_3 + \dots = \sum_r C_r, \text{ etc.} \tag{1.10}$$

where the index r may refer only to a broad separation of contributions into vibrational, electronic, and so on, or to a finer separation into individual vibrational modes (Section 2.6).

Quantities which are not derivatives of $F(T,V)$ are in principle not additive. These include coefficients of thermal expansion as well as compliances and Grüneisen functions. But the thermal pressure coefficient $[\partial P/\partial T]_V$ is additive, and the expansion coefficient can be expressed as

$$\beta = \chi_T [\partial P/\partial T]_V \tag{1.11}$$

and so provided that χ_T is changing with T much more slowly than $[\partial P/\partial T]_V$ we can loosely identify different additive contributions to the thermal expansion.

This is not true of the Grüneisen function. Grüneisen functions can indeed be defined for the separate components of the model, but they are not simply additive. The Grüneisen function for the material is an average weighted by the contribution to the heat capacity of each component:

$$\gamma = \sum_r C_r \gamma_r / \sum_r C_r \tag{1.12}$$

The Grüneisen function for Cu shown in Fig. 1.13 reveals the effect of this weighting at low temperatures when the contribution of C_e becomes an appreciable fraction of C_V, since for Cu γ_e is considerably lower than γ_{vib}.

If the contribution of a component C_r can be scaled to a single characteristic temperature or frequency, its Grüneisen function γ_r is simply a constant parameter, as defined for example in Eq. (1.7).

1.3.3. Vibrational Contributions; Debye Thetas

The Frequency Distribution. We have seen that to a first approximation the vibrational heat capacity of a solid is the sum of contributions from independent harmonic vibrations. The vibrational frequencies may be given as ν (in Hz), as angular frequencies $\omega = 2\pi\nu$ (in rad·s^{-1}), or as their equivalents in meV, etc. (see Table 2.1). For the most part we shall use ω. The number of vibrations with frequencies between ω and $\omega + \delta\omega$ is written as $g(\omega)\delta\omega$, where $g(\omega)$ is called the *frequency distribution* or (less felicitously) the *phonon density of states*. The heat capacity is then

$$C_{vib} = \int g(\omega) c(\hbar\omega/kT) d\omega \tag{1.13}$$

where $c(\hbar\omega/kT)$ is the contribution to C_{vib} of a mode of frequency ω:

$$c(\hbar\omega/kT) = k\frac{x^2}{(e^x - 1)(1 - e^{-x})}, \qquad x = \hbar\omega/kT \tag{1.14}$$

This function has the temperature dependence shown in Fig. 1.1, and rises to a maximum value of k, or to $3R$ for one mole (Einstein solid; see Table C.4).

The shape of $g(\omega)$ is not simple and is different for each solid. It is usually estimated by fitting a lattice dynamical model to experimental data obtained mainly from inelastic neutron scattering (see Section 2.6.2). Examples for two crystals of different simple structures are given in Fig. 1.14. The shape of $g(\omega)$ for Ar is much the same for other rare gas solids, and also for those fcc metals in which nearest neighbor central forces play a dominant rôle. The shape for Si (diamond structure) is fairly similar to those for Ge, for α(grey)-Sn, and for some crystals of zincblende

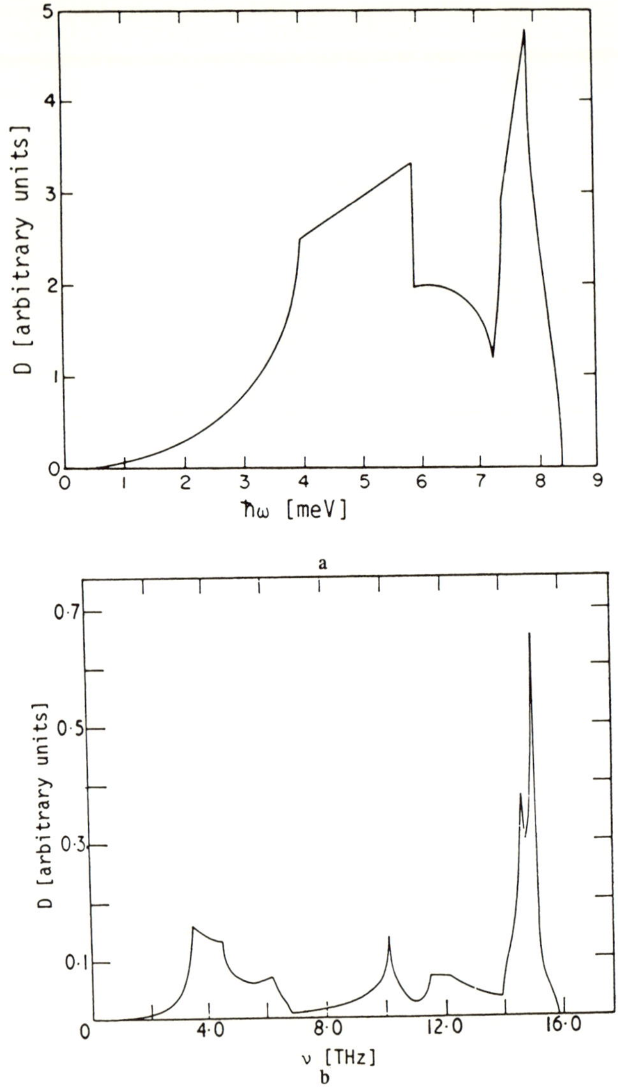

Fig. 1.14. Harmonic frequency distribution $g(\omega)$ (phonon density of states D) for (a) Argon (^{36}Ar) [Fuj74], (b) Silicon [Dol66]. From [Bil79].

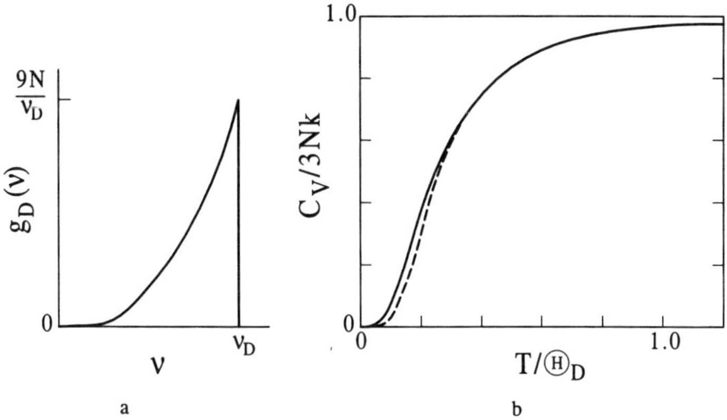

Fig. 1.15. (a) The Debye frequency distribution. The area under the curve is $3N$. (b) —, the Debye heat capacity as a function of T/Θ_D. - - -, the heat capacity of $3N$ Einstein oscillators with $\Theta_E = \sqrt{\frac{3}{5}}\Theta_D$.

structure in which the two atoms are in the same row of the periodic table; but it differs markedly from that for diamond.

The sharp discontinuities in slope (van Hove singularities) are a consequence of the lattice periodicity and are a feature of all crystal frequency distributions.

The Debye Model. Despite their great variety, all $g(\omega)$ for three dimensional crystals have two properties in common: (i) the lowest frequency vibrations are elastic waves of long wave-length, which implies that at low frequencies $g(\omega)$ has the limiting form $\alpha\omega^2$; (ii) the total number of frequencies is $3N$, where N is the number of atoms in the solid. The Debye distribution [Deb12] has the simplest shape with both these properties: the ω^2 dependence is continued over the whole range of frequencies up to a cut-off frequency ω_D, chosen such that there are $3N$ vibrations in total. In terms of the parameter ω_D the Debye distribution is then like that shown in Fig. 1.15(a):

$$g_D(\omega) = 9N\omega^2/\omega_D{}^3 \qquad \text{OR} \qquad g_D(\nu) = 9N\nu^2/\nu_D{}^3 \tag{1.15}$$

where $g_D(\nu)$ is defined such that $g_D(\nu)d\nu$ is the number of frequencies between ν and $\nu + d\nu$.

The temperature dependence of C_V given by Eq. (1.13) for this distribution is shown in Fig. 1.15(b); it scales as T/Θ_D, where $\Theta_D = h\omega_D/k$ is called the Debye temperature. It is usually tabulated as a function of Θ_D/T, either as in Table C.5 for one mole (when $3Nk = 24.94$ J$\cdot$mol$^{-1}\cdot$K^{-1}) or scaled as $C_V/3Nk$:

$$C_V(\text{Debye}) = 3Nkf_D(\Theta_D/T) \tag{1.16}$$

At low temperatures C_V(Debye) tends to zero as T^3, in accord with the behavior of real solids and unlike the exponential behavior of C_V(Einstein) shown for comparison in Fig. 1.15(b) (compare also Tables C.4 and C.5).

The Debye distribution is used widely in solid state theory as an approximation to the true frequency distribution of a solid. Since it is a one-parameter theory, it predicts a constant value $-d\ln\Theta_D/d\ln V$ for the Grüneisen function γ_{vib}. This is a fair approximation for a number of close-packed metals and rare-gas solids but not for more complex solids (Fig. 1.13).

Equivalent Debye Temperatures. Since most materials have C_V curves of rather similar shape, a sensitive means of plotting is needed to bring out their differences. This is usually done by plotting the equivalent Debye temperature Θ^C as a function of temperature. $\Theta^C(T)$ is defined as the Debye temperature Θ_D that would predict the actual value of C_V at temperature T. It can be obtained from Debye tables (e.g., Table C.5) by finding the value of Θ_D/T that gives the experimental $C_V(T)$, and then multiplying by T. If many values are required, automatic computation is more convenient. Plots of Θ^C against T are used extensively in the succeeding chapters (e.g., Figs. 5.4 and 6.1). $\Theta^C(T)$ would of course be constant if $g(\omega)$ were of Debye form, and the shape of its variation with temperature is therefore a characteristic property of the actual frequency distribution of the solid.

Consider for example the two frequency distributions in Fig. 1.14, for which the corresponding $\Theta^C(T)$ are shown in Fig. 1.16. For both Ar and Si the distribution at low frequencies rises above its limiting ω^2 behavior; consequently C_V rises more rapidly than $C_V(Debye)$, and this is shown by a fall in $\Theta^C(T)$ from its initial value Θ_0^C at $T = 0$. For Ar the departure from the Debye distribution is much less severe than for Si, and the total variation of Θ^C until it approaches its high temperature limit Θ_∞^C is fairly small. For Si the initial rise in the distribution above the ω^2 behavior is steep, and there is a big drop in Θ^C in the range above $T = 0$, corresponding to a greatly enhanced heat capacity. But above the first peak the Si distribution is spread out quite thinly until the final peak occurs at much higher frequencies. C_V does not therefore approach $3Nk$ until much higher temperatures are reached, and the high temperature limiting value Θ_∞^C reflects these higher frequencies.

Extreme departures from Debye behavior can occur when there is wide variety of strength of bonding within the crystal, as for example in the layered crystal graphite (Section 5.8.2). The low frequency vibrations involve mainly weak interlayer forces, and despite the low atomic mass Θ_0^C has the moderate value of 413 K. But the high frequency vibrations depending upon the strong intra layer forces are not excited until high temperatures, and at room temperature $\Theta^C \sim 1500$ K and is still rising. Similarly, in molecular crystals the molecules have internal vibrations of much higher frequencies than those involving only the forces between molecules. It is then appropriate to modify the definition of the Debye equivalent temperature by not including these internal vibrations when counting the number of degrees of freedom for the equivalent Debye distribution, thus making this less than $3N$.

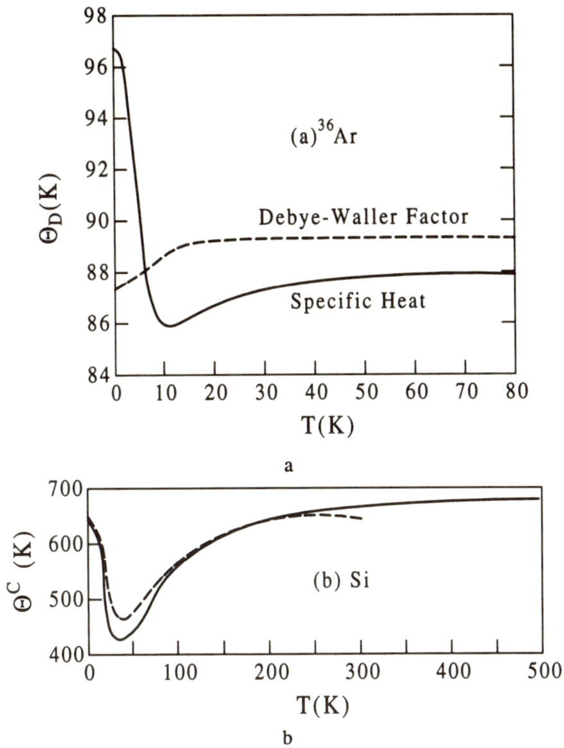

Fig. 1.16. Variation with temperature of Θ^C and (for Ar only) of Θ^M for the distributions $g(\omega)$ of Fig. 1.14. (a)Argon (^{36}Ar) [Fuj74] (b) Silicon: —, from $g(\omega)$; - - -, from experimental heat capacity [Dol66].

The equivalent thetas for the heat capacity most often tabulated are the room temperature value Θ^C_{293} and the low temperature limit Θ^C_0 that determines the coefficient of T^3 in

$$C_{vib} \approx 3Nk(4\pi^4/5)(T/\Theta^C_0)^3 \qquad \text{as} \quad T \to 0 \tag{1.17}$$

Also tabulated sometimes is the estimated limiting value at high temperatures, Θ^C_∞, which is of theoretical importance (Section 2.6). The room temperature values, even if only approximate, can be a valuable guide to the probable behavior of solids at temperatures down to $\Theta/5$ or $\Theta/10$ (see Section 6.2.7).

The representation of experimental C_V values by Θ^C plots can also reveal departure from harmonic behavior at high temperatures. Plots derived for a harmonic vibrational distribution flatten off at high temperatures towards a limiting value Θ^C_∞ — typically for $T \geq 0.3\Theta$. This theoretical behavior is shown in Fig. 1.16(a). But the experimental values for argon corrected to constant volume in Fig. 5.1 show

Θ^C *rising* at high temperatures, indicating a negative anharmonic effect on C_V.* In contrast, for Si a *fall* in Θ^C is observed at high temperatures, indicating a positive anharmonic effect [Flu59].

Other Debye Equivalent Temperatures. Equivalent Debye temperatures can be defined for properties other than the heat capacity — particularly for the entropy (Θ^S) and for the Debye Waller effect (Θ^M). Equivalent thetas for different properties of the same crystal vary differently with temperature (Fig. 1.16(a)), although they are usually of the same order of magnitude (Section 2.6). For this reason the explicit notation Θ^C for the heat capacity is preferable to $\Theta_D(T)$, which is still sometimes used.

The equivalent thetas for the heat capacity and the entropy have the same limiting value at low temperatures, which is therefore often written without superscript:

$$\Theta_0^C = \Theta_0^S = \Theta_0 \tag{1.18}$$

Because the only vibrations contributing to C_V and S at low temperatures are elastic waves, Θ_0 can also be calculated from elastic data (Section 2.8). Values obtained in this way are written Θ_0^{el} to distinguish them from those obtained from calorimetric measurements, written Θ_0^{th}. Whenever possible Θ_0^{el} should be derived from low temperature elastic data, since elastic moduli change with temperature (Fig. 1.3). Values derived from room temperature elastic data should be regarded as a rough approximation for Θ_0; they do *not* give Θ_{293}^C (see also Section 2.6.4). Θ_0^{th} and Θ_0^{el} are usually found to agree within the uncertainty of the measurements (see tabulations [Ale65, Phi71]). Tables of various Debye equivalent thetas are given in Gschneidner's extensive compilation of physical properties of the elements [Gsc64].

1.3.4. Electronic Contributions

The electron theory of metals is a vast subject still in process of development. Here we introduce some simple concepts in common use, which will be discussed further in later chapters. The underlying theory is standard, and can be found in solid state text-books (e.g., [Ash76]).

Independent Particle Model. In a metal the conduction electrons interact with each other as well as with the metallic ions. For many simple metals it is a good first approximation to regard all these interactions as producing an effective potential field in which the electrons move independently, giving a single particle density of electronic states $n(\epsilon)$, where $n(\epsilon)d\epsilon$ is the number of available electron states between ϵ and $\epsilon + d\epsilon$. Electrons are fermions, and multiple occupancy of states is forbidden. At $T = 0$ all the electron states are occupied up to an energy ϵ_{F0} called the *Fermi energy at* $T = 0$, and all higher states are empty. At low temperatures some

*The results in Fig. 5.1 were obtained for natural argon, consisting mainly of ^{40}Ar. The values of Θ^C are consequently lower than for ^{36}Ar.

of these higher states become occupied, giving an electronic heat capacity which is proportional both to T and to the density of states at the Fermi level:

$$C_e = k\left[\frac{1}{3}\pi^2 n(\epsilon_{F0})kT\right] = \Gamma_e T \tag{1.19}$$

The electronic Grüneisen function corresponding to this low temperature limit is then

$$\gamma_e = d\ln[n(\epsilon_{F0})]/d\ln V \tag{1.20}$$

Free Electron Model. The simplest and earliest model for the density of states is to take the electron states as those of a particle of mass m and spin $\frac{1}{2}$ confined to a box of the volume V of the solid. The N conduction electrons then form an ideal Fermi gas (see Section 4.4.1), with

$$n(\epsilon) = \frac{3}{2}N\epsilon^{\frac{1}{2}}/(\epsilon_{F0})^{\frac{3}{2}}, \qquad \epsilon_{F0} = (\hbar^2/2m)(3\pi^2 N/V)^{\frac{2}{3}} \tag{1.21}$$

so that

$$C_e/(Nk) = \frac{1}{2}\pi^2\frac{kT}{\epsilon_{F0}} = \frac{1}{2}\pi^2(T/T_F) \tag{1.22}$$

where the *Fermi temperature* T_F is defined by

$$kT_F = \epsilon_{F0} \tag{1.23}$$

For this model Eq. (1.20) gives for the electronic Grüneisen function $\gamma_e = \frac{2}{3}$, the value for all ideal gases of single particles. This is lower than most experimental values (see, e.g., Table 6.1).

Effective Masses. For the free electron model Eqs. (1.21) and (1.22) show that the electronic heat capacity at low temperatures is proportional to the mass m of an electron. When the electronic heat capacity of a metal differs from the free electron value, an "effective mass" m^* can be defined such that when it is substituted for m in the free electron expression we obtain the correct electronic heat capacity. The ratio m^*/m thus gives the ratio of the actual heat capacity to the free electron value, and is frequently used as a dimensionless measure of the electronic heat capacity. The effective mass is a function of volume, and the electronic Grüneisen parameter is given by

$$\gamma_e = \frac{2}{3} + d\ln m^*/d\ln V \tag{1.24}$$

Effective masses obtained by comparing the predictions of free electron theory with other measurable properties are also commonly used in solid state theory; in general they differ from each other and from that defined above for the heat capacity.

1.3.5. Molecular Gases

The solid state theories mentioned above are relatively simple because crystals are well ordered; even in the free electron gas, where there is no positional order, there is a high degree of order in the momenta. At the other extreme, simplicity also results when in the classical limit an ideal gas becomes highly disordered. This does not happen for an electron gas until very high temperatures are reached ($T \gg T_F$), but for a molecular gas at normal densities the particle states are much closer together and the classical theory of Maxwell, Boltzmann and others holds down to temperatures typically of order 1 K or lower for heavy molecules (Section 2.4.1), giving the familiar results for a monatomic gas:

$$U = \frac{3}{2}NkT, \qquad pV = NkT \tag{1.25}$$

and hence

$$C_V = \frac{3}{2}Nk, \quad C_P = \frac{5}{2}Nk, \quad B_T = P, \quad \beta = (1/T) \tag{1.26}$$

in agreement with the value $\gamma = \frac{2}{3}$ valid at all temperatures. Effects due to the further degrees of freedom of polyatomic molecules are discussed in Section 4.2.2, and those due to departures from ideality due to intermolecular interactions in Sections 4.2.4 and 4.3.

1.4. PLAN OF THIS BOOK

Like the present chapter, the next two chapters are general in the sense that they deal with topics that are relevant to many (and sometimes all) materials. After defining precisely quantitative measures of heat capacity and thermal expansion, Chapter 2 goes on to describe briefly the underlying theoretical framework: first the thermodynamics and statistical mechanics, and then the various types of material to which they are applied, the different types of bonding (ionic, valence, metallic, etc.) giving rise to different types of behavior. Several simple models are then described which have widespread application either directly or indirectly by illustrating concepts important for more complex systems, the aim being to clarify ideas of particular relevance to heat capacity and thermal expansion. We then discuss anisotropic stress and strain, and the thermodynamics of elasticity. The ground is thus prepared for Chapter 3, which discusses methods of measurement and other cryogenic techniques.

Most of the rest of the book deals in turn with different groups of materials. Chapter 4 deals briefly with fluids. Although most materials of cryogenic interest are solids, there are some fluids of great importance. Liquefied gases such as nitrogen, hydrogen and helium are used widely in cooling; their low triple points enable theories of equations of state to be tested up to high reduced temperatures and pressures; and vapors are used to establish the thermodynamic ideal gas scale of temperature. And of course the unique properties of liquid ^{3}He and ^{4}He and their mixtures are of great interest both fundamentally and technically.

Chapter 5 deals with non-metallic solids, including ceramic high temperature superconductors. These are of many different kinds, too many to be covered comprehensively in a book of this type. Emphasis is therefore placed on those materials that are cryogenically important, and on those that are interesting theoretically either because they are well understood or because they present well-defined problems. This is true also of Chapter 6, which deals with metals and semi-metals, but here further theoretical treatment is also needed in order to discuss such wide-spread properties as superconductivity and magnetism.

Composite materials (with polycrystals of a single substance as a special case) are discussed briefly in Chapter 7. Chapter 8, "Cryocrystals, Clathrates and Curiosities," deals with topics which merit inclusion but do not fit conveniently elsewhere. After a brief "Conclusion," there are appendices containing useful information such as methods of thermodynamic manipulation, tables of technical data, Einstein and Debye tables, and a list of commonly used symbols.

Chapter 2

Basic Theory and Techniques

2.1. INTRODUCTION

This chapter summarizes theory used in the discussion of thermodynamic properties, giving references to where more detailed discussion may be found, and elaborating more fully some concepts which are frequently used. The general principles are illustrated first for isotropic behavior. Of these principles, thermodynamics is essential to everyone in that it provides a systematic method of describing, relating and analyzing the bulk properties which form the subject of this monograph. Statistical mechanics relates these properties to atomic and molecular structure, and so forms the basis for their theoretical explanation and prediction. Computational tools developed for this purpose are briefly mentioned, both for the statistical mechanics and for the underlying quantum theory of bonding and cohesion. Some of the applications described are simple, such as ideal gases and Schottky systems; but two others of great general importance, *viz.* cooperative order–disorder effects and vibrational contributions, require longer discussion. So also does the extension of the theory to anisotropic behavior and elasticity.

2.2. THERMODYNAMICS

Like Section 1.3, this section deals with processes that are functions of volume and temperature. The thermodynamics of more general strain (including anisotropic expansion and elasticity) is treated in Section 2.8.

2.2.1. Definitions

Heat Capacity. Heat capacity is defined as the limit of the ratio $\Delta Q/\Delta T$ as $\Delta Q \to 0$, where ΔT is the rise of temperature resulting from an input of heat ΔQ under specified conditions. The heat capacities C_P and C_V, already defined in

Section 1.3.1, can also be expressed as derivatives of the entropy: thus

$$C_P = \left(\frac{\partial H}{\partial T}\right)_P = T\left(\frac{\partial S}{\partial T}\right)_P = -T\left(\frac{\partial^2 G}{\partial T^2}\right)_P \tag{2.1}$$

$$C_V = \left(\frac{\partial U}{\partial T}\right)_V = T\left(\frac{\partial S}{\partial T}\right)_V = -T\left(\frac{\partial^2 F}{\partial T^2}\right)_V \tag{2.2}$$

where U and H are the energy and enthalpy, and F and G the Helmholtz and Gibbs free energies. The estimation of C_P from measurements involving finite intervals is discussed in Sections 1.2 and 3.2. C_V is usually not measured directly, but obtained from C_P by a thermodynamic relation (Eq. 2.10).

The expressions given above can apply to macroscopic systems of any size, but are often taken to refer to molar quantities. We reserve the term *specific heat* for the heat capacity per unit mass or per unit volume, both of which we denote by lower case:

$$c_P = C_P/M, \quad c_V = C_V/M; \quad OR \quad c_P = C_P/V, \quad c_V = C_V/V \tag{2.3}$$

Thermal Expansion. The coefficient of volumetric expansion, already defined in Eq. (1.3), can be expressed in any of the forms

$$\beta = \left(\frac{\partial \ln V}{\partial T}\right)_P = -\left(\frac{\partial \ln \rho}{\partial T}\right)_P = \frac{1}{V}\frac{\partial^2 G}{\partial P \partial T} \tag{2.4}$$

This coefficient is often also denoted by α, but that symbol is more usefully reserved for coefficients of linear expansion, defined by

$$\alpha = \left(\frac{\partial \ln l}{\partial T}\right)_P = \frac{1}{l}\left(\frac{\partial l}{\partial T}\right)_P \tag{2.5}$$

When the expansion is isotropic, $\beta = 3\alpha$.

Data on thermal expansion can be presented in different ways — for example as molar volumes, as dilations $\Delta V/V_0$, or as expansion coefficients. For high precision it may be necessary to make fine distinctions, as for example between β and α as defined thermodynamically above and the quantities β^* and α^* often used as practical definitions:

$$\beta^* = \frac{1}{V_0}\left(\frac{\partial V}{\partial T}\right)_P, \quad \alpha^* = \frac{1}{l_0}\left(\frac{\partial l}{\partial T}\right)_P \tag{2.6}$$

where V_0 and l_0 are usually taken to be the room temperature values of V and l. Details of the treatment of primary dilatometric data are discussed in Section 3.3.1 and in several chapters of [Ho98].

Thermal expansion data are also obtained from the change in crystal lattice parameters measured by Bragg reflection in X-ray or neutron diffraction. Strictly such data are not equivalent to dilatometric data because of crystal imperfections. For example, differences between the volumetric expansion coefficients of the crystallographic unit cell and of the bulk crystal are used to estimate the formation of vacancies (see Section 3.3.2). However, vacancies have a significant effect only near the melting point; at other temperatures the unit cell dimensions change proportionately to those of the macroscopic crystal, and so are equivalent to dilatometric data. Conventions for the nomenclature of crystal axes, and the relation of the change of crystallographic parameters to bulk expansion, are discussed in Appendix A. Analysis of intensities can give also the changing relative positions of atoms within the unit cell (known as internal expansion), but usually with insufficient precision to show perceptible change at low temperatures.

2.2.2. Units and Conversion Factors

Conversion between different energy scales is important for the comparison and interpretation of thermodynamic data. International convention now generally requires the use of SI units, together with allowed related units [Coh87, Nel98]; but cgs and obsolete "practical" units are sometimes found, especially in the older literature. Temperature scales, thermodynamic and practical, are discussed in Section 3.1.

The old unit of heat, the *calorie*, was used in the past in much good calorimetric work. It was defined originally so as to make the specific heat of water at 15° Centigrade equal to 1 $\mathrm{cal \cdot g^{-1} \cdot deg^{-1}}$, but later a *thermochemical calorie* ($\mathrm{cal_{th}}$) was fixed as precisely 4.184 J. Heat capacities are now usually given in molar units of $\mathrm{J \cdot mol^{-1} \cdot K^{-1}}$, or as specific heats in units of $\mathrm{J \cdot g^{-1} \cdot K^{-1}}$ or $\mathrm{J \cdot cm^{-3} \cdot K^{-1}}$.

At the atomic level the electron volt (eV) is often used as a unit of energy, although spectroscopists may also refer to energy differences in terms of the frequency (in Hz) or inverse wave-length (in $\mathrm{cm^{-1}}$) of the equivalent photon. In statistical mechanics we also need to know the temperature range in which the higher energy level becomes appreciably occupied. Table 2.1 gives equivalence factors relating these different energy scales. Thus we can see, for example, that rotational energy levels of molecules, which have microwave spectroscopic transitions of a few $\mathrm{cm^{-1}}$, will contribute to heat capacities at temperatures of a few kelvin and upwards; whereas electronic levels, typically of the order of eV, will usually not contribute at all at low temperatures.

The bulk modulus and other elastic stiffnesses have the dimensions of pressure, for which the SI unit is the pascal:

$$1\ \mathrm{Pa} = 1\ \mathrm{N \cdot m^{-2}} = 10\ \mathrm{dyn \cdot cm^{-2}} \tag{2.7}$$

The unit $\mathrm{dyn \cdot cm^{-2}}$ is now wholly obsolete, but one pre-SI practical unit, the bar, is still acceptable and widely used; 1 bar $= 10^5$ Pa, introduced so that the atmospheric

Table 2.1. Equivalence factors for different energy scales

	J	eV	h×1 THz	hc×1 cm^{-1}	k×1 K
1 eV =	1.602×10^{-19}	1	241.8	8065	11604
h×1 THz =	6.626×10^{-22}	4.136×10^{-3}	1	33.36	47.99
hc×1 cm^{-1} =	1.986×10^{-23}	1.240×10^{-4}	0.02998	1	1.439
k×1 K =	1.381×10^{-23}	8.617×10^{-5}	0.02084	0.6950	1

Also: 1 eV·molecule^{-1} = 96.49 kJ·mol^{-1} = 23.06 kcal$_{th}$·mol^{-1}

pressure is approximately 1 bar. Elastic stiffnesses of solids are typically of the order of 10 to 100 GPa, i.e., 0.1 to 1 Mbar.

2.2.3. Thermodynamic Relations

Methods used for obtaining relationships between thermodynamic quantities are summarized in Appendix B. Here we quote some results widely used in the analysis of thermodynamic data.

The ratio of C_P to C_V is the same as that for B_S to B_T:

$$\frac{C_P}{C_V} = \frac{B_S}{B_T} = \frac{\chi_T}{\chi_S} = 1 + \beta\gamma T \tag{2.8}$$

where γ is the Grüneisen function defined in Eq. (1.5):

$$\gamma = \frac{\beta V}{C_P \chi_S} = \frac{\beta V}{C_V \chi_T} \tag{2.9}$$

Thus

$$C_P = C_V + \frac{\beta^2 V T}{\chi_T}, \qquad \chi_T = \chi_S + \frac{\beta^2 V T}{C_P} \tag{2.10}$$

The following relations are frequently used in the discussion of thermal expansion:

$$\left(\frac{\partial S}{\partial V}\right)_T = \left(\frac{\partial P}{\partial T}\right)_V = \frac{\beta}{\chi_T} = \beta B_T \tag{2.11}$$

Equations (2.11) lead to further expressions for the Grüneisen function γ defined in Section 1.3.1:

$$\gamma = \left(\frac{\partial P}{\partial (U/V)}\right)_V = \frac{V}{C_V}\left(\frac{\partial P}{\partial T}\right)_V = \frac{V}{C_V}\left(\frac{\partial S}{\partial V}\right)_T = -\left(\frac{\partial \ln T}{\partial \ln V}\right)_S \tag{2.12}$$

The last expression in Eq. (2.12) shows that gamma governs the size of temperature changes caused by changing pressure under adiabatic conditions:

$$\left(\frac{\partial T}{\partial P}\right)_S = \gamma T \chi_S \tag{2.13}$$

This provides a direct way of determining γ (Section 3.3.7). The effect has also been used to monitor stress fluctuations in working materials by observing the related temperature fluctuations [Mou78].

Other Grüneisen Functions. Although γ as defined above is the most widely used, other Grüneisen functions may be met in the literature. For example, one is obtained from the volume derivative of the equivalent Debye temperature Θ^C:

$$-[\partial \ln\{\Theta^C(T,V)\}/\partial \ln V]_T$$

This is a good estimate for the γ of Eq. (2.12) when the temperature variation of Θ^C is small, but not when deviations from Debye behavior are large.

Two other thermodynamic Grüneisen functions are used occasionally, called here γ_{EOS} and γ_{SH}. γ_{EOS} is defined from one of the Mie–Grüneisen equations of state, approximations often used by geophysicists and others concerned with high pressure behavior [Eqs. (2.99)]:

$$\gamma_{EOS}(T,V) = \frac{P(T,V) - P(0,V)}{[U(T,V) - U(0,V)]/V} = -\left(\frac{\partial \ln T}{\partial \ln V}\right)_{(F_{th}/T)} \tag{2.14}$$

where $F_{th} = F(T,V) - F(0,V)$, the thermal component of the Helmholtz energy. γ_{SH} was used by Ahlers [Ahl67, Ahl70] in the discussion of C_V data obtained for helium at different fixed volumes, because (unlike γ) it could be obtained from the data without the need to extrapolate the measured heat capacities and their derivatives to $T = 0$:

$$\gamma_{SH} = \frac{(\partial C_V/\partial \ln V)_T}{(\partial C_V/\partial \ln T)_V} = -\left(\frac{\partial \ln T}{\partial \ln V}\right)_{C_V} \tag{2.15}$$

If over the whole range of temperature $\gamma(T,V)$ is a constant depending only on volume, then γ, γ_{EOS} and γ_{SH} are all equal; but if γ varies strongly with temperature they differ markedly.* Relations between the three functions are discussed in [Bar98, Section 1.4.3].

A number of approximations have been proposed which estimate γ from the pressure dependence of the bulk modulus; see [And95a, Section 1.4]. They are based on the theory of lattice vibrations, but since they are themselves thermodynamic expressions they are conveniently described here. Two early estimates, proposed by Slater [Sla39a] and by Dugdale and Macdonald [Dug53], are sometimes still used:

$$\gamma_{\mathrm{sl}} = \frac{1}{2}(dB/dP) - \frac{1}{6}, \qquad \gamma_{\mathrm{d\text{-}m}} = \frac{1}{2}(dB/dP) - \frac{1}{2} \tag{2.16}$$

*In particular, γ_{SH} can become infinite when the heat capacity passes through a maximum or minimum as a function of temperature.

Although useful at the time, these were derived by means of crude assumptions and have long been known to be inaccurate both at low and high temperatures even for the simplest models [Bar57a]; they should now be regarded as of historical interest only. The best estimate of this kind is probably that of Barton and Stacey [Bar85] for the high temperature limit γ_∞ as a function of pressure:

$$\gamma_{\text{ba-s}} = \frac{\frac{1}{2}(\partial B_T/\partial P)_T - \frac{1}{6} - \frac{1}{3}f\left[1 - \frac{1}{3}(P/B_T)\right]}{1 - \frac{4}{3}(P/B_T)} \tag{2.17}$$

where $f = 2.35$. This is based on the concept of pair potentials and includes both bond-stretching and tension effects (see Section 2.6.3). Approximations which make use also of the rigidity modulus and its pressure dependence are discussed by Anderson, see [And95a, Section 1.8].

Pressure and Volume Derivatives. The isothermal pressure derivatives of both the heat capacity and the thermal expansion coefficient can be obtained thermodynamically from the results of measurements carried out at constant pressure:

$$\left(\frac{\partial C_P}{\partial P}\right)_T = -T\left[\frac{\partial}{\partial T}\left(\frac{\partial V}{\partial T}\right)_P\right]_P = -T\left[\frac{\partial(\beta V)}{\partial T}\right]_P = -TV\left[\beta^2 + \left(\frac{\partial \beta}{\partial T}\right)_P\right] \tag{2.18}$$

and

$$\left(\frac{\partial \beta}{\partial P}\right)_T = -\left(\frac{\partial \chi_T}{\partial T}\right)_P \tag{2.19}$$

For the bulk moduli, on the other hand, isothermal pressure derivatives can only be obtained from measurements under varying pressure; they are dimensionless, and can also be expressed as isothermal logarithmic volume derivatives, often denoted by B'_S and B'_T:

$$B'_S = \left(\frac{\partial B_S}{\partial P}\right)_T, \qquad B'_T = \left(\frac{\partial B_T}{\partial P}\right)_T = -\left(\frac{\partial \ln B_T}{\partial \ln V}\right)_T \tag{2.20}$$

Typically they have values between 4 and 6, but decrease appreciably at high pressure [And95a]. As with B_S and B_T, their variation is small at low temperatures; and the difference between them is also small, tending to zero as $T \to 0$. For further discussion see Section 2.7.

In contrast to the isothermal derivatives, the difference between the *isobaric* logarithmic volume derivatives of B_S and B_T is not small. These are the *Anderson-Grüneisen* functions δ_S and δ_T, defined by [Bas68]

$$\delta_S = -\left(\frac{\partial \ln B_S}{\partial \ln V}\right)_P, \qquad \delta_T = -\left(\frac{\partial \ln B_T}{\partial \ln V}\right)_P = \left(\frac{\partial \ln \beta}{\partial \ln V}\right)_T \tag{2.21}$$

They are a different type of function from the Grüneisen functions discussed above; they originate from a parameter used by Grüneisen [Gru12] and later exploited by O. L. Anderson [And66a]. Anderson has pointed out that their small variation at high temperatures make them suitable for extrapolating bulk moduli to higher temperatures. They are not so useful at low temperatures, where both δ_S and δ_T can vary strongly (e.g., [And95a, Fig. 1.1]). At $T \sim \Theta$ the difference between them is given approximately by $\delta_T - \delta_S \sim \gamma$. At low temperatures the difference tends to the limit $\delta_T - \delta_S = (n+1)\gamma_0$, where n is the exponent of T in the limiting form of the heat capacity (1 for metals and 3 for non-metals) and as usual γ_0 is the limiting value of gamma at low temperatures [Bar79].

Measurements under pressure are needed to determine the isothermal volume derivative of the Grüneisen function, which is often described by the dimensionless quantity

$$q = (\partial \ln \gamma / \partial \ln V)_T \tag{2.22}$$

sometimes called the *second Grüneisen function*.* At high temperatures q often has a value of about unity. At low temperatures it may vary strongly, and it becomes infinite whenever gamma changes sign; it is then preferable to use $(\partial \gamma / \partial \ln V)_T$.

Relations between these quantities have been discussed repeatedly in the literature, e.g., [Bas68, Bar79, Bar80, And95a, Bar98].

Electric and Magnetic Fields. The components of electric and magnetic fields are intensive thermodynamic variables additional to temperature and pressure (or stress). For example, under conditions where there is a uniform magnetic field H the differential of the Gibbs free energy ($G = U - TS + PV - \mu_0 H \mathcal{M}$) is

$$dG = -SdT + VdP - \mu_0 \mathcal{M} dH \tag{2.23}$$

where $\mathcal{M}$ is the total magnetic moment of the material. Thermodynamic relations follow by the usual methods; e.g., for first order magnetostriction there is the Maxwell relation

$$\left(\frac{\partial V}{\partial H}\right)_{P,T} = -\mu_0 \left(\frac{\partial \mathcal{M}}{\partial P}\right)_{H,T} \tag{2.24}$$

For more general applications a tensor notation is needed for the stress and strain, and explicit consideration of crystal symmetry [Nye85]; and various subtleties arise [Bor54, Bar98]. We shall not pursue this subject further in this book, except to quote results if needed.

*Other functions have also been given the name of second Grüneisen function, parameter, or constant [Gil56, Dav59, Bar72].

2.2.4. Phase Transitions

Phase transitions have often been discovered through the appearance of unexpected anomalies in measurements of heat capacity or other thermodynamic properties. Indeed, Ehrenfest's historic classification of transitions as first, second, third, etc., is purely thermodynamic. At all phase transitions there is continuity of the Gibbs free energy G across the phase boundary; at a first order transition G is continuous but there are discontinuities in its first derivatives S and V; at a second order transition S and V are continuous but there are discontinuities in the second derivatives of G (and hence in C_P, β, and χ_T); and so on.

Ehrenfest's classification has however proved inadequate for a general description of transitions beyond the first order; for example, superconductors are the only materials which appear to have transitions exhibiting ideal second order behavior. The principal distinction to be drawn is between the first order transitions, which occur between phases of distinct structure and where consequently superheating and supercooling can occur, and other transitions where there is continuity of structure but the transition marks the initiation of some process changing the structure (see the incisive discussion by Pippard in [Pip64, Ch. 9]). These other 'higher order' transitions exhibit many different types of behavior (Section 2.5.4).

First Order Transitions. Consider first a system with two independent intensive variables, P and T. Along the phase boundary in the P–T phase diagram the difference in free energy ΔG between the phases is constant (zero), and so

$$d\Delta G = -\Delta S dT + \Delta V dP = 0 \tag{2.25}$$

The Clapeyron equation for the slope of the transition line follows immediately:

$$\frac{dP}{dT} = \frac{\Delta S}{\Delta V} \tag{2.26}$$

The same method can be applied when there are three (or more) intensive variables, T, P, H say. Using Eq. (2.23) to get $d\Delta G$ we find that on the transition boundary

$$\left(\frac{\partial P}{\partial T}\right)_H = \frac{\Delta S}{\Delta V}, \quad \left(\frac{\partial H}{\partial T}\right)_P = -\frac{\Delta S}{\mu_0 \Delta \mathcal{M}}, \quad \left(\frac{\partial H}{\partial P}\right)_T = \frac{\Delta V}{\mu_0 \Delta \mathcal{M}} \tag{2.27}$$

Second Order Transitions. We consider only two intensive variables T, P. At the transition S and V are continuous, so that along the transition we can equate separately dS and dV to zero. This leads to the two Ehrenfest equations (e.g., [Pip64, Bar98])

$$\frac{dP}{dT} = \frac{1}{VT}\frac{\Delta C_P}{\Delta \beta} = \frac{\Delta \beta}{\Delta \chi_T} \tag{2.28}$$

2.2.5. The Third Law and Its Implications

The Third Law was introduced in Chapter 1 as a consequence of the quantum nature of matter. Here we discuss it strictly from a thermodynamic standpoint, i.e., as a postulate deduced from the observed behavior of macroscopic systems. As such, for many years it was the subject of considerable controversy, due partly to the difficulty of stating it in a form that would be both general and succinct (e.g., [Sim56, Dug96]). Broadly, the essential content of the Law is the existence of an absolute zero for entropy as well as for temperature, which is approached as $T \to 0$ for any system in thermodynamic equilibrium. The Law thus tells us two things: (a) the change in entropy as $T \to 0$ is finite; (b) the change ΔS in entropy is zero between any two equilibrium states of a system at $T = 0$.

Behavior as $T \to 0$. From (a) alone it follows that the heat capacities C_P and C_V tend to zero as $T \to 0$, because otherwise the integrals giving the changes in entropy from $T = 0$ to any other temperature T, *viz.*

$$\Delta S = \int_0^T \frac{C_P(P,T)}{T} dT \qquad \text{or} \qquad \Delta S = \int_0^T \frac{C_V(V,T)}{T} dT \tag{2.29}$$

would be infinite; but to deduce the behavior of other thermodynamic properties the full law is needed. With the aid of a Maxwell relation we prove that $\beta \to 0$ as $T \to 0$, since at $T = 0$

$$\beta = \frac{1}{V}\left(\frac{\partial V}{\partial T}\right)_P = -\frac{1}{V}\left(\frac{\partial S}{\partial P}\right)_T = 0 \tag{2.30}$$

Similarly the thermal pressure coefficient $\left(\frac{\partial P}{\partial T}\right)_V$ can be shown to be zero at $T = 0$. The bulk modulus (and more generally all the elastic stiffnesses) tend to finite limits, but their temperature derivatives tend to zero (e.g., Fig. 1.3):

$$\left(\frac{\partial B_T}{\partial T}\right)_V = -\left[\frac{\partial}{\partial T}\left(\frac{\partial P}{\partial \ln V}\right)_T\right]_V = -V\left[\frac{\partial}{\partial V}\left(\frac{\partial P}{\partial T}\right)_V\right]_T = 0 \tag{2.31}$$

Similar arguments can be applied when instead of P and V we take electric or magnetic conjugate variables, so that for example $\left(\frac{\partial \mathcal{M}}{\partial T}\right)_H$ is zero at $T = 0$, as are also the temperature derivatives of the magnetic permeability χ and the electric permittivity ϵ.

Extension of the Third Law to Sub-Systems. The restriction of the Third Law to systems in equilibrium would forbid any application to substances such as glasses at low temperatures; and indeed their entropy does not approach zero as $T \to 0$. Nevertheless, many conclusions derived from the Third Law are true also for such systems, and they are covered by a revised statement of the Law due to Sir Francis Simon [Sim37], which we give here in the wording of Dugdale in [Dug96, pp. 163-4]: *The contribution to the entropy of a system from each sub-system that is in internal thermodynamic equilibrium vanishes at* $T = 0$. This extended law is applicable to many different types of sub-system. For a glass at low temperatures, the vibrations are in internal equilibrium and contribute to the thermal properties, while the structure is frozen in a disordered arrangement with constant non-zero configurational entropy.

Residual Entropy at $T = 0$. There are many systems which, like a glass, have non-zero entropy at $T = 0$. As such systems are cooled, kT becomes much less than the activation energy of the processes needed to reach thermodynamic equilibrium, so that equilibrium cannot be reached during the time of the experiment (e.g., ice, Fig. 1.12). The approximate temperature at which this occurs will depend upon the rate of cooling, and this may affect the extent to which the system remains disordered at low temperatures. Perhaps the simplest examples of residual entropy are provided by some crystals of asymmetric linear molecules, such as N_2O and CO, for which the activation energy to reverse molecular orientation is much greater than the difference in energy between the initial and final states; molecular orientations then get frozen in a disordered arrangement. If no ordering has occurred before this happens, the residual configurational molar entropy will be

$$S_{\text{config}} = k \ln(2^{N_A}) = R \ln 2 \tag{2.32}$$

as observed for N_2O. However, in CO partial ordering occurs before the remaining disorder becomes frozen in; the residual entropy is appreciably less than $R\ln 2$ and depends on the rate of cooling [Man97, Section 17.1].

Residual entropies are often determined by calculating the absolute entropy of the gaseous phase by the methods of statistical mechanics, using spectroscopic data to obtain the molecular energy states, and then finding the change of entropy on cooling to the solid state by calorimetric methods [Moo62].

The Unattainability of Absolute Zero. Although the state of a system at $T = 0$ is usually the easiest state to treat theoretically, we can deduce from the Third Law that such a state cannot be obtained in the laboratory. Methods for obtaining low temperatures are discussed in detail in many texts (see Section 3.1.1). In essence, reduction of temperature is achieved by changing an experimental variable (such as pressure or magnetic field) under adiabatic conditions, in such a way as to reduce the temperature along the relevant adiabat (Fig. 2.1). If it were possible to reach absolute zero in a finite number of such steps, the final adiabatic step would therefore start in a state with $S > 0$ and end in a state with $S = 0$; but under adiabatic conditions this is forbidden by the Second Law.

It is interesting to note that stating the unattainability of absolute zero is not strictly equivalent to the Third Law, because unattainability would hold for classical as well as for quantum matter [Sim56]. The entropy of all substances would then tend to minus infinity as $T \to 0$, and again the absolute zero would never be reached in a finite number of steps.

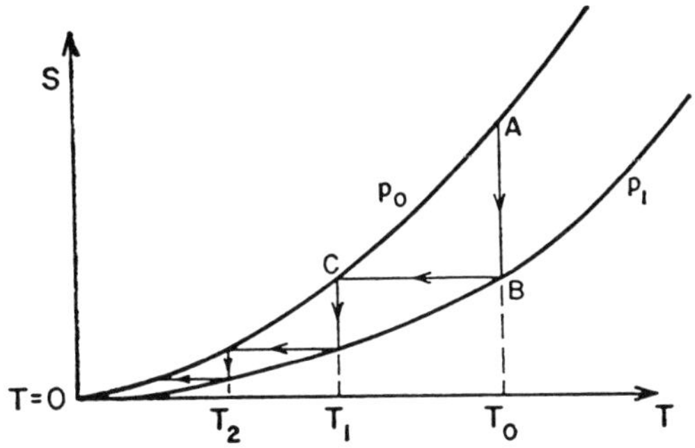

Fig. 2.1. Entropy–temperature diagram of a substance obeying the third law, when the pressure is changed alternately adiabatically and isothermally. The absolute zero is unattainable. From [Dug96, Fig. 22(b)].

2.3. STATISTICAL MECHANICS

2.3.1. Quantum Statistical Mechanics

The nineteenth century work of Maxwell, Boltzmann and Gibbs treated the statistical behavior of large numbers of atoms and molecules obeying the laws of classical mechanics. Despite its great success in accounting for the First and Second Laws of thermodynamics, and also for the behavior of dilute gases, it failed to account for the equilibrium between a solid body and its surrounding radiation field; and of course it was this that led to Planck's original quantum hypothesis. As we have seen in Chapter 1, quantum theory is essential for understanding the Third Law, including the behavior of the heat capacity and thermal expansion at low temperatures.

There are very many books on statistical mechanics, with different applications and different levels of theory. Among these are the classic text by Tolman [Tol38], and [Hil60, Ric67, Fey72, Gop74].

The Boltzmann Factor and the Partition Function. Consider a physical system with possible quantum states i, having energies E_i, in thermal equilibrium with surroundings at temperature T. The probability $\mathcal{P}_i$ that the system is in state i is proportional to the *Boltzmann factor*, $exp(-E_i/kT)$, and is given by

$$\mathcal{P}_i = \frac{\exp(-E_i/kT)}{Z} \tag{2.33}$$

where Z is the *partition function* or *sum over states* (German *Zustands-summe*)

$$Z = \sum_i \exp(-E_i/kT) \tag{2.34}$$

The Helmholtz energy and the entropy are then given by

$$F = -kT \ln Z, \qquad S = -k \sum_i \mathcal{P}_i \ln \mathcal{P}_i \tag{2.35}$$

The expression for the entropy is quite general. In particular, it reduces to the familiar $S = k \ln \Omega$ when there is a finite number Ω of possible states, all equally probable, since then $\mathcal{P}_i = 1/\Omega$ for all i.

For a macroscopic system the energies E_i are functions of the volume, $E_i(V)$, or more generally of the strain (see Section 2.8). Expressions for other thermodynamic quantities are found by differentiating F with respect to T and V, and are functions of averages over the states i weighted by $\mathcal{P}_i$. With the general notation

$$\langle X \rangle = \sum_i \mathcal{P}_i X_i \tag{2.36}$$

the pressure is given by

$$P = -\sum_i \mathcal{P}_i E_i' = -\langle E' \rangle \tag{2.37}$$

and

$$B_T = \frac{1}{\chi_T} = V[\langle E'' \rangle - \{\langle E'^2 \rangle - \langle E' \rangle^2\}/kT] \tag{2.38}$$

$$C_V = \{\langle E^2 \rangle - \langle E \rangle^2\}/kT^2 = \langle (\Delta E)^2 \rangle / kT^2 \tag{2.39}$$

$$(\partial P/\partial T)_V = (\partial S/\partial V)_T = -\{\langle E'E \rangle - \langle E' \rangle \langle E \rangle\}/kT^2 \tag{2.40}$$

The thermal expansion coefficient β is then obtained by using Eq. (2.11), and the Grüneisen function is

$$\gamma(T, V) = -V \frac{\langle E'E \rangle - \langle E' \rangle \langle E \rangle}{\langle E^2 \rangle - \langle E \rangle^2} \tag{2.41}$$

If all the energies E_i scale with volume in the same way, so that they are all proportional to a single characteristic energy E_c, this reduces to a single Grüneisen parameter

$$\gamma(T, V) = \gamma_c = -(d \ln E_c / d \ln V) \tag{2.42}$$

The above equations show that $\mathcal{P}_i$, S and C_V depend only on the intervals between the energies E_i, which may be determined spectroscopically. Spectroscopic measurements under pressure give also the volume derivatives of energy intervals, and hence in principal the data needed to derive γ and β.

Application: the Two State Schottky System. A simple but important illustration of these general results is provided by a system which has only two possible quantum states — for example, the magnetic states of a nucleus of spin $\frac{1}{2}$ in a magnetic field, with energies $\epsilon_1 = 0$, $\epsilon_2 = \Delta\epsilon$. This is a special case of the general class of *Schottky* systems discussed in Section 2.5.3. For $T \ll \Delta\epsilon$, the system will be in the lower energy state, so that $\mathcal{P}_1 = 1$ and $\mathcal{P}_2 = 0$; and the entropy is $-k\ln 1 = 0$. For $T \gg \Delta\epsilon$, there are 2 possible states of equal probability $\frac{1}{2}$, and the entropy is $k\ln 2$. At intermediate temperatures the probabilities for the system to be in the lower and upper states are as shown in Fig. 2.2:

$$\mathcal{P}_1 = \frac{1}{1+e^{-x}} = \frac{e^x}{1+e^x}, \qquad \mathcal{P}_2 = \frac{e^{-x}}{1+e^{-x}} = \frac{1}{1+e^x} \tag{2.43}$$

where $x = \Delta\epsilon/kT$ and $1+e^{-x}$ is the partition function. Other thermodynamic functions follow immediately from Eqs. (2.33)–(2.41); in particular the entropy and heat capacity are

$$s_{sch} = k\left[\ln(1+e^{-x}) + \frac{x}{1+e^x}\right], \qquad c_{sch} = k\frac{x^2}{(e^x+1)(1+e^{-x})} \tag{2.44}$$

where the lower case symbols denote properties of a microscopic subsystem rather than of bulk material. The formal similarity of the expression for the heat capacity to that in Eq. (1.14) for a harmonic oscillator is a good aid to memory, but the plus signs in the brackets lead to the behavior shown in Fig. 2.2(c), very different from that of a harmonic oscillator (Fig. 1.1). In the high temperature limit $x \to 0$, and c_{sch} tends to zero as $\frac{1}{4}k(\Delta\epsilon/kT)^2$.

Additive Contributions. According to statistical mechanics, the additivity of different contributions to thermodynamic functions has its origin in the additivity of different contributions to the energies of excited quantum states. For example, in α-nickel sulphate (see Fig. 1.5) the excitation of the magnetic energy levels is to a very good approximation independent of the lattice vibrations, and at low temperatures a total excited state of the crystal is specified by giving both its vibrational state v and its magnetic state x. The energy is

$$E_{v,x} = E_g + E_v + E_x \tag{2.45}$$

where E_g is the energy of the electronic ground state. The partition function then factories:

$$Z = \sum_{v,x} e^{-(E_g+E_v+E_x)/kT} = e^{-E_g/kT} Z_{vib} Z_m \tag{2.46}$$

where Z_{vib} and Z_m are the vibrational and magnetic partition functions, giving the Helmholtz energy as

$$F = -kT\ln Z = E_g + F_{vib} + F_m \tag{2.47}$$

Separate contributions to P, S, C_V, etc., follow by differentiation (e.g., Section 5.11.1). Another example is provided by excited states of molecules, which can be labelled by their electronic, vibrational and rotational states (Section 4.2), although there is significant interaction between rotations and vibrations.

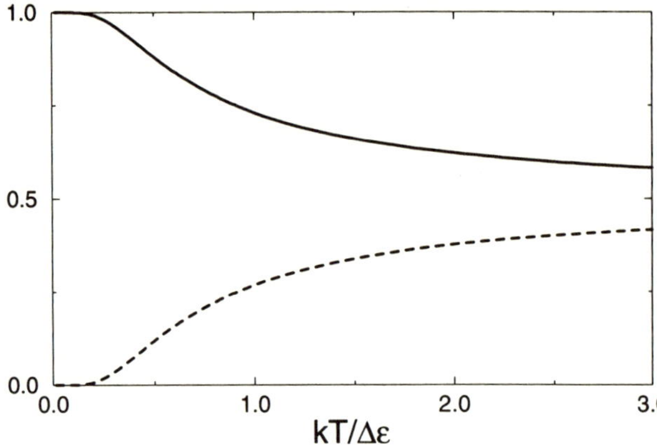

a

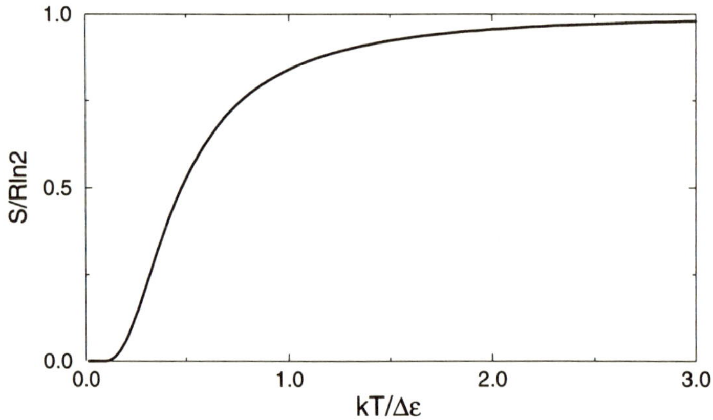

b

Fig. 2.2. Properties of a mole of identical Schottky systems with two non-degenerate levels (see text): (a) occupation of levels $\mathcal{P}_1$ (solid curve) and $\mathcal{P}_2$ (dashed curve); (b) entropy; (c) heat capacity.

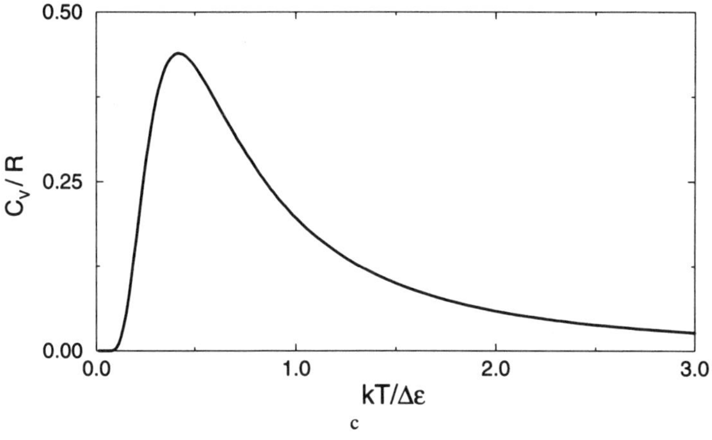

Fig. 2.2. *(Continued)*.

Grand Partition Function. The method of averaging just described is that of the *canonical ensemble*, applicable to a system of fixed volume and composition in contact with a heat bath at temperature T. Statistical mechanics can also be applied to systems under different conditions. For example, the *grand canonical ensemble* is applicable to a system of fixed volume in contact both with a heat bath and with a reservoir of particles of chemical potential μ, so that it can exchange particles with the environment as well as energy. The states of the system then have energies $E_{N,i}(V)$, where the subscripts indicate the ith quantum state of the system when it contains N particles. The statistical probability of this state is then

$$P_{N,i} = \frac{\exp\{(N\mu - E_{N,i})/kT\}}{\Xi} \tag{2.48}$$

where μ is the chemical potential (per particle) in the environment and Ξ is the *grand partition function*, defined by

$$\Xi = \sum_{N=0}^{\infty} \sum_{i} \exp\{(N\mu - E_{N,i})/kT\} \tag{2.49}$$

From Ξ is obtained the virial PV expressed as a function of T and μ:

$$PV = kT \ln \Xi \tag{2.50}$$

Other thermodynamic properties follow by differentiation, since

$$d(PV) = SdT + PdV + Nd\mu \tag{2.51}$$

For some systems the grand canonical ensemble is easier to apply than the canonical ensemble — notably to systems of non-interacting particles, such as the quantum ideal gas (Section 4.4.1) and the independent particle model for electronic properties of solids. We do not need to consider the whole system simultaneously; we can treat each particle quantum state j with energy ϵ_j as an independent sub-system in an environment of temperature T and electron chemical potential μ, so that in the formalism N becomes the number of electrons in state j. Since electrons are fermions there are only two possibilities: j is occupied, giving $N = 1$ and energy $E = \epsilon_j$; or it is unoccupied, giving $N = 0$ and $E = 0$. The grand partition function is then simply $1+\exp\{(\mu-\epsilon_j)/kT\}$, and the probability of occupation is given by the mean value of N:

$$p = \frac{\exp\{(\mu-\epsilon_j)/kT\}}{1+\exp\{(\mu-\epsilon_j)/kT\}} = \frac{1}{\exp\{(\epsilon_j-\mu)/kT\}+1} \tag{2.52}$$

This is the Fermi–Dirac distribution function referred to in Section 4.4.1.

The treatment of independent boson systems is similar, except that each particle state can be multiply occupied: N now takes any value between 0 and ∞. The grand partition function is then

$$\Xi = \sum_{N=0}^{\infty} \exp\{N(\mu-\epsilon_j)/kT\} = \frac{1}{1-\exp\{(\mu-\epsilon_j)/kT\}} \tag{2.53}$$

giving a mean occupation number

$$n_j = \langle N \rangle = \sum_{N=1}^{\infty} N \exp\{N(\mu-\epsilon_j)/kT\}/\Xi \tag{2.54}$$

which reduces to

$$n_j = \frac{1}{\exp\{(\epsilon_j \quad \mu)/kT\}-1} \tag{2.55}$$

This is the Bose–Einstein distribution referred to in Section 4.4.1.

Use of Quantum Operators. The theory given above is expressed in terms of the energies E_i of the solutions of the Schrödinger equation for the system; but these are not always known. Complex systems are therefore often treated in an equivalent but more general formulation which expresses the sums over i in Eqs. (2.34)–(2.37) as traces of quantum mechanical operators (e.g., [Bar74b]); such traces are invariant whichever complete orthonormal set of wave functions are used. For example, the partition function Z defined in Eq. (2.34) can also be written as

$$Z = \mathcal{T}r\{\exp(-\hat{H}/kT)\} = \sum_j < \phi_j|\exp(-\hat{H}/kT)|\phi_j > \tag{2.56}$$

where $\hat{H}$ is the Hamiltonian energy operator and the quantum states $|\phi_j>$ do not have to be the energy states of the system but can be taken to be any complete set of orthonormal states that is convenient to use; e.g., for an anharmonic crystal the $|\phi_j>$ can be taken to be the harmonic vibrational states, whose properties are well known.

2.3.2. Classical Statistical Mechanics

Classical statistical mechanics is valid in the limit of high temperatures. The mechanical state of a classical system is specified by the positions x, y, z and momenta p_x, p_y, p_z of all the N particles, equivalent to a point in $6N$-dimensional *phase space* [Tol38, Cal60]. The statistical probability that the system is in a state within some given region of phase space is determined by a *probability density* $P(\mathbf{r}_1, \cdots, \mathbf{r}_N; \mathbf{p}_1, \cdots, \mathbf{p}_N)$, which in the *canonical ensemble* is proportional to the Boltzmann factor $\exp(-E/kT)$:

$$P(\mathbf{r}_1, \cdots, \mathbf{p}_N) = \frac{\exp\{-E(\mathbf{r}_1, \cdots, \mathbf{p}_N)/kT\}}{\int_{-\infty}^{\infty} d\mathbf{r}_1 \cdots \int_{-\infty}^{\infty} d\mathbf{p}_N \exp\{-E(\mathbf{r}_1, \cdots, \mathbf{p}_N)/kT\}} \tag{2.57}$$

where the integral is over the whole of phase space. The expressions in Eqs. (2.37)–(2.41) remain valid, the averages being now quotients of integrals. A purely classical theory cannot give absolute values for the entropy and free energy, but the correspondence principle of quantum mechanics implies that at sufficiently high temperatures it agrees with quantum theory if the density of quantum states in phase space for a system of N identical particles is taken to be $h^{-3N}/N!$. The partition function is then

$$Z = \frac{h^{-3N}}{N!} \int_{-\infty}^{\infty} d\mathbf{r}_1 \cdots \int_{-\infty}^{\infty} d\mathbf{p}_N \exp\{-E(\mathbf{r}_1, \cdots, \mathbf{p}_N)/kT\} \tag{2.58}$$

Since

$$E = \sum_{i=1}^{N} \frac{|\mathbf{p}_i|^2}{2m} + \Phi(\mathbf{r}_1, \cdots, \mathbf{r}_N) \tag{2.59}$$

where m is the mass of a particle and Φ is the potential energy, the integration over the momenta can be done analytically to give

$$Z = \left(\frac{2\pi mkT}{h^2}\right)^{\frac{3N}{2}} \frac{1}{N!} \int_{-\infty}^{\infty} d\mathbf{r}_1 \cdots \int_{-\infty}^{\infty} d\mathbf{r}_N \exp\{-\Phi(\mathbf{r}_1, \cdots, \mathbf{r}_N)/kT\} \tag{2.60}$$

The entropy is given by

$$S = \langle E \rangle / T + k \ln Z \tag{2.61}$$

2.3.3. Computational Methods

Quantum Calculations. Both quantum and classical statistical mechanics are widely used in the calculation of thermodynamic properties from models of physical systems. The most direct applications of quantum statistical mechanics are to models for which the quantum states can be classified and their energies calculated for use in the equations of Section 2.3.1, with or without algebraic summation of analytic expressions. Examples include Schottky systems (Section 2.5.3), vibrating crystals in the harmonic approximation (Section 2.6.2) and dilute molecular gases (Section 4.2). The computational task is thus two-fold: to solve the Schrödinger equation, at least approximately; and to perform the required summations over all states. For all except these simple separable systems this can be a formidable task.

Classical Calculations. The validity of classical mechanics at sufficiently high temperatures is widely exploited. The classical integrals are sometimes easier to evaluate, either analytically or numerically, than the corresponding quantum sums; for example, they enable the virial expansions for the behavior of imperfect gases to be expressed in terms of integrals over clusters of interacting molecules (Section 4.2.4). But except in special cases like these, where the integrals over phase space can be reduced to the calculation for small clusters, numerical computation can be carried out only for comparatively small systems. For these surface effects would be important, but in the study of bulk matter they are avoided mathematically by employing a cyclic boundary condition, as first used by Born in the theory of crystal vibrations. The finite system under study is repeated periodically over all space, with the effect that, for example, material near the left boundary of the system is in direct interaction with material near the right boundary. As the size of the system increases, its properties approach those of the bulk material.

To be practicable, the integration over phase space requires some method of avoiding the vast regions which have very low probability. Two techniques, *Monte Carlo* (MC) and *Molecular Dynamics* (MD), are widely used for this purpose.

MC methods use techniques for random successive sampling of phase space which are systematically biassed against regions of low probability density. The earliest and best known of these is that of Metropolis *et al.* [Met53] for canonical ensemble averaging; but many others have been developed for a variety of ensembles, and applied to many different types of model systems, including fluids, interfaces and strongly anharmonic solids. Details and discussions of the accuracy and reliability of the methods are available in several texts (e.g., [All87, Fre96]).

In an MD simulation, initial positions and velocities of all the particles are chosen compatible with the desired macroscopic conditions, and the forces on each atom calculated from the model potential function. Newtonian mechanics is then used to deduce the development of the system over a short time step, after which the forces are recalculated and the process repeated. By taking a large number of such steps we may follow the development of the system over time. Thermodynamic properties are then estimated by averaging over time, on the *ergodic hypothesis* that all significant regions of phase space are covered statistically during the progress of the calculation. Mechanical properties such as energy and momentum are given by direct averages; statistical properties such as entropy and free energy are deduced from fluctuations of the system and obtained less accurately. The method is also used to obtain time correlation functions required for the calculation of spectroscopic properties, and to follow the kinetics of non-equilibrium processes. Details can again be found in the texts referenced above.

Advantages of classical calculations are: (i) the classical states of any model system are known, enabling the methods to be applied immediately to a wide variety of systems, including highly disordered materials (fluids and solids) and strongly anharmonic solids; (ii) the results obtained for the disposition and mutual behavior of the atoms and molecules can be presented graphically in ways that are easy to interpret. Disadvantages are: (i) the theory is invalid at temperatures below which quantum effects are dominant (although small quantum effects may be treated as a perturbation); (ii) the number of independent particles considered is much smaller (typically $10^2 - 10^5$) than that in bulk materials, often causing results to depend on sample size; (iii) important regions of phase space may be inadvertently neglected; (iv) getting even modest precision may be expensive in computer resources.

2.4. BONDING AND INTERATOMIC POTENTIALS

2.4.1. The Separation of Vibrational and Electronic Effects

In principle, thermodynamic properties can be calculated exactly by applying statistical mechanics to the system of nuclei and electrons constituting each material, but in practice approximations must be made (except perhaps for dilute gases composed of small molecules). A good approximation may lead to a simplified model in terms of which the general behavior of the material can be understood.

The Born–Oppenheimer approximation, which leads to the separation of electronic and nuclear motions, is of this type. The nuclei are considered to move subject to an effective potential energy Φ which is a function of the positions of all the nuclei, obtained by solving the electronic Schrödinger equation for each set of nuclear positions; thus both the kinetic energy and the potential energy of the electrons contribute to the effective potential energy seen by the nuclei. In principle, there is a different Born–Oppenheimer potential energy function for each electronic energy state. This is important in molecular spectroscopy, as can be seen in the vibrational fine structure of electronic molecular spectra. For the statistical mechanics of materials, however, we usually need consider only the electronic ground state. This is obviously a good approximation for most insulators, where higher electronic states are not excited at temperatures of interest. Even for metals the excitation of electronic levels is usually found to make little difference to the effective potential in which the nuclei move, allowing us to treat separately the electronic and vibrational contributions to the energy and hence to the thermodynamic properties (see Sections 1.3.4 and 6.1.1). When this is not so, phenomena are usually discussed in terms of interaction between separately conceived electronic and vibrational systems (see Ch. 6). The small energies associated with nuclear magnetic dipoles and electric quadrupoles are similarly treated as separate systems.

Even with the Born–Oppenheimer separation, the electronic structure of materials presents formidable theoretical problems which are currently the subject of widespread research. A modern introduction to this vast subject is provided by two recent undergraduate texts, *Electronic Structure of Materials* by Adrian Sutton [Sut93] and *Bonding and Structure of Molecules and Solids* by David Pettifor [Pet95], and by some of the general reviews in the centenary volume *Electron* [Spr97a].

2.4.2. *Ab Initio* Calculations

Within the Born–Oppenheimer approximation, potential energy functions can in principle be calculated *ab initio*, i.e., quantum mechanically without resort to fitting of parameters to empirical data. In recent years the increasing computational power available has made this practicable, at least for simple systems. For example, the frequencies of vibrations with wave numbers of high symmetry have been calculated for a number of crystals, including tetrahedral semi-conductors [Yin82, Kin90], using the *frozen phonon* method in which the change of electronic energy associated with a vibrational displacement is calculated quantum mechanically. Again, *ab initio* methods have been used to calculate directly the forces on the atoms at each step of a molecular dynamics computation [Car85].

A short account of some of the approximations used and results obtained is given in [Bar98, Section 1.7.7.5].

2.4.3. Models of Bonding

Even for quite simple systems *ab initio* calculations can be expensive in computer resources, and for complex systems they become unrealistic. It is therefore often necessary or desirable to work with *ad hoc* potentials for different types of material, with adjustable parameters that in early work were always adjusted to fit experimental properties but are now frequently fitted to *ab initio* results for selected atomic displacements (e.g., [Fra98]). Such a model can often encapsulate the most essential features of a material, and so give immediate insight into the processes underlying their thermodynamic properties. On the other hand, caution is needed; it should not be assumed that a model which gives a good account of known experimental properties will necessarily predict unknown properties correctly.

The simplest models are those in which the atoms interact only in pairs through short-range potentials $\phi(r)$; these potentials give rise to *central forces*, i.e., forces which act along the lines between the atomic nuclei. This is a good model for rare gas solids and fluids, especially when it is modified to take account also of much weaker many-body interactions [Kle76]. It has also been fairly successful in accounting for the phonon dispersion curves of simple metals, both fcc and bcc, indicating the importance of central force interactions between neighboring atoms in these materials; although there are usually serious departures from experiment at small wave vectors for some of the acoustic branches. Since acoustic frequencies in the long-wave limit are determined by the elastic stiffnesses, this discrepancy indicates a failure of the model to account fully for the elastic properties. In metals positive ions are bonded together by the sea of non-localized electrons extending

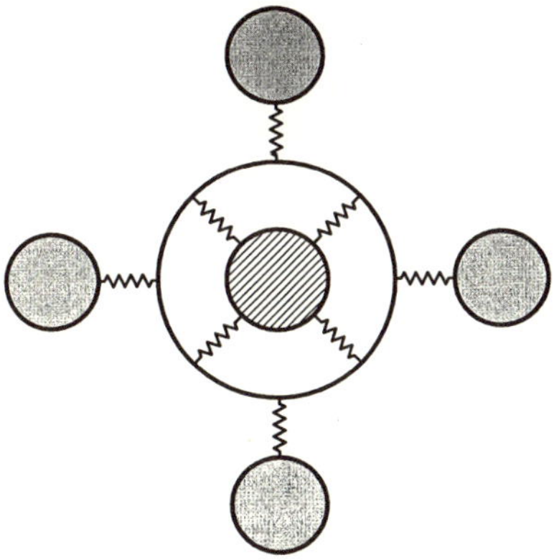

Fig. 2.3. Schematic representation of a simple shell model in which only the anions have shells. From [Coc73, Fig. 7.8].

throughout the crystal, as described in Section 6.1, and this cannot be represented solely by effective pair potentials.

Molecular crystals to some extent resemble rare gas crystals; the intermolecular interaction is much weaker than the bonding within the molecules, which can therefore to a first approximation be treated as rigid. However, molecules are not spherical but "knobbly" in their short range interactions, and depending on their symmetry may possess permanent electric dipoles or quadrupoles giving rise to longer range interactions. To a greater or less extent molecules are also "wobbly," so that there can be appreciable interaction between the crystal vibrations and the internal molecular vibrations.

Models for ionic solids include both long-range Coulombic forces and short-range forces. *Rigid ion* models give a surprisingly good account of many crystal properties, and are still widely used because of their simplicity and ready applicability to disordered and other complex systems; but they obviously cannot take account of the polarisability of the ions and the interaction of this with the vibrations. Various models which allow the ions to deform have therefore been designed [Har79, Bil79, Mad96]. Of these the shell model is the best known, which gives a simple mechanical representation of ion distortion: each polarizable ion is represented by a massive charged core surrounded by a massless charged spherical shell, which interacts through a short-range potential with neighboring shells (and in some models with neighboring cores also). Polarization arises from the displacement of the shell relative to the core, and can be affected both by the local electric field and by short-range forces exerted by neighboring ions (Fig. 2.3). The applicability of ionic models is extended further by using models which allow changes of size and shape of the ions.

In covalent crystals, among which are the diamond structure elements and many organic materials, the bonding is strongly directional, and all models employ non-central forces of some kind. These are often provided by adding many-body potentials (particularly three-body) to pair-potential models. For crystals (and molecules) of definite structure, short-range *valence force fields* are much used: the potential energy is expressed as a sum of second order terms in small changes δr and $\delta\theta$ in the lengths of valence bonds and in the angles between them. Where there is similar bonding in different substances, as for example in many organic materials, the same force field may be applicable. The number of possible parameters is increased greatly when anharmonic effects are calculated, since this requires the inclusion of third order terms; but often only the third order terms in the δr are included. For glasses and other disordered structures other types of many-body potential are used, because the greatly varying local arrangement of the atoms invalidates the use of a single valence-force field.

In some materials there is strong covalent bonding in some directions and weaker, less directional, bonding in others — for example in polymers, where there is covalent bonding along the polymer chain only, or in crystals such as graphite, where layers of covalently bonded atoms interact with much weaker Van der Waals forces.

In other materials the bonding is intermediate between ionic and covalent. Thus the compounds XY of zinc-blende structure (similar to diamond but with each atom bonded to one of different type) provide examples of differing ionicity, which J·C Phillips [Phi73] has classified by a numerical *ionicity factor* f_i on a scale varying from 0 to 1 (e.g., Table 5.4). Silica provides another example: the tetrahedral surroundings of the silicon atoms and the two-fold coordination of the oxygen atoms point to the importance of covalency, although exclusively ionic models have been reasonably successful in predicting experimental properties. In materials with multiatomic ions both covalent and ionic bonding coexist: for example, ammonium salts are strongly ionic, but internally the NH_4^+ ion is covalently bonded (Section 8.2.4).

A fuller discussion of potential models is given elsewhere [Bar98]. For many materials containing atoms in the lower part of the periodic table the bonding does not fall completely into any of the simple categories listed above, and it becomes hard to design suitable *ad hoc* models for them.

2.5. SOME MODEL SYSTEMS

2.5.1. Ideal Gases

Ideal gas behavior occurs in the limit of infinite dilution, when interaction between gas molecules may be neglected. It is one of the simplest systems to treat by statistical mechanics, because at virtually all temperatures of interest the translational positions and momenta of the molecules may be treated in the classical high temperature limit, and are purely random; the other degrees of freedom can be treated separately for each molecule. Because the motion of the molecular centers of mass is independent of all the other molecular degrees of freedom, the free energy separates into two components, one translational and the other non-translational. Both components depend on the temperature, and so contribute to the entropy and heat capacity; but only the translational component depends on the volume and contributes to the pressure $P = -(\partial F/\partial V)_T$. It is for this reason that the equation of state of all classical ideal gases is the same as that of a monatomic gas (Section 1.3.5): $PV = NkT$, where N is the total number of molecules in the gas.

The total heat capacity of an ideal molecular gas is obtained by adding the contributions of the non-translational molecular degrees of freedom to the monatomic values of $C_V = \frac{3}{2}Nk$, $C_P = \frac{5}{2}Nk$, giving a much richer behavior, sometimes with subtle effects from quantum statistics (Section 4.2.3).

2.5.2. Ideal Crystals

The ordered periodic structure of an ideal crystal is at the other extreme from the random structure of an ideal gas, but again makes for simplicity in the theoretical treatments described in later sections. The periodicity aids the analysis both of vibrational behavior (Section 2.6), and of electronic structure and excitations. If there are localized non-interacting systems they are repeated identically throughout the crystal, giving rise to simple Schottky behavior (Section 2.5.3). If however such systems interact with each other, complex patterns of behavior result which are more difficult to treat theoretically (Section 2.5.4).

2.5.3. Schottky Systems

A Schottky system is localized, interacting only weakly with other degrees of freedom; and each system has only a small number of accessible energy states. The general results of Eqs. (2.33)–(2.41) can be applied immediately. For a system with n energy levels ϵ_l with degeneracy g_l, Eq. (2.33) gives for the probability that level l is occupied

$$f_l = \frac{g_l \exp(-\epsilon_l/kT)}{\sum_{j=1}^{n} g_j \exp(-\epsilon_j/kT)} \tag{2.62}$$

A system with only two, non-degenerate levels, separated by an interval $\Delta\epsilon = \epsilon_2 - \epsilon_1$, has already been discussed in Section 2.3.1. Contributions of the system to the entropy and heat capacity are shown in Fig. 2.2. In general, the detailed behavior of a Schottky system depends upon the number of energy levels, their degeneracies and the spacing between them. The Schottky systems in α-$NiSO_4.6H_2O$ (see Section 1.1) have three levels, all non-degenerate, giving a high temperature entropy of $R\ln 3$; and an analysis of the heat capacity (Fig. 1.5) has shown that the two higher levels are relatively close together at 4.48 ± 0.07 and 5.05 ± 0.07 cm^{-1} above the lowest level, giving a higher peak in C_{sch} than that shown in Fig. 2.2(c) for the system with only two levels of equal degeneracy. All Schottky systems have a heat capacity that rises exponentially at sufficiently low temperatures and falls off as T^{-2} at sufficiently high

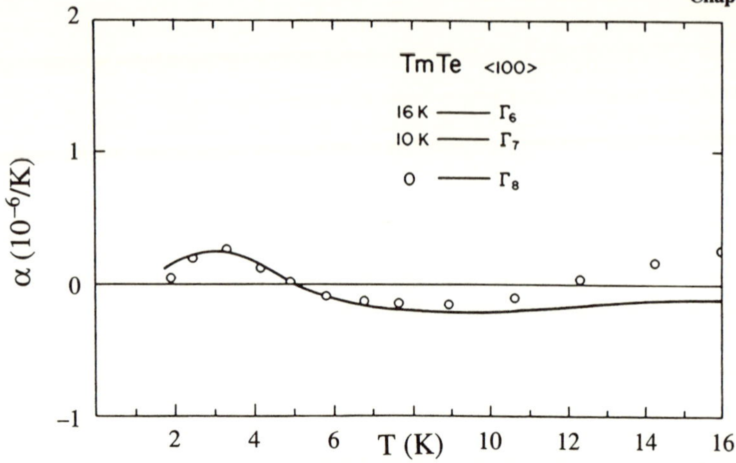

Fig. 2.4. Linear thermal expansion $\alpha/[10^{-6}\mathrm{K}^{-1}]$ of TmTe. Circles are experimental values; the full line is the calculated Schottky contribution with $\gamma_{\{\Gamma_7\}} = 1.3$, $\gamma_{\{\Gamma_6\}} = -1.5$. The Γ_n label the symmetry species of the levels [Ott77].

temperatures. Often the Schottky peak occurs at such low temperatures that only the high temperature tail is detected experimentally (e.g., Fig. 1.8).

Schottky contributions to the thermal expansion are determined by the dependence of the energy intervals on volume. For a two-level system, which has only the one interval $\Delta\epsilon$, there is a single Grüneisen parameter $\gamma_{sch} = -d\ln\Delta\epsilon/d\ln V$; the Schottky anomaly in the thermal expansion is then similar in shape to that in the heat capacity, its sign and magnitude depending on γ_{sch}. This remains true for multilevel systems if the different energy intervals change with volume by the same factor and so have a common Grüneisen parameter, as in the three-level system of TmSb, for which the Grüneisen parameter and consequent Schottky expansion are negative (Fig. 5.39). More complex behavior is seen for those multilevel systems which have energy intervals with different Grüneisen parameters; Eq. (2.41) then gives the thermodynamic Grüneisen function as

$$\gamma_{sch}(T,V) = \frac{\langle\gamma E^2\rangle - \langle\gamma E\rangle\langle E\rangle}{\langle E^2\rangle - \langle E\rangle\langle E\rangle} \tag{2.63}$$

where the denominator is proportional to the Schottky heat capacity. The thermal expansion of TmTe shows such behavior (Fig. 2.4). Between 2 and 10 K it is dominated by the contribution from a 3-level Schottky system with a positive γ for the lower excited level and a negative γ for the higher excited level [Ott77].

Schottky systems in anisotropic crystals behave similarly to those in isotropic systems, except that the energy levels are now functions of all the independent strain coordinates. For example, in an axial crystal with two independent dimensions a and

c, a two-level Schottky system has two Grüneisen parameters, defined by

$$\gamma_{\perp} = -\frac{1}{2}\left(\frac{\partial \Delta\epsilon}{\partial \ln a}\right)_c, \qquad \gamma_{\parallel} = -\left(\frac{\partial \Delta\epsilon}{\partial \ln c}\right)_a \tag{2.64}$$

2.5.4. Cooperative Order–Disorder Systems

The Ising Model. At the opposite extreme to having localized Schottky systems contributing independently to the free energy, we now turn to materials in which it is the interactions between systems that determine the behavior. To illustrate this, we take the spin $\frac{1}{2}$ Ising model of ferromagnetism, extensively studied over decades because of its simplicity in conception and difficulty in solution [Dom96]; although (unlike the Heisenberg model of Section 6.4.1) it probably does not represent very closely the magnetism of any real solid [deJ74]. It consists of a periodic array of spins, each of which can be in one of two states, up or down. It is postulated that there is an energy of interaction between each pair of nearest neighbors (representing the quantum mechanical exchange effect) which depends upon whether their spins are like, $\uparrow\uparrow$, or unlike, $\uparrow\downarrow$; but any direct interaction between the magnetic dipoles associated with the spins is neglected. Ferromagnetism is then favored if $\Delta\epsilon \equiv \epsilon_{\uparrow\downarrow} - \epsilon_{\uparrow\uparrow} > 0$, antiferromagnetism if $\Delta\epsilon < 0$. Because $\Delta\epsilon$ is the only energy parameter, S and C_V are functions of $kT/\Delta\epsilon$. Changing $\Delta\epsilon$ translates horizontally plots of S and C_V against $\ln T$, but otherwise does not alter them. For the same reason there is only one Grüneisen parameter, $-(d\ln\Delta\epsilon/d\ln V)$, and $\beta(T)$ will always be proportional to $C_V(T)$.

In the ferromagnet, when $kT \ll \Delta\epsilon$ all the spins are aligned in one direction: we say that there is complete *long range order*. Conversely, when $kT \gg \Delta\epsilon$ the directions of the spins are random, and disorder is complete. As T increases from low temperatures, some spins will reverse, although at first most spins will be in the original direction: there is then partial long-range order. The existence of reversed spins then makes it less energetically unfavorable to reverse other spins; so as the temperature is increased further, the loss of long-range order becomes more rapid. Finally a critical temperature T_c is reached beyond which there remains no long range order: knowledge of the spin directions in one part of the crystal no longer enables us to predict anything about their behavior in a distant part of the crystal. On the other hand, there is still some *short range order* at temperatures above T_c, because on average each spin still has more like neighbors than unlike. Figure 2.5 illustrates the effect this behavior has on the heat capacity of two models, one two-dimensional and one three-dimensional.

Consider the heat capacity of the three-dimensional model, with spins on a face-centered cubic lattice. The long range order parameter f is defined for a mole of N_A spins by

$$f = (|N_{\uparrow} - N_{\downarrow}|)/N_A \tag{2.65}$$

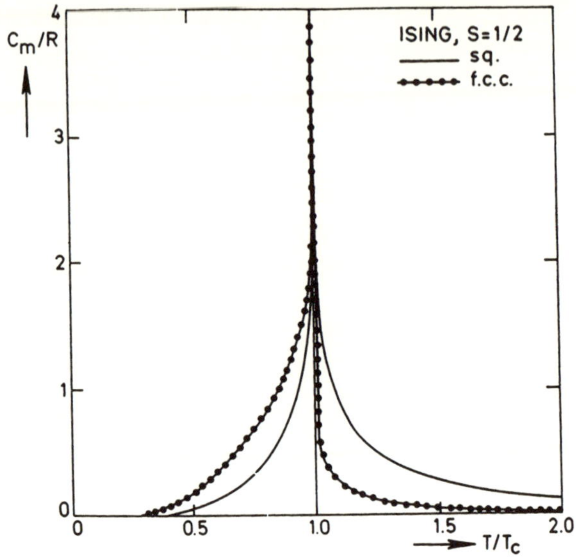

Fig. 2.5. Magnetic heat capacity of Ising spins: •-•-•, on a fcc lattice; —, on a two dimensional square lattice. From [deJ74, after C. Domb].

Loss of order implies an increase in entropy, which in the completely random high temperature limit has the same value as that of the two-state Schottky systems:

$$\Delta S = k \ln(2^{N_A}) = R \ln 2 \tag{2.66}$$

However, the way in which the system passes from complete order to complete disorder is quite different from the smooth passage of the Schottky systems. The heat capacity, given by $C_V = T(\partial S/\partial T)_V$, has a sharp peak at the transition temperature T_c; this is often called a *lambda* peak, because for some systems it resembles the Greek letter λ. This peak is associated with the catastrophic loss of long range order, and is characteristic of cooperative phase transitions; above $T_c, f = 0$, but there is still a contribution to the heat capacity due to progressive loss of short range order.

Mean Field Theory. Precise calculations on order–disorder models are not simple. For example, the Ising model requires simultaneous consideration of the disposition of spins throughout the macroscopic crystal; this cannot be done analytically (except in two dimensions), and results to the accuracy of those shown in Fig. 2.5 are obtained only after considerable computation. Approximations have therefore been devised, and of these the simplest is the *mean field approximation*, in which the potential field seen by each unit is replaced by an average taken over the whole system. In the Ising model, for example, the distribution of neighbors about an up spin (or a down spin) varies; but in the mean field approximation we assume that the field seen by each spin is that corresponding to the average excess of up neighbors over down, zf, where z is the total number of neighbors of each spin. The change of energy in a flip from down to up is then $zf\Delta\epsilon$, and for self consistency the resulting Boltzmann factor must give the ratio of up to down spins:

$$\exp(-zf\Delta\epsilon/kT) = (1+f)/(1-f) \tag{2.67}$$

This equation is easily solved numerically for different values of $\Delta\epsilon/T$, giving a result differing considerably from that shown in Fig. 2.5. The critical temperature T_c marking the disappearance of long range order is too high by nearly 20%, and the peak in C_V is finite in height and markedly different in shape. In particular, the mean field approximation neglects any additional local correlation between spins; and so there is no short range order and consequently no tail in C_V above T_c.

Real Materials. Order–disorder transitions occur in many materials, both at cryogenic and at higher temperatures. The detailed behavior varies widely, and effects are seen not only in heat capacity but also in thermal expansion and other properties. In alloys such as β-brass (CuZn), where the order is in the arrangement of the different types of atom on the crystal lattice, the order–disorder contribution to the heat capacity is very similar to that of the Ising model, with a large peak making an additional contribution to the high temperature entropy of $R\ln 2$. On the other hand, in real ferromagnetic and antiferromagnetic materials there are significant magnetic contributions to C_V and β at lower temperatures, both when the relevant spins are localized and when they are itinerant (Section 6.4), and the final peak is considerably smaller than an Ising peak (e.g., Cr in Fig. 6.15). In molecular crystals and liquid crystals the order is in the orientation of the molecules, and successive transition temperatures can occur as order is lost for different orientational degrees of freedom (e.g., HBr in Fig. 8.3). In 'simple' type I superconductors, where the order is only in the momenta of the relevant particles or particle pairs, the peak has no tail above T_c, since the concept of short-range order is not relevant here (Fig. 6.19).

Both real materials and theoretical models have been studied intensely over many years [deJ74, Dom96], particularly the variation of their properties immediately below and above T_c, mainly with the aim of elucidating the nature of the transitions for different systems and the behavior of different properties in the neighborhood of T_c ('critical exponents'). Further references are given in Section 5.11.1.

2.5.5. Glasses

Unlike a crystal, a glass is frozen in a random structure, and so there is no unique model on an atomic scale; studies must be done on individual random assemblies of atoms. Early work involved the laborious construction of random configurations consistent with an appropriate type of bonding, but this can now be done in a way analogous to the experimental formation of glasses, by using configurations obtained from computer simulations of the liquid material. Such models are of limited size, typically containing 10^3 to 10^5 atoms, obeying a periodic boundary condition. With suitable intermolecular potentials they can give quite a good account of thermodynamic properties at high and intermediate temperatures. But the models cannot easily be used to interpret the striking experimental results obtained at very low temperatures (Section 5.7), because the periodic boundary condition applied to samples of small size prevents the study of the effect of random structure both on very low frequency vibrations and on the distribution of energy intervals in tunnelling centers.

Although most glasses are insulators, electronic properties may be studied in metallic and semi-conducting glasses [Cus87].

2.6. LATTICE VIBRATIONS

2.6.1. General

In all solid materials vibrations contribute to the heat capacity and thermal expansion. The theory of vibrations in solids, called *lattice dynamics*, has been developed in great detail, from the work of Born and his colleagues onwards; it is described in many texts [Bor54, Bla55, deL56, Mar71, Hor74, Ven75]. In this section we fill out and make more precise some of the concepts already introduced in Chapter 1, in preparation for later discussion of specific materials.

2.6.2. Harmonic Theory

The Harmonic Approximation. The starting point of lattice dynamics is the existence of a potential energy, as given by the Born–Oppenheimer approximation (Section 2.4.1). This potential can be expressed as a Taylor series in the displacements of the atoms from their mean positions:

$$\Phi = \Phi_L + \Phi_1 + \Phi_2 + \Phi_3 + \Phi_4 + \cdots \qquad (2.68)$$

Here Φ_L is the potential energy of the static lattice with the atoms in their mean positions, and the Φ_n are the sums of all the terms of the nth order in the displacements. The harmonic approximation is to neglect all terms beyond the second order in the displacements. The motion can then be resolved into the superposition of a set of independent *normal modes* j, with angular frequencies $\omega_j = 2\pi\nu_j$ [Gol50]. In a bulk solid these form the continuous frequency distribution $g(\omega)$ defined in Section 1.3.3.

At this point, some clarification is needed. (i) For a purely harmonic solid the Taylor series would terminate at Φ_2. But real solids are never purely harmonic; indeed, apart from one-dimensional models it is mathematically virtually impossible for a purely harmonic lattice to exist [Bar57a]. (ii) It is easy to show (e.g., [Bar98]) that if a purely harmonic solid could exist, its frequencies would have no dependence on volume or strain. There would then be no thermal expansion, and no temperature dependence of the elasticity; both these are essentially anharmonic effects. (iii) There are many independent anharmonic terms in the expansion of the potential energy function, and the anharmonicity cannot be specified by a single parameter. (iv)Although no real solid is ever strictly harmonic, the harmonic approximation is a good approximation for the vibrational heat capacities of most solids, especially at low temperatures.

The Vibrational Spectrum. The structure of a crystal is described by its periodic *lattice*, consisting of points given by the vectors

$$\mathbf{x}(l) = l_1\mathbf{a}_1 + l_2\mathbf{a}_2 + l_3\mathbf{a}_3 \tag{2.69}$$

where l_1, l_2, l_3 are integers, together with a set of vectors $\mathbf{x}(\kappa)$ determining the positions $\mathbf{x}(l) + \mathbf{x}(\kappa)$ of the n atoms (the *basis*) in each cell of the lattice. The translational symmetry enables the normal vibrations to be classified by wave vectors $\mathbf{q}$, such that the phase difference between any two cells separated by a vector $\mathbf{x}(l)$ is $\mathbf{q}\cdot\mathbf{x}(l)$ (Fig. 2.6a,b); in modes for which $\mathbf{q} = 0$ the atoms in each cell are in phase with the corresponding atoms in all other cells (Fig. 2.6c). This phase factor is periodic in $\mathbf{q}$-space (*reciprocal space*), so that the same mode can be ascribed to different values of $\mathbf{q}$ (Fig. 2.7). Wave vectors need therefore be taken only over a finite region of reciprocal space. This is usually chosen to be the *First Brillouin Zone* (FBZ), which comprises those independent $\mathbf{q}$ that are nearest to the origin of reciprocal space (e.g., [Kit76, Ch. 2]). For a crystal of volume V, the allowed $\mathbf{q}$ are uniformly distributed throughout the FBZ with a density $V/(2\pi)^3$. In diagrams the point at the zone center ($\mathbf{q} = 0$) is usually labelled Γ, and referred to in speech as the "Gamma point." First Brillouin Zones for many lattices are given in the appendices of [Hel81].

The periodicity in $\mathbf{q}$-space is described by the *reciprocal lattice*, which is the set of all points giving the same phase factors as the Gamma point. These are given by $n_1\mathbf{b}_1 + n_2\mathbf{b}_2 + n_3\mathbf{b}_3$, where n_1, n_2, n_3 are integers and the *reciprocal lattice vectors* $\mathbf{b}_j$ are related to the $\mathbf{a}_i$ by

$$\mathbf{a}_i \cdot \mathbf{b}_j = 2\pi\delta_{ij} \tag{2.70}$$

The position of a $\mathbf{q}$-vector referred to the reciprocal lattice is given by reduced dimensionless coordinates $\zeta \equiv (\zeta_1, \zeta_2, \zeta_3)$, so that

$$\mathbf{q} = \zeta_1\mathbf{b}_1 + \zeta_2\mathbf{b}_2 + \zeta_3\mathbf{b}_3 \tag{2.71}$$

and the cell phase factors are $2\pi\sum_i \zeta_i l_i$. Strictly it is ζ rather than $\mathbf{q}$ that should be used to label a given normal mode of vibration, because (except at the Γ-point) $\mathbf{q}$ changes when the crystal is strained so as to keep the phase factor for each cell unaltered. This must be remembered when Grüneisen parameters are derived for individual normal modes. Details may be found in the references in Section 2.6.1, and in [Bar98].

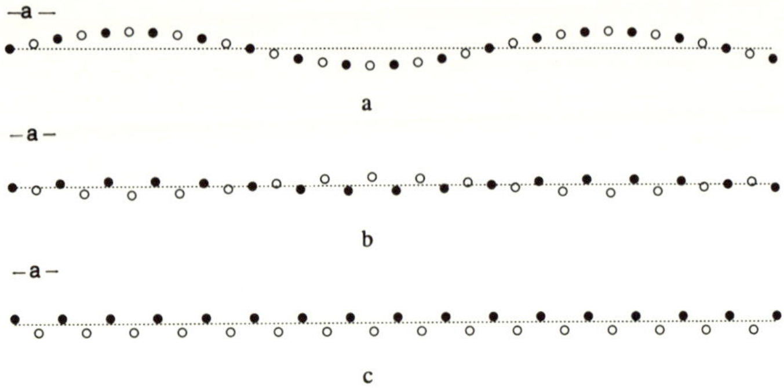

Fig. 2.6. Atomic displacements for transverse vibrations in a diatomic linear chain (schematic): •, heavier atom; ∘, lighter atom. (a) $q = 0.1(2\pi/a)$, acoustic branch; (b) $q = 0.1(2\pi/a)$, optical branch; (c) $q = 0$, optical branch.

In a three dimensional crystal, for each $\mathbf{q}$ there are $3n$ modes, labelled $\mathbf{q}s$, with frequencies $\omega_{\mathbf{q}s}$. As $\mathbf{q}$ varies, these form $3n$ *branches* of the vibrational spectrum $(s = 1, 3n)$. Of these, three are called *acoustic branches*, because as $\mathbf{q} \to 0$ the modes become macroscopic elastic sound waves; the rest are called *optical branches* because their limiting frequencies as $\mathbf{q} \to 0$ can often be measured by infra-red or Raman spectroscopy. At $\mathbf{q} = 0$ the acoustic modes become translations of the whole crystal, with zero frequency, but the optical modes remain vibrations of finite frequency (Fig. 2.6c). Frequencies of modes for which $\mathbf{q} \neq 0$ are usually measured by inelastic neutron scattering [Dol74]. Plots of $\omega_{\mathbf{q}s}$ against q for different directions of $\mathbf{q}$ are called dispersion curves. Figures 2.8 and 2.9 show dispersion curves in directions of high symmetry for argon and silicon. With only one atom per cell, argon has three acoustic branches; however, along the [001] and [111] directions the two transverse branches are degenerate. Silicon, with two atoms per cell, has in addition three optical branches; the marked flattening of the dispersion curves for the transverse acoustic modes is responsible for the low frequency peak in $g(\omega)$ centered on 4 THz and hence for the deep minimum in $\Theta^C(T)$ (Figs. 1.14 and 1.16). Dispersion curves of many other solids are given by Bilz and Kress [Bil79] for non-metals and in Landolt–Börnstein for metals [Sch81] and alloys [Kre83].

Surfaces and Imperfections. The periodicity of a crystal is broken at the surface, and so no real crystal has strict translational symmetry. In modelling crystals for

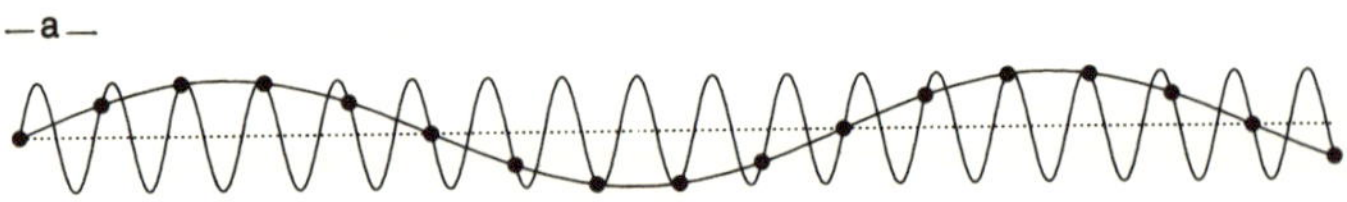

Fig. 2.7. Two different wave vectors, $q = 0.1(2\pi/a)$ and $q = 1.1(2\pi/a)$, describing the same vibration.

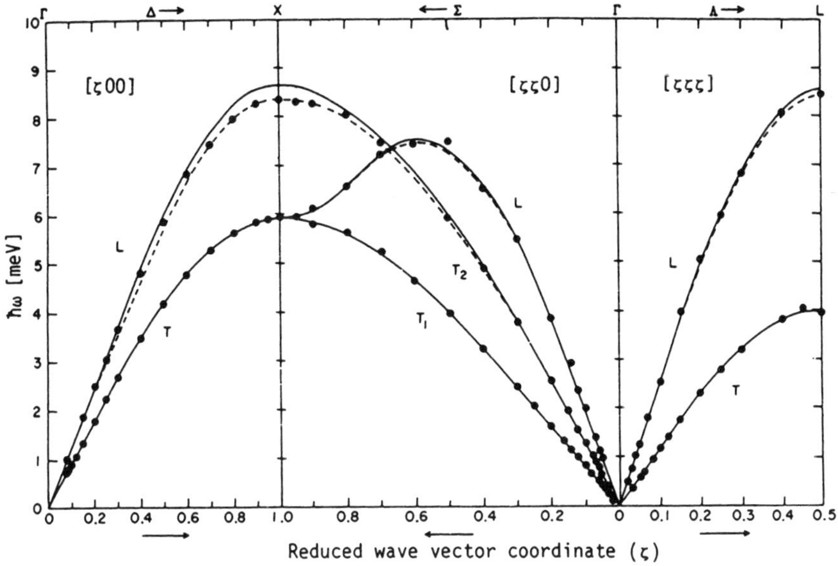

Fig. 2.8. Phonon dispersion curves for argon. Full and dashed curves are computed from two different models fitted to the experimental points. From [Bil79], who give original sources.

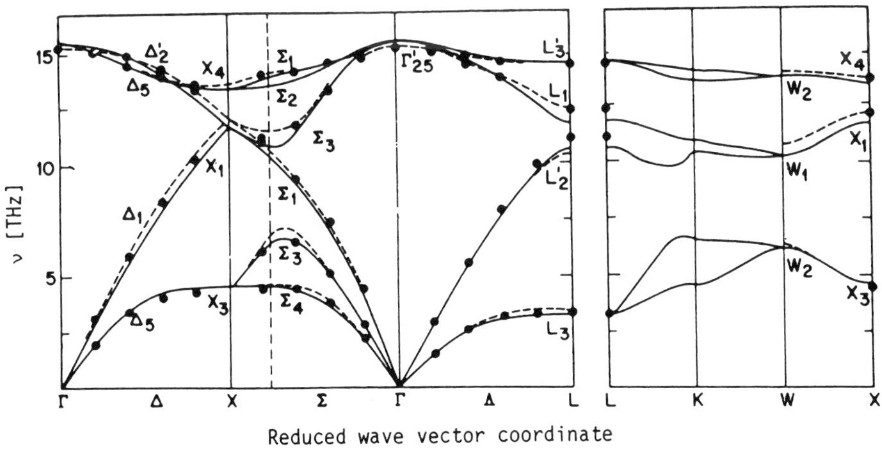

Fig. 2.9. Phonon dispersion curves for silicon. Δ, Σ, Λ indicate propagation in ⟨100⟩, ⟨110⟩, ⟨111⟩ directions respectively. Full and dashed curves are computed from two different models fitted to the experimental points. From [Bil79], who give original sources.

calculating bulk properties this difficulty is avoided by imposing a cyclic boundary condition, as described in Section 2.3.3. Surface vibrations therefore require separate theoretical treatment (e.g., [Mar71, Ch. IX] and [Wal75]). Their effect on heat capacity has been measured by comparing data for bulk crystals with data for fine powders of large surface area (e.g., [Dug54, Pat55]).

Periodicity is also destroyed by internal imperfections, whether due to the presence of impurity atoms (Section 5.12) or to structural defects such as vacancies, interstitials, dislocations etc. These again require special theory (e.g., [Mar71, Ch. VIII] and [Tay75]).

Moments $<\omega^n>$ of the Vibrational Spectrum; $\Theta_D(n)$. Some theoretical expressions for thermodynamic and other properties involve the moments of the vibrational spectrum, which are averages of powers of the frequencies defined by

$$<\omega^n>=\frac{1}{3N}\sum_{\omega_j\neq 0}\omega_j^n=\frac{\int \omega^n g(\omega)d\omega}{\int g(\omega)d\omega} \tag{2.72}$$

here brackets of the type $<\cdots>$ are used to distinguish averages over the frequency distribution from the thermal averages $\langle\cdots\rangle$. Provided that we exclude the zero frequency modes, which correspond not to vibrations but to displacements of the whole body, Eq. (2.72) can be used in computer calculations for values of n greater than -3; for $n\leq -3$, the integral in the numerator diverges and the moments $<\omega^n>$ become infinite. Expressions for the entropy and free energy involve also the geometric mean frequency ω_g, given by

$$\omega_g=\exp(<\ln\omega>) \tag{2.73}$$

To compare the moments with each other, we define equivalent Debye frequencies $\omega_D(n)$ and temperatures $\Theta_D(n)$ (written by some authors Θ_n), corresponding to the Debye distribution which gives the same nth moment as the spectrum under study. They are given by

$$\Theta_D(n)=\frac{\hbar\omega_D(n)}{k}=\frac{\hbar}{k}\left(\frac{n+3}{3}<\omega^n>\right)^{\frac{1}{n}} \tag{2.74}$$

As $n\to 0$ and as $n\to -3$, $\Theta_D(n)$ tends to values

$$\Theta_D(0)=e^{\frac{1}{3}}\left(\frac{\hbar\omega_g}{k}\right),\qquad \Theta_D(-3)=\Theta_0^C \tag{2.75}$$

the latter value being determined by the coefficient of ω^2 in the low frequency expansion for $g(\omega)$ in Eq. (2.86).* The $\omega_D(n)$ or $\Theta_D(n)$ are used also in the analysis of experimental data (Section 2.6.5).

*The symbol Θ_0 is used by D. C. Wallace (see, for example, [Wal72, Wal92]) to denote $\exp[-(1/3)]\Theta_D(0)=\hbar\omega_g/k$.

Both $\omega_D(n)$ and $\Theta^C(T)$ result from comparing properties of $g(\omega)$ with those of the Debye distribution. Thus for silicon the steep initial rise in $g(\omega)$ above its initial T^2 dependence (Fig. 1.14) is reflected by initial falls in both $\Theta^C(T)$ (Fig. 1.16) and $\omega_D(n)$ (Fig. 2.10), which then rise again as the high frequency vibrations take effect. For large n, $\omega_D(n)$ approaches the maximum frequency of the distribution, ω_{max}. Some values of $\Theta_D(n)$ give high or low temperature limiting values for $\Theta^C(T)$ and $\Theta^M(T)$, the latter being the equivalent Debye temperature for the Debye–Waller effect [Sal65]:

$$\Theta_D(2) = \Theta^C_\infty, \qquad \Theta_D(0) = \Theta^S_\infty, \qquad \Theta_D(-3) = \Theta^C_0 = \Theta^S_0,$$
$$\Theta_D(-2) = \Theta^M_\infty, \qquad \Theta_D(-1) = \Theta^M_0 \tag{2.76}$$

Thus values of $\omega_D(n)$ obtained by analyzing the heat capacity (Section 2.6.5) can be correlated with Debye Waller data [Sal65, Bar77b]. An accurate estimate of the zero point energy [see Eq. (2.83)] is given by

$$E_Z = \frac{9}{8}Nk\Theta_D(1) \tag{2.77}$$

Heat Capacity, Entropy and Helmholtz Energy. The $\omega_{\mathbf{q}s}$ comprise all the vibrational frequencies ω_j and so determine the frequency distribution $g(\omega)$ as defined in Section 1.3.3.* For example, Fig. 1.14 shows $g(\omega)$ for Ar and Si, obtained from force-constant models fitted to neutron scattering data. Given the frequency distribution, Eq. (1.13) can be used to obtain the heat capacity, and other thermodynamic functions can be obtained similarly. But in practice it is more direct to calculate thermodynamic functions, and also the moments $<\omega^n>$, by integrating over the FBZ. For example,

$$C_V = \sum_j c(\hbar\omega_j/kT) = [V/(2\pi)^3] \int_{FBZ} d\mathbf{q} \sum_s c(\hbar\omega_{\mathbf{q}s}/kT) \tag{2.78}$$

where the function $c(x)$ is given in Eq. (1.14). Similar expressions may be written for F_{vib}, S, and U_{vib}, where the vibrational contributions for each mode are respectively

$$f = kT\left[\frac{1}{2}x + \ln(1-e^{-x})\right] \tag{2.79}$$

$$s = k\left[\frac{x}{e^x-1} - \ln(1-e^{-x})\right] \tag{2.80}$$

$$u = kT\left[\frac{1}{2}x + \frac{x}{e^x-1}\right] \tag{2.81}$$

*Formally $g(\omega) = \sum_j \delta(\omega - \omega_j)$, where $\delta(x)$ is the Dirac delta function.

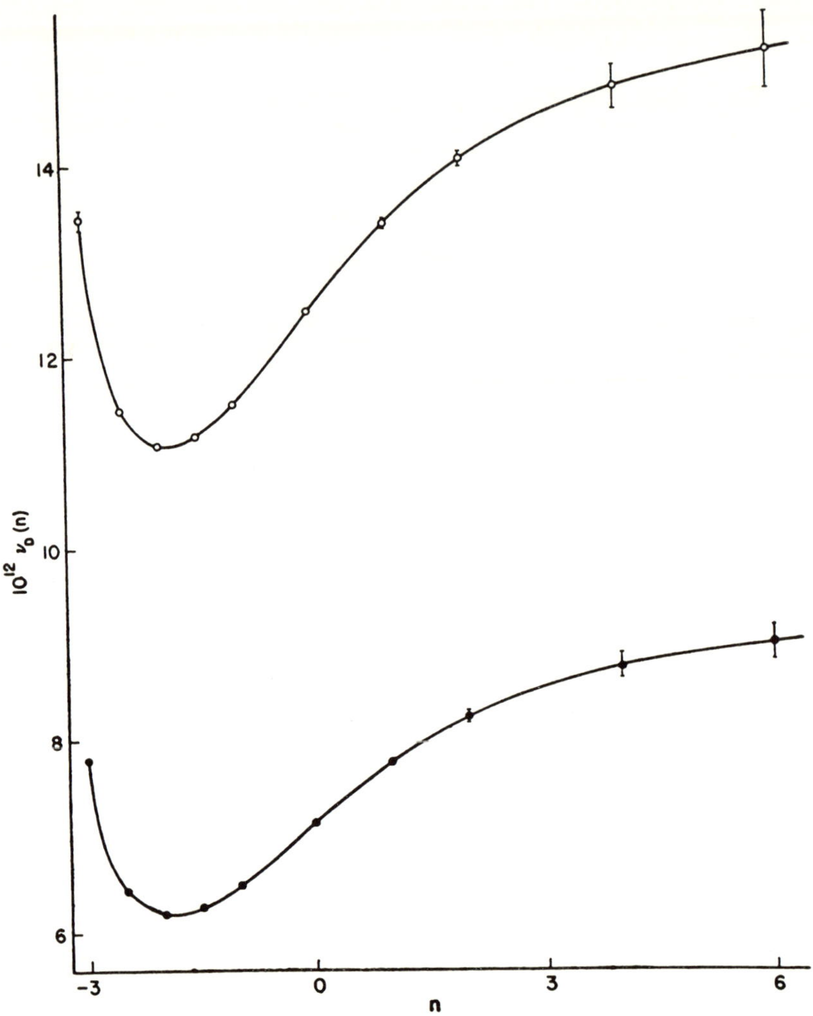

Fig. 2.10. $\nu_D(n) = \omega_D(n)/2\pi$ as a function of n for Si (upper curve) and Ge (lower curve), from analysis of thermodynamic data [Flu59].

Here $x = \hbar\omega/kT$, so that $kT\frac{1}{2}x$ is the zero point energy $\frac{1}{2}\hbar\omega$. The thermal components of f and u are

$$f_{th} = kT\ln(1-e^{-x}) \quad \text{and} \quad u_{th} = kT[x/(e^x-1)] \tag{2.82}$$

Behavior at Low Temperatures. As $T \to 0$, $x = \hbar\omega/kT \to \infty$; thus while $c(x)$ and $s(x) \to 0$, $u(x) \to \frac{1}{2}\hbar\omega$. The zero point energy of the solid is therefore

$$E_Z = \sum_j \frac{1}{2}\hbar\omega_j \tag{2.83}$$

For the Debye model $E_Z = \frac{9}{8}Nk\Theta_D$, where N is the total number of atoms in the solid; and this is usually a good approximation for simple solids if for Θ_D we take Θ_∞^C, the equivalent temperature for the heat capacity at the high temperature limit [Dom52]. Eq. (2.77) is more accurate, but requires more detailed analysis of the heat capacity.

For temperatures so low that only elastic waves contribute to S and C_V, i.e., acoustic modes with wave vectors $\mathbf{q}$ very near the zone center, Debye's theory (first applied only to an elastically isotropic solid) is generalized to take account of the dependence of sound velocities on direction of wave propagation. The T^3-dependence as $T \to 0$ is given by an integral over all directions [Bor54, Eq. 6.3]:

$$\frac{C_{vib}}{V} \simeq \frac{3S_{vib}}{V} \simeq k\left\{\frac{\pi}{30}\left(\frac{k}{\hbar}\right)^3 \int \sum_{s=1}^{3}[v_s(\hat{\mathbf{q}})]^{-3}d\Omega\right\}T^3 \tag{2.84}$$

where Ω is an element of solid angle, $\hat{\mathbf{q}}$ is a unit vector giving the direction of propagation, and the $v_s(\hat{\mathbf{q}})$ are velocities of sound for the three different wave polarizations given by continuum elasticity theory (Section 2.8.7). The corresponding Debye equivalent temperature when all degrees of atomic freedom are taken into account is

$$\Theta_0^{el} = \frac{h}{k}\left\{\frac{\mathrm{v}_a}{9}\int \sum_{s=1}^{3}[v_s(\hat{\mathbf{q}})]^{-3}d\Omega\right\}^{-\frac{1}{3}} \tag{2.85}$$

where v_a is the mean volume per atom and the superscript ^{el} denotes that the value is obtained from elastic data. For an elastically isotropic solid the integral reduces to $4\pi(v_L^{-3}+2v_T^{-3})$, where the subscripts denote longitudinal and transverse polarization.

For molecular crystals and other complex solids a smaller number of degrees of freedom may be used for the equivalent Debye spectrum (see Section 1.3); if so, the quotient $\mathrm{v}_a/9$ occurring in Eq. (2.85) must be increased in proportion. Implementation of Eqs. (2.84) and (2.85) is discussed in Section 2.9.

Various methods used for estimating from elastic data the behavior of S and C_V above the T^3 region are reviewed elsewhere [Bar80, Bar98]; they are all approximations and their degree of reliability depends upon the type of material. Precise theory

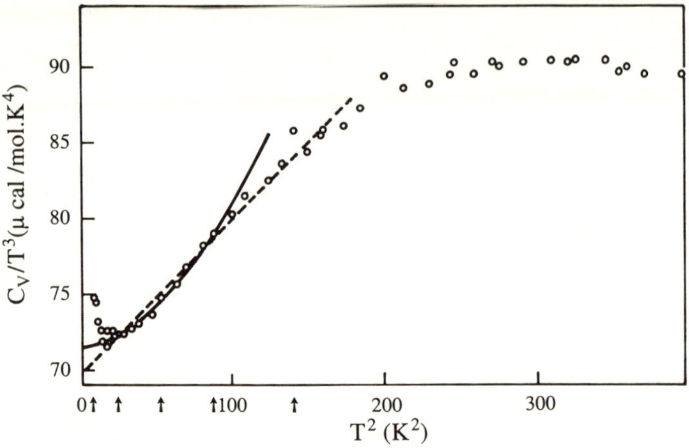

Fig. 2.11. C_V/T^3 plotted against T^2 for KCl. The dashed line indicates a spurious $AT + BT^3$ fit. Arrows denote temperatures $\Theta_0/100$, $2\Theta_0/100$, etc. [Bar57c].

requires lattice dynamics, which takes account of the discrete atomic structure of the solid. At low frequencies this gives series expansions of the form

$$g(\omega) = a\omega^2 + b\omega^4 + c\omega^6 + \cdots \tag{2.86}$$

leading to low temperature expansions for C_{vib} of the form

$$C_{vib} = AT^3 + BT^5 + CT^7 + \cdots \tag{2.87}$$

Plots of C_V/T^3 (or C/T for metals) against T^2 give smooth curves enabling the first two or three coefficients to be estimated from good experimental data (Fig. 2.11). In particular, the coefficients A give calorimetric Debye temperatures Θ_0^C which are usually found to be in good agreement with Θ_0^{el}.

High Temperature Behavior. For many solids C_V and Θ^C approach their high temperature limits below room temperature, *viz.* in the cryogenic region (e.g., Fig. 1.2). In the harmonic approximation the high temperature behavior is obtained by expanding the harmonic expressions for these quantities as power series in inverse powers of T. When T is large, $x = \hbar\omega/kT$ is small, and the functions c, s, u and f of Eqs. (1.14) and (2.79)–(2.81) can be expanded in powers of x; for example,

$$u = kT\left[1 + \frac{B_2}{2!}x^2 - \frac{B_4}{4!}x^4 + \frac{B_6}{6!}x^6 - \cdots\right] \tag{2.88}$$

where $B_2, B_4, B_6, \cdots$ are the Bernoulli numbers $\frac{1}{6}, \frac{1}{30}, \frac{1}{42}, \cdots$ (e.g., [Jef50]). Summing over the frequency distribution then gives the Thirring expansions for the bulk properties [Thi13, Bar57b], which can be used in the analysis of experimental data (see

Section 2.6.5):

$$U_{vib} = 3NkT\left[1 + \frac{B_2}{2!}\left(\frac{\hbar}{k}\right)^2 \frac{<\omega^2>}{T^2} - \frac{B_4}{4!}\left(\frac{\hbar}{k}\right)^4 \frac{<\omega^4>}{T^4} + \cdots\right] \tag{2.89}$$

$$S_{vib} = 3Nk\left[1 - \ln\frac{\hbar\omega_g}{kT} + \frac{B_2}{2(2!)}\left(\frac{\hbar}{k}\right)^2 \frac{<\omega^2>}{T^2} - \frac{3B_4}{4(4!)}\left(\frac{\hbar}{k}\right)^4 \frac{<\omega^4>}{T^4} + \cdots\right] \tag{2.90}$$

$$C_{vib} = 3Nk\left[1 - \frac{B_2}{2!}\left(\frac{\hbar}{k}\right)^2 \frac{<\omega^2>}{T^2} + \frac{3B_4}{4!}\left(\frac{\hbar}{k}\right)^4 \frac{<\omega^4>}{T^4} - \cdots\right] \tag{2.91}$$

An expansion for $(\Theta^C)^2$ can be derived from Eq. (2.91):

$$(\Theta^C)^2 = (\Theta^C_\infty)^2\left\{1 - A\left(\frac{\Theta^C_\infty}{T}\right)^2 + B\left(\frac{\Theta^C_\infty}{T}\right)^4 - \cdots\right\} \tag{2.92}$$

where the harmonic high temperature limiting value of Θ^C is

$$\Theta^C_\infty = \frac{\hbar}{k}\left(\frac{5<\omega^2>}{3}\right)^{\frac{1}{2}} \tag{2.93}$$

and

$$A = \frac{3}{100}\left(\frac{<\omega^4>}{<\omega^2>^2} - \frac{25}{21}\right) \tag{2.94}$$

$$B = \frac{1}{1400}\left\{\left(\frac{<\omega^6>}{<\omega^2>^3} - \frac{125}{81}\right) - 100A\right\} \tag{2.95}$$

We may note that in the expressions for U and C_V all terms involving $\hbar$ tend to zero as $T \to \infty$, leaving only the terms $3NkT$ and $3Nk$ given by classical statistical mechanics. In particular, the zero-point energy E_Z does not appear in the expansion for U, which at high temperatures is asymptotic to $3NkT$ (and not to $3NkT + E_Z$) as shown in Fig. 2.12. In contrast, the absolute value of the entropy at high temperatures depends on $\hbar$ and the frequencies, and is thus a quantum property.

We may also note that Θ^C departs from its limiting value as T^{-2}, as does C_V. However, the coefficients of the first few terms are relatively much smaller than those for C_V, so that $\Theta^C(T)$ starts to approach Θ^C_∞ at lower temperatures.

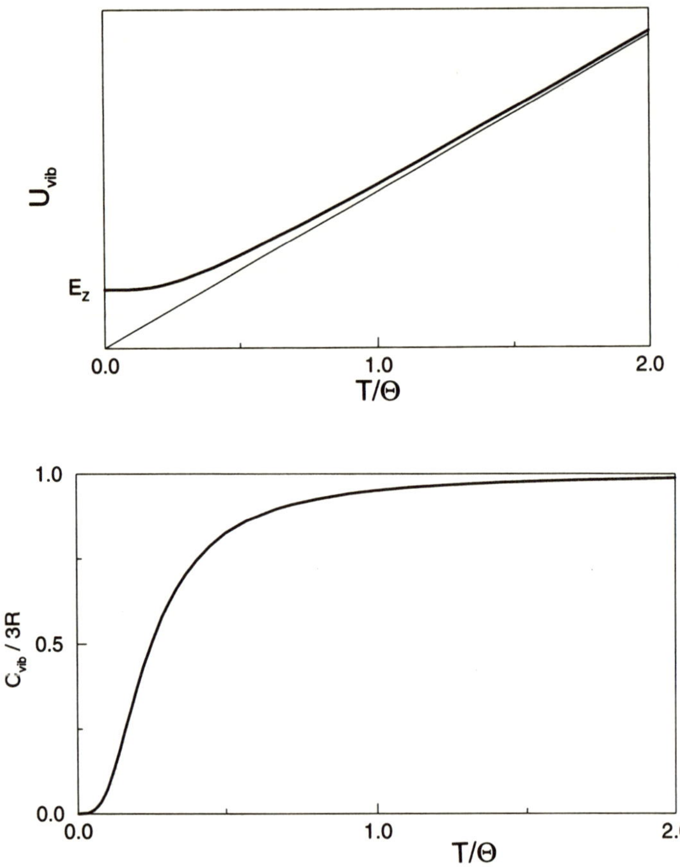

Fig. 2.12. U and C_V as functions of T for a harmonic crystal (schematic).

2.6.3. Anharmonic Theory — Quasiharmonic Approximation

At low enough temperatures the vibrational amplitudes in most solids are small, and the anharmonic part of the potential energy can be treated as a perturbation; it is only for a few 'quantum solids,' where the zero-point energy becomes comparable to the cohesive energy, that this is not so. To the first order of perturbation theory, the anharmonic potential has two effects: (i) interaction between different normal vibrations; (ii) volume (and strain) dependence of the normal mode frequencies. The first of these provides a mechanism for thermal resistance (e.g., [Ber76b]); the second affects the thermodynamic properties, giving rise to thermal pressure and hence to thermal expansion, as well as to vibrational contributions to the bulk modulus and other elastic stiffnesses. Although there is no exact relation between the two effects, approximate quantitative correlations can often be found (e.g., [Whi89b]). There are no first order anharmonic contributions to the expressions for F, S and C_V.

The quasi-harmonic approximation takes the dependence of the harmonic frequencies on volume or strain into account, but neglects all other anharmonic effects. In calculating thermal stress and hence the Grüneisen function, it is equivalent to first order perturbation theory (e.g., [Bar74b]). Since the normal mode contributions to F are additive, the heat capacity and Grüneisen function are as given in Section 1.3.2, with the subscript $_r$ replaced by mode labels $_j$ or $_{\mathbf{q},s}$. The vibrational Grüneisen function is thus the average of all the $\gamma_j = -d\ln\omega_j/d\ln V$ weighted by the heat capacities c_j of each mode:

$$\gamma_{vib}(V,T) = \frac{\sum_j \gamma_j c_j}{\sum_j c_j} = \frac{\sum_j \gamma_j (\hbar\omega_j/kT)^2 e^{\hbar\omega_j/kT}/(e^{\hbar\omega_j/kT}-1)^2}{\sum_j (\hbar\omega_j/kT)^2 e^{\hbar\omega_j/kT}/(e^{\hbar\omega_j/kT}-1)^2} \tag{2.96}$$

The coefficient of thermal expansion is given by

$$\beta_{vib} = \frac{\chi_T}{V}\sum_j \gamma_j c_j = \frac{\chi_T}{V}\sum_j \gamma_j \frac{k(\hbar\omega_j/kT)^2}{(e^{\hbar\omega_j/kT}-1)(1-e^{-\hbar\omega_j/kT})} \tag{2.97}$$

Similarly, the equation of state can be written in one of the forms

$$PV = -V\Phi_L'(V) + \sum_j \gamma_j u_j = -VU_0'(V) + \sum_j \gamma_j u_{th\,j} \tag{2.98}$$

where Φ_L is the static lattice energy, U_0 is the internal energy $\Phi_L + E_Z$ at $T=0$, u_j is the total vibrational energy of the mode j and $u_{th\,j}$ is the thermal energy (Eq. 2.82). Taking the same value γ for all the γ_j leads to the Mie–Grüneisen equations of state:

$$PV = -V\Phi_L'(V) + \gamma U_{vib} = -VU_0'(V) + \gamma U_{th} \tag{2.99}$$

Thus in the quasi-harmonic approximation the thermodynamic function γ_{EOS} defined in Section 2.2.3 is an average of the γ_j weighted by the $u_{th,j}$. It tends to the same low and high temperature limits as γ (γ_0 and γ_∞), but reaches γ_∞ more slowly.

The bulk modulus B_T can be obtained by differentiating P as given by the first of Eqs. (2.98). It involves the second derivatives of the frequencies.

Anisotropic Expansion. Of the non-cubic crystals for which low temperature data is commonly available, axial crystals (tetragonal, hexagonal, trigonal) have two independent expansion coefficients (written $\alpha_\perp, \alpha_\parallel$), and orthorhombic have three (written α_a, α_b, α_c or α_1, α_2, α_3); of the others, monoclinic have four and triclinic six. The thermodynamics of anisotropic expansion (see Section 2.8) therefore has to consider stress, strain and elasticity rather than only pressure, volume and compressibility. The quasiharmonic theory then requires mode Grüneisen parameters $\gamma_{\lambda,j}$ for each independent strain coordinate η_λ. For axial crystals

$$\gamma_{\perp,j} = -\frac{1}{2}\left(\frac{\partial \ln \omega_j}{\partial \ln a}\right)_c, \qquad \gamma_{\parallel,j} = -\left(\frac{\partial \ln \omega_j}{\partial \ln c}\right)_a \tag{2.100}$$

where the factor $\frac{1}{2}$ arises because altering the crystallographic parameter a affects both dimensions perpendicular to the axis; and for orthorhombic crystals

$$\gamma_{1,j} = -\left(\frac{\partial \ln \omega_j}{\partial \ln a}\right)_{b,c}, \quad \gamma_{2,j} = -\left(\frac{\partial \ln \omega_j}{\partial \ln b}\right)_{a,c}, \quad \gamma_{3,j} = -\left(\frac{\partial \ln \omega_j}{\partial \ln c}\right)_{a,b} \tag{2.101}$$

Averages of these Grüneisen parameters weighted by the mode contributions to the heat capacity give thermodynamic anisotropic Grüneisen functions [cf. Eq. (2.96)]

$$\gamma_\lambda = \sum_j \gamma_{\lambda,j} c_j \Big/ \sum_j c_j \tag{2.102}$$

for use in the thermodynamic equations of Section 2.8.4.

Expansion Behavior at Low and High Temperatures. At low temperatures only long wavelength acoustic modes are excited and the thermal expansion has a temperature dependence like that of the heat capacity [see Eq. (2.87)]:

$$\beta_{vib} = b_3 T^3 + b_5 T^5 + b_7 T^7 + \cdots \tag{2.103}$$

The Grüneisen function is given by a power series in T^2, in which the first term γ_0 is related to that in the series for the heat capacity by

$$\gamma_0 = \frac{1}{3} d \ln A / d \ln V = -d \ln \Theta_0 / d \ln V \tag{2.104}$$

At high temperatures Eqs. (2.96)–(2.98) can be expanded as series in inverse powers of T^2. The limiting value of the expression for $\gamma(V,T)$ as $T \to \infty$ is written $\gamma_\infty(V)$; it is the arithmetic mean of all the γ_j. Higher terms in the expansion involve the weighted means

$$\gamma(n) = \sum_j \gamma_j \omega_j^n \Big/ \sum_j \omega_j^n = -d \ln \omega_D(n) / d \ln V \tag{2.105}$$

Techniques for Computing Low Temperature Behavior. In calculations on theoretical models the squares of the frequencies $(\omega_{\mathbf{q},s})^2$ for each value of **q** are obtained as the eigenvalues of a *dynamical matrix* $\mathbf{D}(\mathbf{q})$, and the associated mode Grüneisen parameters can be derived from the volume or stress derivatives of $\mathbf{D}(\mathbf{q})$ by perturbation theory (e.g., [Wal72, Kan95, Tay97a]). Thermodynamic properties are then obtained by integrating over the FBZ or other equivalent region in **q**-space. At most temperatures the required accuracy is given by integration grids which have typically 100 to 10000 points in the whole FBZ.

But vibrational effects at very low temperatures depend mainly on acoustic modes with **q** close to the Γ-point, and to study them much finer grids are required in this region. To extend such grids over the entire zone would be prohibitively wasteful, and so some procedure is needed to allow the use of progressively finer grids as the Γ-point is approached. One simple iterative method starts with a grid adequate for intermediate and high temperatures. In the first iteration the integration over an inner region with linear dimensions half those of the whole zone is recalculated with a finer mesh which has the same number of points in the inner region as used originally for the whole zone. In the second iteration an inner region of the first inner region is treated similarly, and so on. At each step there is an eightfold increase in the density of points in the innermost region. The number of iterations needed to obtain convergence — typically three to six — depends upon the lowest temperature for which precise results are required. A complete calculation shows the approach to the Debye limit, as checked by an independent calculation of Θ_0^{el}.

Fortunately this iteration need be done only once. At low temperatures the thermal expansion is small, and to calculate the dynamical matrix, frequencies and mode Grüneisen parameters we can use the geometry at $T = 0$. Once these are found, properties at each low temperature can be calculated by simultaneous integration over the FBZ, so that only one set of iterations of the integration grid is needed. To find the equilibrium geometry at $T = 0$ the total static and zero-point vibrational energy must be minimized, but since the zero-point energy is dominated by higher frequencies this does not require fine grids near the Γ-point. Similarly at higher temperatures, where the effect of thermal expansion on the dynamical matrix is appreciable, the total free energy must be minimized with respect to volume or strain for each T, necessitating repeated calculation and diagonalization of the changing dynamical matrix; but this does not require fine grids at the zone center.

Central Force Mechanisms in Thermal Expansion. Equation (2.97) shows that thermal expansion will be positive or negative depending upon whether positive or negative γ_j predominate in the weighted average. Positive γ_j arise chiefly because a typical interatomic potential has the asymmetric shape shown in Fig. 2.13a, so that the force constant affecting vibrations along the line joining the atoms decreases with increasing distance. This mechanism for positive expansion due to vibration along the line of atomic centers can also be seen in two other equivalent ways: (i) it is easier for the atoms to move further from each other than nearer to each other; (ii) if the volume and hence the distances between the mean positions of the atoms are kept fixed when the atoms vibrate, the average force between the atoms will be repulsive, leading to a positive thermal pressure (Fig. 2.13a). This is sometimes called the *bond stretching mechanism*, because it causes the mean distance between atoms to increase.

Negative γ_j arise chiefly because of vibrations perpendicular to the line of centers. The tension between the atoms, and hence the frequencies of transverse vibrations, increase with interatomic distance, as in a stretched banjo string. This mechanism too can be seen also in an equivalent way, by considering the thermal stresses caused by the vibrations: if the mean atomic positions are kept fixed, transverse displacements about these positions increase the mean distance between the atoms, causing a tension tending to pull them towards each other (Fig. 2.13b) and so to restore the mean interatomic distance to its original value. This is sometimes called the *tension mechanism.* In the quasiharmonic approximation each mechanism contributes additively to the Grüneisen parameters and hence to the thermal expansion, so that their relative importance may be easily calculated for theoretical models (e.g., [Bru98]).

In most crystal vibrations the displacements have components both along and perpendicular to the lines joining neighboring atoms. Both mechanisms then operate.* Moreover, displacements that are roughly along one line of centers can be roughly perpendicular to another. Usually the bond-stretching mechanism of Fig. 2.13a has the larger effect, and the thermal expansion is positive. But those modes which have on average larger components of vibration normal to lines of centers tend to have lower frequencies, and so are preferentially excited at low temperatures. For this reason γ_{vib} often decreases at low temperatures [Bla57, Bar57a]. The effect is particularly marked for crystals of open structure, where there can be modes for which the relative atomic displacements are predominantly transverse to the lines of centers; among the simplest examples are the rocksalt and zincblende structures (see Sections 5.3 and 5.5). For some very open structures β can be negative even at room temperature.

In non-cubic crystals the same mechanisms can give rise to a wide variety of behavior, as in the layered and chain structures discussed in Sections 5.8 and 5.9. This has been exemplified by a model primitive rhombohedral lattice, in which the ratio of the interaction between planes perpendicular to the trigonal axis to that within the planes could be altered [Bar74a]. A small ratio gave layered crystals, in which the low frequency modes were polarized largely perpendicular to the planes; so that at low temperatures the expansion along the axis was relatively large and positive due to bond-stretching, while that perpendicular to the axis was small, and negative due to the tension effect and a large negative cross compliance. In contrast, a large ratio gave linked polymer chains, with the opposite behavior.

*There is also a third mechanism tending to rotate the line of centers away from the direction of vibration, but its net effect is small or zero except for strongly anisotropic non-cubic crystals such as polyethylene [Bru98].

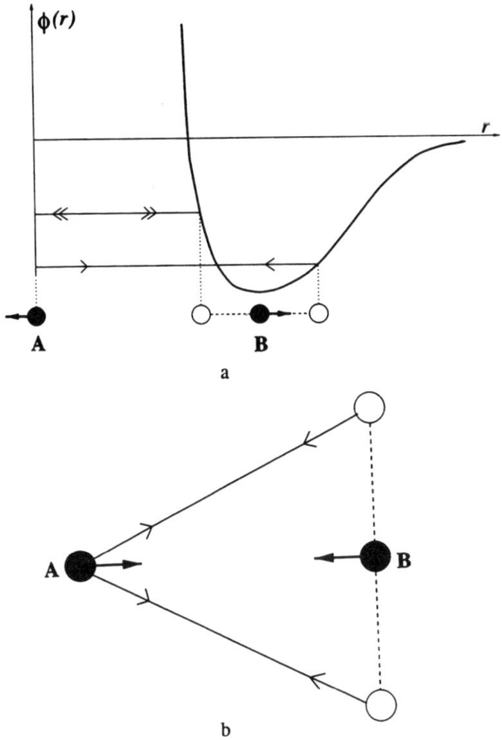

Fig. 2.13. Central force mechanisms in thermal expansion (schematic). Dashed lines show amplitude of motion of atom B with respect to atom A. (a) *Bond stretching:* displacements along the "bond" direction produce a mean repulsive force, because the potential well is asymmetric; (b) *Tension:* displacements perpendicular to the "bond" direction produce a mean attractive force [Bru98].

2.6.4. Strongly Anharmonic Vibrations

For most solids that have been studied the quasi-harmonic approximation gives a good account of the thermodynamic properties at low temperatures, although it breaks down progressively as the amplitude of lattice vibrations increases with temperature. The temperatures at which this becomes serious vary for different substances and the pressures at which they are held; there is no universal rule. For example the approximation is totally invalid at all temperatures for solid helium except at high pressures (Section 5.2), and also for the rotational motions of molecules in solid H_2 and CH_4 (Ch. 8); but for tightly bound crystals such as the ceramic oxides it is useful well above room temperature. Calculations on the thermodynamic behavior of models of alkali halides and heavy rare gas solids under zero pressure have shown that noticeable deviations from quasi-harmonic behavior may begin at about 0.2 T_m, where T_m is the melting point (e.g., [Gly71]); but the approximation can remain useful to considerably higher temperatures, especially when there is cancellation between different anharmonic terms.

Several of the theoretical methods of treating anharmonic effects are summarized briefly in [Bar98], and described in detail in various chapters of [Hor74, Hor75, Kle76, Kle77]. Successive higher orders of perturbation theory (PT) extend the range to higher temperatures, and the variational method of the *self-consistent* (SC) phonon theory to yet higher temperatures, especially when it is combined with PT to give the *improved self-consistent* (ISC) phonon theory. These methods, together with some cell models, have more recently been further tested and compared with each other and with classical Monte Carlo methods by E. R. Cowley, G. K. Horton and colleagues, again using simple models for rare gas solids and ionic solids (e.g., [Shu85, Gon88, Cow90]; see also Fig. 2.14).

At temperatures where quantum effects are small classical Monte Carlo and Molecular Dynamics are often used, because of their comparative simplicity (Section 2.3.3). Usually it is sufficient to use quasiharmonic theory at low temperatures, and check its validity at higher temperatures by comparison with MC or MD results, interpolating between them. There are also some methods of quantum simulation valid over the whole temperature range [Hor95], such as the Feynman path integral (PI) method which can be evaluated by techniques mathematically equivalent to classical Monte Carlo [McG95]. Figure 2.14 illustrates the power of this method, and also shows that for the more strongly anharmonic (lighter) rare gas solids it becomes progressively less reliable to interpolate between quasiharmonic theory at low temperatures and classical theory at high temperatures without recourse to further theory. A faster approximate path integral technique, the *improved effective potential* (IEP) method, has been found useful for quantum solids such as neon and helium [Cow95, Hor96, Aco96].

According to both PT and SC theory the entropy (but not the heat capacity) is given correctly by the harmonic expression, but with shifted frequencies equal to those given by spectroscopic measurements, including neutron diffraction. We thus have a temperature-dependent 'renormalized' spectrum. Even as $T \to 0$ there are still some anharmonic effects due to zero-point energy, but S and C_V still have a T^3 dependence related to the elasticity in the same way as for a harmonic solid. On the other hand room temperature elastic constants can be used only to calculate the ω^2 term in the room temperature renormalized spectrum, which does *not* correspond exactly to any measured heat capacity.

2.6.5. Analysis of Thermodynamic Data

The most usual way of treating vibrational data is to fit to them the parameters of a suitable lattice dynamical model, and then to use the fitted model to derive other information not given directly by the experimental data. But also some properties of the frequency distribution of a solid can be derived to a good accuracy from its measured thermodynamic properties without the use of specific models.

The analysis of thermodynamic data is discussed in several books and papers, including [Bar57b, Bar64, Wal72, Yat72, Bar82, Gui91, Wal92]. We have already

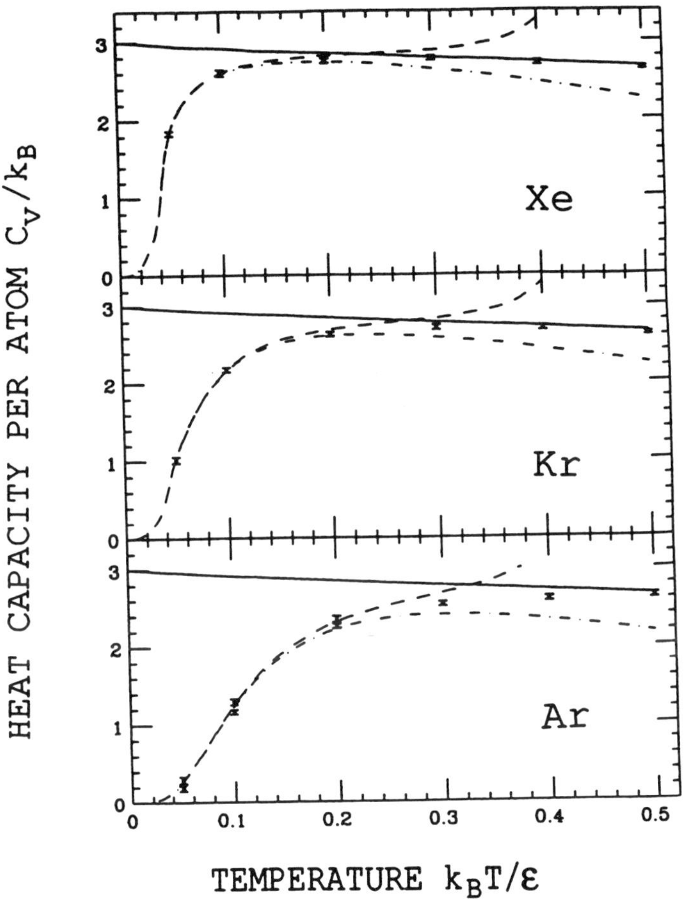

Fig. 2.14. Heat capacities of nearest neighbor central force models for rare gas solids according to different approximations (see text). •, first order PI; ×, second order PI; —, smoothed classical MC; - - -: lower curves, PT with lowest order anharmonic terms; upper curves, PT with next order terms added [Liu91].

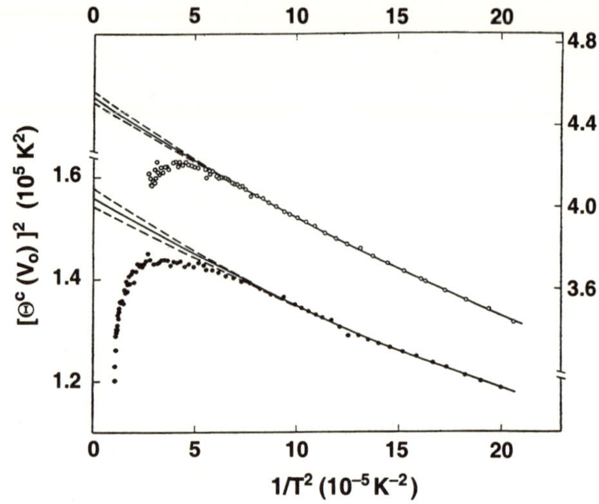

Fig. 2.15. Estimation of coefficients in the high temperature expansion for Θ^C: $(\Theta^C)^2$ plotted against T^{-2} for Si (upper curve) and Ge (lower curve) [Flu59].

seen that at low temperatures by plotting C_V/T^3 (for metals C_V/T) against T^2 we can estimate the first two or three coefficients in the low frequency expansion [Eq. (2.86)]. The $g(\omega)$ to which this corresponds is that of the renormalized spectrum at $T = 0$, which takes into account any anharmonic effects due to the zero-point energy. To derive further information about this spectrum from higher temperature measurements the data must first be corrected to the volume V_0 at $T = 0$, with the aid either of thermodynamic expressions for the volume derivatives of heat capacity, entropy, thermal expansion, etc. (see Section 2.2.3 and [Wal72]) or of appropriate Grüneisen parameters (see [Bar57b, Bar64]). However, since Grüneisen parameters are based on quasiharmonic theory, they should not be used in the fully anharmonic analysis of high temperature data.

Quasiharmonic analysis. The volume-corrected data is fitted to the quasiharmonic high temperature expansions for the thermodynamic properties or for Debye temperatures and Grüneisen functions derived from them. This can be done in various ways. Figure 2.15 shows how $\Theta_\infty^C(V_0)$ has been estimated for Si and Ge, by plotting $[\Theta^C(V_0)]^2$ against T^{-2}. Although experimental uncertainty and increasing anharmonic effects at higher temperatures limit the precision of the extrapolated intercepts at $T^{-2} = 0$, Θ_∞^C is obtained to within about 1%. Good estimates can also be obtained of the slopes, from which $< \omega^4 >$ and hence $\omega_D(4)$ are derived, and even $\omega_D(6)$ can be roughly estimated.

The most accurate information, however, is obtained from data at intermediate temperatures. For all values of n (including non-integral) in the range $-3 < n < 0$ the moments of the distribution are given by integrals over the harmonic heat capacity:

$$(\hbar/k)^n < \omega^n > = \frac{1}{3Nk}\int_0^\infty (C_V/T^{1-n})dT \Big/ \Gamma(2-n)\zeta(1-n) \tag{2.106}$$

where $\Gamma(x)$ is the mathematical Gamma function, $\zeta(x)$ is the Riemann zeta function [Abr65], and the high temperature part of the integral is calculated using the high temperature (harmonic) expansions already obtained. In addition, ω_g and the first moment $< \omega >$ are obtained from the high temperature expansions for S and U. Results for Si and Ge, correlated in $\omega_D(n)$ plots, are shown in Fig. 2.10. This type of analysis can be extended to the thermal expansion to give values of $\gamma(n)$ [Bar64], and similarly $\gamma_\lambda(n)$ for anisotropic solids [Bar67, Bar82].

Fully anharmonic analysis. A fully anharmonic analysis usually refers to data obtained above cryogenic temperatures, and detailed discussion is beyond the scope of this monograph. Wallace [Wal72] has shown how a plot of $(C_V - 3Nk)/T$ against T^{-3} can be extrapolated to give the coefficient of a linear term in T in C_V, corresponding to the lowest order non-zero term in classical perturbation theory; while $< \omega^2 >$ is given to a good accuracy by the slope. Such plots also demonstrate the breakdown of perturbation theory at higher temperatures. It is useful in this range to work at least partly with the entropy, since this corresponds to the temperature dependent renormalized spectrum, about which information may be gained from neutron diffraction. In conductors there are also electronic contributions, for which reliable band-structure calculations may sometimes be available. For further discussion see [Gui91, Eri92, Wal92, Whi93c].

2.7. APPROXIMATE EQUATIONS OF STATE

For many solids data are not available over all temperatures and pressures of interest, nor can they be calculated reliably from theoretical models. But examination of the experimental and theoretical data that is available has led to working rules for estimating equations of state of a wide range of materials, especially in the high temperature limit relevant to geophysics and many technical applications.

2.7.1. Behavior for $T \geq \Theta$

When $T \geq \Theta$, to a first approximation $C_V \simeq 3Nk$ (the Dulong and Petit value) and γ has reached its high temperature limiting value γ_∞. The Mie–Grüneisen equation of state is then

$$P(T,V) \simeq -\Phi_L'(V) + \frac{\gamma_\infty(V)}{V} 3NkT \tag{2.107}$$

and the thermal pressure coefficient is

$$\left(\frac{\partial P}{\partial T}\right)_V = \beta B_T \simeq 3Nk\frac{\gamma_\infty(V)}{V} \tag{2.108}$$

which is a function of volume but not of temperature.

To this Swenson [Swe68] added the further approximation, based on empirical observation on many cubic solids, that at high temperatures B_T is a function of volume only; this gives [cf. Eq. (2.31)]

$$\left(\frac{\partial B_T}{\partial T}\right)_V = -\left[\frac{\partial}{\partial \ln V}\left(\frac{\partial P}{\partial T}\right)_V\right]_T \simeq 0 \tag{2.109}$$

Together with Eqs. (2.107), this implies that the thermal pressure coefficient [in Eq. (2.108)] is independent of both volume and temperature, and that γ_∞/V is also constant; and that B_T can be identified with the bulk modulus of the static lattice:

$$q_\infty = (\partial \ln \gamma_\infty/\partial \ln V)_T \simeq 1, \qquad \text{and} \qquad B_T \simeq V\Phi_L'' \tag{2.110}$$

The constancy of $(\partial P/\partial T)_V$ implies that all isothermal equations of state are parallel:

$$P(T,V) \simeq P(T^*,V) + \beta(T^*,V^*)B_T(T^*,V^*)(T-T^*) \tag{2.111}$$

Because of its simplicity this approximation is widely used. Its reliability in the light of more extensive data than those used by Swenson is discussed by O. L. Anderson [And95a], who points out that it will be valid when the difference between the dimensionless functions δ_T and B_T' is small, since

$$\left(\frac{\partial B_T}{\partial T}\right)_V = \beta B_T(B_T' - \delta_T) \tag{2.112}$$

Although (like Swenson) he finds significant departures in observed behavior from that predicted by the approximation, he concludes that it is a good first approximation over a wide range of pressures for many different types of solid.

Equations of State. To complete the information required to estimate the equation of state at all pressures, Swenson used the Murnaghan equation of state at temperature T^*:

$$P(T^*,V) \simeq \frac{B_T(T^*,V^*)}{B_T'(T^*,V^*)}\left[\left(\frac{V^*}{V}\right)^{B_T'(T^*,V^*)} - 1\right] \tag{2.113}$$

based on the assumption that the dimensionless derivative $B_T' = (\partial B_T/\partial P)_T$ is independent of volume. At higher pressures than those Swenson considered, observed

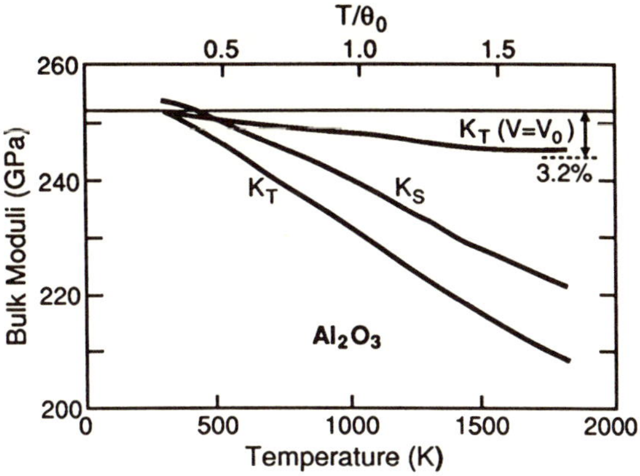

Fig. 2.16. Bulk moduli $K_T(T,V_0)$, $K_T(T,P=0)$, $K_S(T,P=0)$ as functions of temperature for Al_2O_3. From [And95a], noting his use of K rather than B.

behavior diverges from the Murnaghan equation, and a more sophisticated approximation is needed. Perhaps the best known and most successful is the Vinet "universal" equation of state, which may be written in the analytic form [Vin87a, Vin87b]

$$P(T^*,V) = \frac{3B(T^*,V^*)}{X^2}(1-X)\exp[\eta(T^*,V^*)(1-X)] \tag{2.114}$$

where

$$X = \left(\frac{V}{V^*}\right)^{\frac{1}{3}} \quad \text{and} \quad \eta(T^*,V^*) = \frac{3}{2}[B'(T^*,V^*)-1] \tag{2.115}$$

2.7.2. Behavior at Low Temperatures

At $T=0$ the quasi-harmonic expressions for the vibrational free energy and pressure are

$$F_{vib}(0,V) = E_Z = \frac{9}{8}R\Theta_D(1), \quad P_{vib} = \frac{\gamma(1)}{V}\frac{9}{8}R\Theta_D(1) \tag{2.116}$$

Neglecting any volume dependence of $\gamma(1)/V$, we find [Swe68]

$$B_T(0,V) = B_S(0,V) \simeq B_L(V) + \frac{[\gamma(1)]^2}{V}\frac{9}{8}R\Theta_D(1) \tag{2.117}$$

The vibrational term is positive, explaining why on cooling at constant volume B_T is observed to rise above its high temperature value to a new limit as $T \to 0$ (Fig. 2.16).

2.8. ANISOTROPIC STRAIN AND STRESS: ELASTICITY

2.8.1. Introduction

For thermal properties of materials under isotropic pressure, the same thermodynamics applies to both fluids and solids; V is the only relevant geometric variable. But unlike liquids, solids can sustain anisotropic stress; ideally they cannot flow, and neighboring atoms remain so permanently even when the solid is put under stress.* In particular, in a crystal under anisotropic stress the unit cell retains its identity, but changes its shape as well as its volume. Six parameters are needed to define its dimensions — for example, the lengths of the edges a, b, c and the angles $\hat{\alpha}, \hat{\beta}, \hat{\gamma}$ between them.

Depending on the specific application, we have therefore to generalize the thermodynamics of Section 2.2 to take account of up to six geometrical degrees of freedom. For example, we might replace the one parameter V by the six unit cell parameters, and it may sometimes be appropriate to do this. However, the stresses conjugate to these parameters are not in general simply related to the macroscopic applied stress, and in any case such a theory could not be applied to non-crystalline solids. For these and other reasons, standard elasticity theory is developed in terms not of unit cell parameters but of macroscopic parameters describing the distortion of the material from some configuration chosen as reference. Many applications are concerned only with the limit of infinitesimal strains from a state originally at zero pressure, and it is then quite straightforward to define stresses, strains and the related (second order) elastic coefficients with little fear of confusion (e.g., [Nye85]); with these we can generalize most of the relations of Section 2.2.3 to take account simultaneously of the different independent strain coordinates. Later in this section, however, we shall have to go further. We have seen that at very low temperatures the vibrational free energy depends on the frequencies of long wave acoustic phonons, and these in turn depend on the second order elastic coefficients. The low temperature thermal expansion therefore depends on the strain-derivatives of these frequencies, and so involves higher order elasticity. There is an extensive literature on this (later work includes [Thu64, Thu65a, Thu65b, Bru64, Bru65, Bru67, Wal70, Wal72, Bar98]), but the complexity of the subject makes much of it difficult to read. Here we shall be concerned only with the minimum needed for present purposes.

2.8.2. Stress and Strain

The Cauchy Stress. The best known measure of stress is the *Cauchy stress tensor*, $\sigma_{\alpha\beta}$, defined such that if dA is an element of surface area within the solid separating regions labelled I and II, the force exerted across dA by region II on region I is given by vector components

$$F_\alpha = \sigma_{\alpha\beta} n_\beta dA \tag{2.118}$$

*We shall not here be concerned with visco-elastic substances.

where **n** is a unit vector normal to dA in the direction I to II; here and from now on we use the Einstein convention that a repeated suffix indicates summation (in this case $\Sigma_{\beta=1}^{3}$). For example, a positive value of σ_{11} indicates a tension along the direction of the x-axis, and a negative value indicates a compression. Off-diagonal elements, and also differences between the diagonal elements, indicate shear stress. For a solid under hydrostatic pressure P, all the off-diagonal elements vanish and $\sigma_{\alpha\beta} = -P\delta_{\alpha\beta}$; here $\delta_{\alpha\beta}$ is the Kronecker delta, which is unity when α and β are the same and zero when they are different.

Normally (for exceptions see [Nye85]) the stress tensor is symmetric ($\sigma_{\alpha\beta} = \sigma_{\beta\alpha}$), and so has only six independent elements. In the abbreviated notation of Voigt these are written with a single subscript, as σ_λ, where $\lambda = 1, \cdots, 6$:

$$\sigma_1 = \sigma_{11}, \quad \sigma_2 = \sigma_{22}, \quad \sigma_{33} = \sigma_3$$

$$\sigma_4 = \sigma_{23} = \sigma_{32}, \quad \sigma_5 = \sigma_{31} = \sigma_{13}, \quad \sigma_6 = \sigma_{12} = \sigma_{21} \tag{2.119}$$

Strain Coordinates. Strain coordinates describe the distortion of a material from the chosen reference configuration. They too may be expressed either as components of a tensor or in a Voigt abbreviated notation. Two sets in common use are the *infinitesimal strain coordinates*, written as $e_{\alpha\beta}$ or e_λ, which are sufficient for many applications (e.g., [Nye85]); and the *Lagrange finite strain coordinates*, written as $\eta_{\alpha\beta}$ or η_λ, which determine uniquely any state of strain however large, and so are widely used in the treatment of higher order elasticity. To the first order in the strain the two sets are the same, but they differ to higher orders.

In this book we shall use mainly the Lagrange coordinates, making it clear when use of the infinitesimal coordinates would give different results. To define them, we choose a set of rectilinear Cartesian axes in the reference state, usually determined by crystal symmetry. In a state of uniform strain, lines that were straight in the reference state remain straight in the strained state. In particular, a unit cube in the reference state, with edges $\mathring{\boldsymbol{\ell}}_1$, $\mathring{\boldsymbol{\ell}}_2$, $\mathring{\boldsymbol{\ell}}_3$ parallel to the coordinate axes, becomes in the strained state a parallelepiped with edges $\boldsymbol{\ell}_1, \boldsymbol{\ell}_2, \boldsymbol{\ell}_3$ that in general are no longer of unit length nor at right angles to each other. The *Lagrange finite strain tensor* can then be defined by

$$\eta_{\alpha\beta} = \frac{1}{2}(\boldsymbol{\ell}_\alpha \cdot \boldsymbol{\ell}_\beta - \mathring{\boldsymbol{\ell}}_\alpha \cdot \mathring{\boldsymbol{\ell}}_\beta) \tag{2.120}$$

Its significance is that the square of the length of any vector $\mathring{\mathbf{r}}$ in the reference state is changed from $\mathring{r}^2$ to $\mathring{r}^2 + 2\eta_{\alpha\beta}\mathring{r}_\alpha\mathring{r}_\beta$ in the strained state.

Like $\sigma_{\alpha\beta}$ this tensor is symmetric, with only six independent elements. In the abbreviated Voigt notation, strain coordinates η_λ are defined as follows:

$$\eta_1 = \eta_{11}, \quad \eta_2 = \eta_{22}, \quad \eta_3 = \eta_{33}, \quad \eta_4 = 2\eta_{23}, \quad \eta_5 = 2\eta_{13}, \quad \eta_6 = 2\eta_{12} \tag{2.121}$$

The factor 2 is introduced here for the off-diagonal elements of the strain tensor to make summation and differentiation equivalent in the two notations. To the first order in the strain the finite strain coordinates have a simple geometrical meaning: η_1, η_2, η_3 are the dilations $\delta l/l$ along the three coordinate axes, and $-\eta_4, -\eta_5, -\eta_6$ are the changes $\delta\hat{\theta}_{23}, \delta\hat{\theta}_{31}, \delta\hat{\theta}_{12}$ in the angles between the straight lines originally along the positive coordinate axes.

The Lagrange strain tensor is also often defined equivalently in terms of the linear transformation specifying the position $\mathbf{x}(\mathring{\mathbf{x}})$ in the strained state of each point of the body originally at $\mathring{\mathbf{x}}$ in the reference state:

$$x_\alpha = (\delta_{\alpha\beta} + u_{\alpha\beta})\mathring{x}_\beta \tag{2.122}$$

This transformation takes account of rotation of the body as well as homogeneous strain, and so in general all nine elements of the tensor $u_{\alpha\beta}$ are independent. In terms of the $u_{\alpha\beta}$, Eq. (2.120) gives the Lagrangian tensor as

$$\eta_{\alpha\beta} = \frac{1}{2}(u_{\alpha\beta} + u_{\beta\alpha} + u_{\gamma\alpha}u_{\gamma\beta}) \tag{2.123}$$

The infinitesimal strain tensor is then defined by omitting the second order terms in this expression:

$$e_{\alpha\beta} = \frac{1}{2}(u_{\alpha\beta} + u_{\beta\alpha}) \tag{2.124}$$

and is thus the symmetric part of $u_{\alpha\beta}$. Voigt coordinates e_λ are defined in an analogous way to the η_λ. Infinitesimal rotation of the body is described by the antisymmetric part $\omega_{\alpha\beta}$ of $u_{\alpha\beta}$:

$$\omega_{\alpha\beta} = \frac{1}{2}(u_{\alpha\beta} - u_{\beta\alpha}) \tag{2.125}$$

2.8.3. Elastic Stiffnesses $c_{\lambda\mu}$ and Compliances $s_{\lambda\mu}$

There are many different ways of treating elasticity, and of defining elastic coefficients. Some of these will be discussed in Section 2.8.5. But for the present we shall be concerned only with the "stress–strain" coefficients* obtained from the dependence of the Cauchy stress on the strain when the instantaneous state of the system is taken as reference configuration; whether we use η_λ or e_λ then makes no difference to first order derivatives. The isothermal and adiabatic *stiffnesses* are generalizations of B_T and B_S, defined by

$$c^T_{\lambda\mu} = \left(\frac{\partial\sigma_\lambda}{\partial\eta_\mu}\right)_{\eta',\omega,T}, \qquad c^S_{\lambda\mu} = \left(\frac{\partial\sigma_\lambda}{\partial\eta_\mu}\right)_{\eta',\omega,S} \tag{2.126}$$

*They have also been called "effective," and "physical."

where the subscript η' denotes that all strain coordinates except η_μ are kept constant during differentiation, and the subscript ω denotes that the body does not rotate. Similarly, compliances are generalizations of χ_T and χ_S:

$$s^T_{\lambda\mu} = \left(\frac{\partial\eta_\lambda}{\partial\sigma_\mu}\right)_{\eta',\omega,T}, \qquad s^S_{\lambda\mu} = \left(\frac{\partial\eta_\lambda}{\partial\sigma_\mu}\right)_{\eta',\omega,S} \tag{2.127}$$

The stiffness and compliance matrices are reciprocal, in the sense that

$$s^T_{\lambda\mu}c^T_{\mu\nu} = c^T_{\lambda\mu}s^T_{\mu\nu} = \delta_{\lambda\nu}; \qquad s^S_{\lambda\mu}c^S_{\mu\nu} = c^S_{\lambda\mu}s^S_{\mu\nu} = \delta_{\lambda\nu} \tag{2.128}$$

the repeated suffices now imply the summation $\sum_{\mu=1}^{6}$.

The stiffnesses tell us how each stress coordinate changes when one strain coordinate is changed, while all the other strain coordinates are kept at zero. Conversely, the compliances tell us how each strain coordinate responds to an applied stress: thus $s_{21}\sigma_1$ is the dilation in the y-direction resulting from a stress σ_1 stretching the material in the x-direction, while all other stresses are kept constant. The number of independent coefficients depends upon symmetry. If the stress is isotropic, $c_{\lambda\mu} = c_{\mu\lambda}$ and $s_{\lambda\mu} = s_{\mu\lambda}$, reducing the number of independent coefficients to twenty-one. Crystal symmetry further reduces this number [Nye85]; for example, cubic crystals have only three (c_{11}, c_{12}, c_{44}).

For solids under isotropic pressure the *directional compressibilities* χ_λ provide another useful generalization of χ. They tell us how the volume responds to a change in the single stress coordinate σ_λ, and also how the strain coordinate η_λ responds to a change in pressure:

$$\chi^T_\lambda = \left(\frac{\partial \ln V}{\partial\sigma_\lambda}\right)_{\sigma',T} = -\left(\frac{\partial\eta_\lambda}{\partial P}\right)_T \tag{2.129}$$

Adiabatic directional compressibilities are defined similarly. χ_λ can be expressed as the sum of three compliances:

$$\chi^T_\lambda = s^T_{\lambda 1} + s^T_{\lambda 2} + s^T_{\lambda 3} \tag{2.130}$$

In turn the total compressibility is

$$\chi_T = -\left(\frac{\partial \ln V}{\partial P}\right)_T = \sum_{\lambda=1}^{3}\chi^T_\lambda = \sum_{\lambda=1}^{3}\sum_{\mu=1}^{3} s^T_{\lambda\mu} \tag{2.131}$$

The reciprocal concept of bulk modulus is experimentally not so useful for solids with symmetry lower than cubic, since specifying the volume leaves the shape undetermined. However, for processes carried out under hydrostatic pressure B_T may be defined as the reciprocal of the compressibility:

$$B_T = -\left(\frac{\partial P}{\partial \ln V}\right)_{T\ \text{(isotropic stress)}} \tag{2.132}$$

Elasticity of Isotropic Materials. We may ask how these quantities are related to Young's modulus E, Poisson's ratio σ, and other coefficients used by geophysicists and engineers to describe the elastic properties of an isotropic material. Here all directions are equivalent, so that $c_{11} = c_{22} = c_{33}$, $c_{12} = c_{13}$, etc.; also $c_{44} = \frac{1}{2}(c_{11} - c_{12})$, etc. Other stiffnesses, such as c_{14}, c_{15} and c_{45}, are all zero. Young's modulus refers to the relation between stress and strain along a single axis while all other *stresses* are kept constant; and so although it has the dimensions of a stiffness, it is actually a reciprocal compliance: $E = (s_{11})^{-1}$. Similarly, Poisson's ratio is a ratio of two compliances: $\sigma = -s_{12}/s_{11}$. On the other hand, the bulk and rigidity moduli are both true stiffnesses: $B = (c_{11} + 2c_{12})/3$, and $G = (c_{11} - c_{12})/2 = c_{44}$. So also are the Lamé coefficients, λ and μ, given by $\lambda = c_{12}$ and $\mu = c_{44}$. In terms of E and σ, $B = E/\{3(1-2\sigma)\}$ and $G = E/\{2(1+\sigma)\}$.

2.8.4. Thermodynamic Relations

We can now generalize the results of Section 2.2.3. C_V and C_P are replaced by heat capacities at constant strain and stress:

$$C_\eta = T\left(\frac{\partial S}{\partial T}\right)_\eta, \quad C_\sigma = T\left(\frac{\partial S}{\partial T}\right)_\sigma \tag{2.133}$$

where C_σ becomes C_P when the stress is isotropic. Thermal expansion coefficients and Grüneisen functions are defined by

$$\alpha_\lambda = \left(\frac{\partial \eta_\lambda}{\partial T}\right)_{\sigma,\omega}, \quad \gamma_\lambda = -\left(\frac{\partial \sigma_\lambda}{\partial (U/V)}\right)_{\eta,\omega} \tag{2.134}$$

Other expressions for γ_λ are

$$\gamma_\lambda = -\frac{V}{C_\eta}\left(\frac{\partial \sigma_\lambda}{\partial T}\right)_{\eta,\omega} = -\left(\frac{\partial \ln T}{\partial \eta_\lambda}\right)_{\eta',\omega,S} \tag{2.135}$$

Consideration of first increasing the temperature at constant strain and then allowing the stress to relax isothermally leads to a generalization of $\beta = \gamma(C_V/V)\chi_T$:

$$\alpha_\lambda = \frac{C_\eta}{V} s^T_{\lambda\mu}\gamma_\mu = \frac{C_\sigma}{V} s^S_{\lambda\mu}\gamma_\mu \tag{2.136}$$

where the second equality can be derived by thermodynamic manipulation. Anisotropic thermal expansion is thus a result of the interplay of the thermal stress coefficients (proportional to the γ_λ) and the elastic compliances [Mun68]. Reciprocal relations give the γ_λ in terms of the stiffnesses and expansion coefficients:

$$\gamma_\lambda = \frac{V}{C_\eta} c^T_{\lambda\mu}\alpha_\mu = \frac{V}{C_\sigma} c^S_{\lambda\mu}\alpha_\mu \tag{2.137}$$

In terms of all these quantities relations have been found between C_σ and C_η, and between isothermal and adiabatic elastic coefficients. For the heat capacities we have the two equivalent relations

$$C_\sigma = C_\eta(1+\alpha_\lambda\gamma_\lambda T), \quad C_\sigma = C_\eta + VTc^T_{\lambda\mu}\alpha_\lambda\alpha_\mu \tag{2.138}$$

and for the stiffnesses and compliances

$$c^S_{\lambda\mu} = c^T_{\lambda\mu} + \frac{TC_\eta}{V}\gamma_\lambda\gamma_\mu, \quad s^S_{\lambda\mu} = s^T_{\lambda\mu} - \frac{TV}{C_\sigma}\alpha_\lambda\alpha_\mu \tag{2.139}$$

Crystals of Axial and Orthorhombic Crystals. Much work, both experimental and theoretical, has been done on crystals of high symmetry, for which the above equations take simple explicit forms. For axial crystals (tetragonal, trigonal and hexagonal) there are only two independent coefficients of expansion, perpendicular and parallel to the axis, and similarly two Grüneisen functions, related by:

$$\alpha_\perp = \frac{C_\eta}{V}[(s^T_{11}+s^T_{12})\gamma_\perp + s^T_{13}\gamma_\parallel] = \frac{C_\sigma}{V}[(s^S_{11}+s^S_{12})\gamma_\perp + s^S_{13}\gamma_\parallel] \tag{2.140}$$

$$\alpha_\parallel = \frac{C_\eta}{V}[2s^T_{13}\gamma_\perp + s^T_{33}\gamma_\parallel] = \frac{C_\sigma}{V}[2s^S_{13}\gamma_\perp + s^S_{33}\gamma_\parallel] \tag{2.141}$$

and

$$\gamma_\perp = \frac{V}{C_\eta}[(c^T_{11}+c^T_{12})\alpha_\perp + c^T_{13}\alpha_\parallel] = \frac{V}{C_\sigma}[(c^S_{11}+c^S_{12})\alpha_\perp + c^S_{13}\alpha_\parallel] \tag{2.142}$$

$$\gamma_\parallel = \frac{V}{C_\eta}[2c^T_{13}\alpha_\perp + c^T_{33}\alpha_\parallel] = \frac{V}{C_\sigma}[2c^S_{13}\alpha_\perp + c^S_{33}\alpha_\parallel] \tag{2.143}$$

Similarly

$$C_\sigma = C_\eta(1 + 2\alpha_\perp\gamma_\perp T + \alpha_\parallel\gamma_\parallel T) \tag{2.144}$$

and so on for the rest of Eqs. (2.137)–(2.139).

For orthorhombic crystals there are three independent expansion coefficients and Grüneisen functions, related by

$$\alpha_1 = \frac{C_\eta}{V}[s^T_{11}\gamma_1 + s^T_{12}\gamma_2 + s^T_{13}\gamma_3] = \frac{C_\sigma}{V}[s^S_{11}\gamma_1 + s^S_{12}\gamma_2 + s^S_{13}\gamma_3] \tag{2.145}$$

$$\alpha_2 = \frac{C_\eta}{V}[s^T_{12}\gamma_1 + s^T_{22}\gamma_2 + s^T_{23}\gamma_3] = \frac{C_\sigma}{V}[s^S_{12}\gamma_1 + s^S_{22}\gamma_2 + s^S_{23}\gamma_3] \tag{2.146}$$

$$\alpha_3 = \frac{C_\eta}{V}[s^T_{13}\gamma_1 + s^T_{23}\gamma_2 + s^T_{33}\gamma_3] = \frac{C_\sigma}{V}[s^S_{13}\gamma_1 + s^S_{23}\gamma_2 + s^S_{33}\gamma_3] \tag{2.147}$$

A model of orthorhombic polyethylene [Bru98] provides a good example of the interplay of thermal pressure and compliance. There is a strong negative cross compliance s_{12} perpendicular to the polymer chains, so that a small change in the anisotropy of the Grüneisen tensor with temperature leads to a much larger change in that of the thermal expansion. Examples for axial crystals may be found in [Mun69, Mun72] and in Chs. 5 and 6.

2.8.5. Thermodynamic Stiffnesses $C_{\lambda\mu}$

The Thermodynamic Stress t_λ. In general the Cauchy stress is not thermodynamically conjugate to any set of strain coordinates. The stress conjugate to η_λ is

$$t_\lambda = \frac{1}{\overset{\circ}{V}}\left(\frac{\partial U}{\partial \eta_\lambda}\right)_{\eta',S} = \frac{1}{\overset{\circ}{V}}\left(\frac{\partial F}{\partial \eta_\lambda}\right)_{\eta',T} \tag{2.148}$$

where $\overset{\circ}{V}$ is the volume in the reference configuration. t_λ is equal to σ_λ only in the reference configuration, or when all stress coordinates are zero. The general relation between the thermodynamic and Cauchy stresses is discussed briefly in [Bar98].

The advantage of using t_λ is that it is a purely thermodynamic variable, independent of whether the material has been rotated. Since t_λ is defined by Eq. (2.148) for all states of strain, a full set of thermodynamic relationships can be developed straightforwardly in a systematic manner, to any order of differentiation, and related later to the Cauchy stress if required (e.g., [Bar98]). Here we shall be concerned only with the elastic properties, so that we can relate to each other three different sets of second order stiffnesses all commonly found in current literature, and go on to discuss their stress-dependence in terms of higher order thermodynamic stiffnesses.

Second Order Stiffnesses. Thermodynamic stiffnesses are usually written in upper case. The second order stiffnesses are

$$C^S_{\lambda\mu} = \left(\frac{\partial t_\lambda}{\partial \eta_\mu}\right)_{\eta',S} = \frac{1}{\overset{\circ}{V}}\left(\frac{\partial^2 U}{\partial \eta_\lambda \partial \eta_\mu}\right)_{\eta',S}$$

$$C^T_{\lambda\mu} = \left(\frac{\partial t_\lambda}{\partial \eta_\mu}\right)_{\eta',T} = \frac{1}{\overset{\circ}{V}}\left(\frac{\partial^2 F}{\partial \eta_\lambda \partial \eta_\mu}\right)_{\eta',T} \tag{2.149}$$

Unless the stress is zero, these stiffnesses are different from the stress–strain $c_{\lambda\mu}$ defined in Section 2.8.3 even in the reference configuration; and they are different again from another set of stiffnesses sometimes used in theoretical modelling, *viz.* those obtained from second order derivatives of F or U with respect to the infinitesimal strain coordinates e_λ. The relations between these three sets of second order stiffnesses (in the reference configuration) take a simple form when the solid in the reference configuration is under hydrostatic pressure $\mathring{P}$; thus for the adiabatic stiffnesses

$$\mathring{C}_{11} = \mathring{c}_{11} + \mathring{P} = \frac{1}{\mathring{V}} \left(\frac{\partial^2 U}{\partial e_1{}^2} \right)_{e',\omega,S} + \mathring{P} \tag{2.150}$$

$$\mathring{C}_{12} = \mathring{c}_{12} - \mathring{P} = \frac{1}{\mathring{V}} \left(\frac{\partial^2 U}{\partial e_1 \partial e_2} \right)_{e',\omega,S} \tag{2.151}$$

$$\mathring{C}_{44} = \mathring{c}_{44} + \mathring{P} = \frac{1}{\mathring{V}} \left(\frac{\partial^2 U}{\partial e_4{}^2} \right)_{e',\omega,S} + \frac{1}{2}\mathring{P} \tag{2.152}$$

When the stress $\mathring{\sigma}_\lambda$ in the reference state is anisotropic, the relation between the $\mathring{C}_{\lambda\mu}$ and the $\mathring{c}_{\lambda\mu}$ can be written as

$$\mathring{c}_{\lambda\mu} = \mathring{C}_{\lambda\mu} + P_{\lambda\mu} \tag{2.153}$$

where $P_{\lambda\mu}$ is the matrix

$$\begin{pmatrix} \mathring{\sigma}_1 & -\mathring{\sigma}_1 & -\mathring{\sigma}_1 & 0 & \mathring{\sigma}_5 & \mathring{\sigma}_6 \\ -\mathring{\sigma}_2 & \mathring{\sigma}_2 & -\mathring{\sigma}_2 & \mathring{\sigma}_4 & 0 & \mathring{\sigma}_6 \\ -\mathring{\sigma}_3 & -\mathring{\sigma}_3 & \mathring{\sigma}_3 & \mathring{\sigma}_4 & \mathring{\sigma}_5 & 0 \\ -\mathring{\sigma}_4 & 0 & 0 & \frac{1}{2}(\mathring{\sigma}_2 + \mathring{\sigma}_3) & \frac{1}{2}\mathring{\sigma}_6 & \frac{1}{2}\mathring{\sigma}_5 \\ 0 & -\mathring{\sigma}_5 & 0 & \frac{1}{2}\mathring{\sigma}_6 & \frac{1}{2}(\mathring{\sigma}_3 + \mathring{\sigma}_1) & \frac{1}{2}\mathring{\sigma}_4 \\ 0 & 0 & -\mathring{\sigma}_6 & \frac{1}{2}\mathring{\sigma}_5 & \frac{1}{2}\mathring{\sigma}_4 & \frac{1}{2}(\mathring{\sigma}_1 + \mathring{\sigma}_2) \end{pmatrix}$$

We note that although $C_{\lambda\mu} = C_{\mu\lambda}$, $c_{\lambda\mu} = c_{\mu\lambda}$ only for a solid under hydrostatic pressure, when Eq. (2.153) reduces to Eqs. (2.150)–(2.152). An expression for the elements of $P_{\lambda\mu}$ in tensor notation is given in Eq. (2.161).

Thermodynamic compliance matrices $S_{\lambda\mu}$ are reciprocal to the stiffness matrices $C_{\lambda\mu}$.

Higher Order Stiffnesses. Higher order thermodynamic stiffnesses are derivatives of lower order stiffnesses with respect to the η_λ. They may be adiabatic, such as

$$C^S_{\lambda\mu\nu} = \left(\frac{\partial C^S_{\lambda\mu}}{\partial \eta_\nu} \right)_{\eta',S} = \left(\frac{\partial^2 t_\lambda}{\partial \eta_\mu \partial \eta_\nu} \right)_{\eta',S} = \frac{1}{\mathring{V}} \left(\frac{\partial^3 U}{\partial \eta_\lambda \partial \eta_\mu \partial \eta_\nu} \right)_{\eta',S} \tag{2.154}$$

isothermal, or "mixed." The most important example of "mixed" is given by the third order stiffnesses

$$C^{ST}_{\lambda\mu\nu} = \left(\frac{\partial C^S_{\lambda\mu}}{\partial \eta_\nu} \right)_{\eta',T} = \left[\frac{\partial}{\partial \eta_\nu} \left(\frac{\partial t_\lambda}{\partial \eta_\mu} \right)_{\eta',S} \right]_{\eta',T} \tag{2.155}$$

which are determined experimentally by ultrasonic measurements under varying stress. Relations between pure and mixed third order stiffnesses are discussed by Skove and Powell [Sko67].

2.8.6. Tensor Notation

Despite the convenience and simplicity of Voigt notation, a full tensor notation is needed for some topics — for example, the propagation of elastic waves. The second order stiffnesses then appear as fourth rank tensors; e.g.,

$$c^S_{\alpha\beta\gamma\delta} = \left(\frac{\partial \sigma_{\alpha\beta}}{\partial \eta_{\gamma\delta}}\right)_{\eta',S} \tag{2.156}$$

At this point it must be made clear what is meant by differentiating partially with respect to the elements of a symmetric tensor, and in particular how $\eta_{\alpha\beta}$ is considered to be altered while $\eta_{\beta\alpha}$ is kept constant. The convention is that the function to be differentiated is first expressed in a form symmetric to the interchange of $\eta_{\alpha\beta}$ and $\eta_{\beta\alpha}$, and then all nine elements are treated as independent during differentiation. With this convention the stiffnesses $c_{\alpha\beta\gamma\delta}$ equal the corresponding Voigt stiffnesses; e.g.,

$$c_{1123} = c_{1132} = c_{14} \tag{2.157}$$

However, the compliance elements have a factor of one half for each off-diagonal $\alpha\beta$ or $\gamma\delta$; e.g.,

$$s_{1111} = s_{11}, \quad s_{1122} = s_{12}, \quad s_{1123} = \frac{1}{2}s_{14}, \quad s_{1213} = \frac{1}{4}s_{65} \tag{2.158}$$

Also, while all the $\gamma_{\alpha\beta}$ equal the corresponding γ_λ, the thermal expansion coefficients have similar relationships to those between $\eta_{\alpha\beta}$ and η_λ, i.e., $\alpha_{11} = \alpha_1$ but $\alpha_{12} = \frac{1}{2}\alpha_6$. The reciprocal relationship between stiffnesses and compliances then becomes

$$c_{\alpha\beta\gamma\delta}s_{\gamma\delta\xi\eta} = s_{\alpha\beta\gamma\delta}c_{\gamma\delta\xi\eta} = \frac{1}{2}(\delta_{\alpha\xi}\delta_{\beta\eta} + \delta_{\alpha\eta}\delta_{\beta\xi}) \tag{2.159}$$

All the thermodynamic equations of Sections 2.8.3–2.8.4 can now be written in tensor form. For example, Eq. (2.136) becomes

$$\alpha_{\alpha\beta} = \frac{C_\eta}{V}s^T_{\alpha\beta\gamma\delta}\gamma_{\gamma\delta} = \frac{C_\sigma}{V}s^S_{\alpha\beta\gamma\delta}\gamma_{\gamma\delta} \tag{2.160}$$

The difference $P_{\lambda\mu}$ between the thermodynamic and stress–strain stiffnesses in the reference configuration can be expressed in tensor notation as

$$P_{\alpha\beta\gamma\delta} = \frac{1}{2}(\delta_{\alpha\gamma}\mathring{\sigma}_{\beta\delta} + \delta_{\beta\delta}\mathring{\sigma}_{\alpha\gamma} + \delta_{\alpha\delta}\mathring{\sigma}_{\beta\gamma} + \delta_{\beta\gamma}\mathring{\sigma}_{\alpha\delta}) - \delta_{\gamma\delta}\mathring{\sigma}_{\alpha\beta} \tag{2.161}$$

Cauchy Relations. Here we digress from thermodynamics to describe simple relations between elastic stiffnesses which are often used as a test for the predominance of central forces. They apply strictly only to static models in which each atom is a center of inversion symmetry, so that all interatomic distances change under strain uniformly, and when the only interactions are pair potentials. By writing these potentials in the form $f(r^2)$, and remembering that in the strained state r^2 is increased from $\mathring{r}^2$ by $2\eta_{\alpha\beta}\mathring{r}_\alpha\mathring{r}_\beta$, it follows that for each pair of atoms

$$\partial^2 f(r^2)/\partial\eta_{\alpha\beta}\partial\eta_{\gamma\delta} = \mathring{r}_\alpha\mathring{r}_\beta\mathring{r}_\gamma\mathring{r}_\delta f''(\mathring{r}^2)$$

and hence that the value of $C_{\alpha\beta\gamma\delta}$ is unchanged when the indices $\alpha, \beta, \gamma, \delta$ are permuted; for example, $C_{2233} = C_{2323}$. This gives six relations in Voigt notation,

$$\begin{aligned} &C_{23} = C_{44}, \qquad C_{13} = C_{55}, \qquad C_{12} = C_{66}, \\ &C_{14} = C_{56}, \qquad C_{25} = C_{46}, \qquad C_{36} = C_{45} \end{aligned} \tag{2.162}$$

Similarly, Cauchy relations for higher order stiffnesses follow from permuting the Cartesian indices in $C_{\alpha\beta\ldots}$. However, the best known application is to the second order stiffnesses of unstressed cubic crystals, for which the thermodynamic and stress–strain stiffnesses are equivalent and there is but one relation $c_{12} = c_{44}$. As a test for the dominance of central forces, it is best applied at low temperatures, where the crystal approximates more closely to a static model.

2.8.7. Velocities of Elastic Waves

During wave propagation the local strain-rotation tensor $u_{\alpha\beta}$ is given at any point of the body by

$$u_{\alpha\beta} = \frac{\partial x_\alpha}{\partial \mathring{x}_\beta} \tag{2.163}$$

Elastic waves normally propagate adiabatically, and the equations of motion in the harmonic limit of small amplitudes involve the adiabatic second order derivatives of the energy with respect to the $u_{\alpha\beta}$, written $A^S_{\alpha\beta\gamma\delta}$, which are simply related to the thermodynamic stiffnesses $C^S_{\alpha\beta\gamma\delta}$:

$$A^S_{\alpha\beta\gamma\delta} = \left(\frac{\partial^2 U}{\partial u_{\alpha\beta}\partial u_{\gamma\delta}}\right)_{u',S} = C^S_{\alpha\beta\gamma\delta} + \delta_{\alpha\gamma}\mathring{\sigma}_{\beta\delta} \tag{2.164}$$

It is the $A^S_{\alpha\beta\gamma\delta}$ that determine the elastic wave velocities. Note that they are not symmetric with respect to separate interchange of α with β and γ with δ, and so cannot be expressed in Voigt notation.

For propagation in the direction of the unit vector $\hat{\mathring{\mathbf{q}}}$, the wave-form is $\exp i[q(\hat{\mathring{\mathbf{q}}}\cdot\mathring{\mathbf{x}}) - \omega t]$. The three eigenvalues of the matrix

$$M_{\alpha\gamma} = \frac{1}{\rho}A^S_{\alpha\beta\gamma\delta}\hat{\mathring{q}}_\beta\hat{\mathring{q}}_\delta \tag{2.165}$$

then give the squares of the three possible wave velocities, and the normalized eigenvectors $\hat{\mathbf{w}}$ give the corresponding directions of polarization. Eqs. (2.164) and

(2.165) are valid for arbitrary stress in the reference state. When the stress is isotropic, $A^S_{\alpha\beta\gamma\delta}$ can be replaced in Eq. (2.165) by the stress–strain stiffness $c^S_{\alpha\beta\gamma\delta}$, because the difference between them is antisymmetric with respect to interchange of β and δ.

Measurement of wave velocities in suitably chosen directions can thus be used to determine the second order elastic stiffnesses. For cubic symmetry we have

$$\rho M_{11} = c_{11}(\hat{q}_1)^2 + c_{44}[(\hat{q}_2)^2 + (\hat{q}_3)^2], \text{ etc.,}$$
$$\rho M_{12} = (c_{12} + c_{44})\hat{q}_1\hat{q}_2, \quad \text{etc.} \tag{2.166}$$

For propagation in the [100] direction this becomes a diagonal matrix, giving a longitudinal velocity $(c_{11}/\rho)^{1/2}$ and two equal transverse velocities $(c_{44}/\rho)^{1/2}$, as described in Section 3.4.4. Application to different symmetries is discussed in [Wal70, Thu64].

For propagation of waves in piezoelectric media, explicit account must be taken of the electric field arising from polarization (see [Bor54, Section 32]).

2.9. CALCULATION OF Θ_0^{el}, γ_0^{el} AND $\gamma_{\lambda,0}^{el}$ FROM ELASTIC DATA

Θ_0^{el} can be computed from the integral of Eq. (2.85) with velocities obtained from the eigenvalues of the matrix $M_{\alpha\gamma}$ in Eq. (2.165). Before powerful electronic computers became widely available various ingenious methods were employed for this integration (see [Ale65]), but now it can be readily performed to any degree of required precision with suitable variables such as the spherical polar coordinates θ and ϕ, in terms of which

$$\hat{\mathbf{q}} = (\sin\theta\cos\phi, \sin\theta\sin\phi, \cos\theta), \quad d\Omega = \sin\theta\, d\theta\, d\phi \tag{2.167}$$

A standard computer subroutine can be used to obtain the eigenvalues and eigenvectors of the matrix $M_{\alpha\gamma}$ for each direction of $\hat{\mathbf{q}}$ used in the integration.

The integration can also be done to a good accuracy by using the set of 489 directions uniformly distributed over 1/48 of the unit sphere given by Overton and Schuch [Ove65]. This is directly applicable to cubic crystals, and can be extended to lower symmetries by appropriate permutation of the direction cosines.

Numerical tables for Θ_0^{el} have been published by de Launay [deL54, deL56] covering a wide range of stiffness ratios for non-piezoelectric cubic crystals, and by Wolcott [Wol59] covering some ratios for hexagonal crystals.

The Grüneisen functions at $T = 0$ can be expressed as derivatives of Θ_0, and so

$$\gamma_0^{el} = -\frac{d\ln\Theta_0^{el}}{d\ln V}, \qquad \gamma_{\lambda,0}^{el} = -\left(\frac{\partial\ln\Theta_0^{el}}{\partial\eta_\lambda}\right)_{\eta'} \tag{2.168}$$

In principle there are two different but equivalent ways of evaluating them from elastic properties.

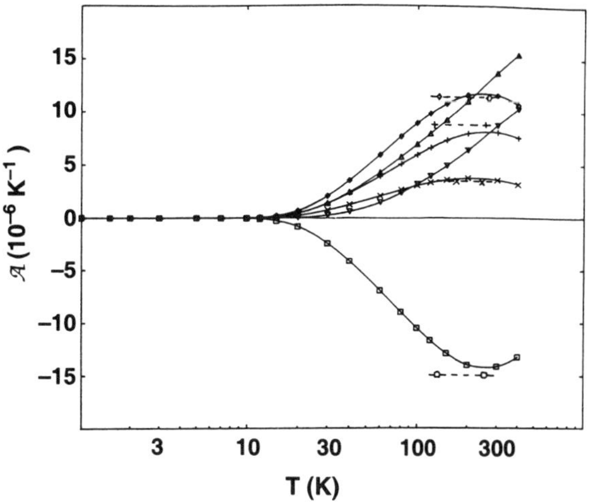

Fig. 2.17. Internal and macroscopic expansion coefficients for a central force model of α-quartz (see text). —, model; - - -, approximate experimental internal coefficients [LeP80]. $\triangle$, $\alpha_\perp$; ∇, $\alpha_\parallel$; $+$, $\mathcal{A}_u$; $\times$, $\mathcal{A}_x$; $\square$, $\mathcal{A}_y$; $\diamond$, $\mathcal{A}_z$. From [Bar87].

Derivatives of Θ_0^{el} can be obtained by brute force calculation of Θ_0^{el} for neighboring strains, or (for cubic crystals) by using de Launay's tables [Dan62a]. This method has often been used to obtain γ_0^{el} from pressure derivatives of the stress–strain stiffnesses $c_{\lambda\mu}$, particularly for cubic crystals. Application to piezoelectric crystals is straightforward if the volume-dependence of relevant properties is known [Han74, Bar77a]. It can also be used for theoretical models.

The second method averages mode Grüneisen parameters of elastic waves over all polarizations and directions of propagation, weighted by inverse third powers of their velocities; thus

$$\gamma_0^{el} = \frac{\int \sum_{s=1}^{3}\{[v_s(\hat{\mathbf{q}})]^{-3}\gamma(\hat{\mathbf{q}},s)\}d\Omega}{\int \sum_{s=1}^{3}[v_s(\hat{\mathbf{q}})]^{-3}d\Omega}, \qquad \gamma_{\lambda,0}^{el} = \frac{\int \sum_{s=1}^{3}\{[v_s(\hat{\mathbf{q}})]^{-3}\gamma_\lambda(\hat{\mathbf{q}},s)\}d\Omega}{\int \sum_{s=1}^{3}[v_s(\hat{\mathbf{q}})]^{-3}d\Omega} \tag{2.169}$$

The mode Grüneisen parameters are derived from the thermodynamic elastic stiffnesses. In tensor notation, Wallace gives for an unstressed crystal at $T = 0$ [Wal70]

$$\gamma_{\alpha\beta}(\hat{\mathbf{q}},s) = -w_\alpha w_\beta - \frac{1}{2\rho v^2}\sum_{\zeta=1}^{3}\sum_{\eta=1}^{3}\hat{q}_\zeta\hat{q}_\eta\Big[C_{\zeta\eta\alpha\beta} + \sum_{\gamma=1}^{3}\sum_{\delta=1}^{3}C_{\zeta\gamma\eta\delta\alpha\beta}w_\gamma w_\delta\Big] \tag{2.170}$$

where $\mathbf{w}$ is the normalized polarization vector for the wave and v is its velocity; and the general expression for $\gamma(\hat{\mathbf{q}},s)$ is

$$\gamma(\hat{\mathbf{q}},s) = B\sum_{\alpha=1}^{3}\sum_{\beta=1}^{3}\sum_{\xi=1}^{3}s_{\alpha\beta\xi\xi}\gamma_{\alpha\beta}(\hat{\mathbf{q}},s) \tag{2.171}$$

See [Wal70] for further discussion, and [Bru67] for application to crystals of high symmetry.

2.10. INTERNAL STRAIN

In many crystals, the positions of atoms within the unit cell are not wholly determined by symmetry, and are specified in crystallography by one or more additional parameters. These parameters describe the state of *internal strain*, and they (or some equivalent set of internal coordinates) can be treated thermodynamically on the same footing as the macroscopic strain coordinates. This is particularly useful in the study of theoretical models, where it is usually simplest to calculate a generalized Helmholtz energy as a function of both internal and macroscopic strain. The subsequent derivation of purely macroscopic properties is discussed in [Bar71, Bar98], and illustrated for a model of tellurium by Gibbons [Gib73].

Internal expansion coefficients are temperature derivatives of internal strain coordinates. They can be calculated for theoretical models either by minimization of free energies at different temperatures with respect to internal and external strains or from analytical expressions (e.g., [Tay97a, Kan95, Bru98]). Experimentally they are usually determined from the analysis of intensities in x-ray diffraction, which for most crystals is not sufficiently sensitive to detect changes in internal strain below about 100 K (see Chapter 3). But when they are available they provide additional checks on the validity of theoretical models of more complex crystals, as exemplified by the four independent internal expansion coefficients of α-quartz (Fig. 2.17).

Because straining a crystal can lower its symmetry, modelling elasticity usually requires a fuller set of internal strain coordinates than is provided by the crystallographic parameters. A comprehensive discussion of internal elasticity is given in tensor notation by Cousins [Cou78].

Chapter 3

Measurement Techniques

3.1. GENERAL PRINCIPLES

3.1.1. Introduction

Measurements of C_P and expansivity have many requirements in common: accurate measurement of temperature and temperature difference, temperature control, and thermal isolation. They differ in that measurement of C_P requires accurate knowledge of heat input ΔQ to determine

$$C_P = (dQ/dT)_P = (\Delta Q/\Delta T)_{\Delta T \to 0}$$

while α (or β) requires accurate measurement of dimensional changes Δl (or ΔV) to determine

$$\beta = (\partial \ln V/\partial T)_P = (\Delta V/V\Delta T)_{\Delta T \to 0}$$

Another difference is that measurements of expansion may be taken during either heating or cooling, which is useful when studying phase transitions and hysteresis.

Measurements of elastic moduli usually place less stringent requirements on temperature measurement and stability, since the moduli rarely change rapidly with T except at a phase transition. But they usually require precise measurement of another parameter, travel time and/or frequency of an ultrasonic wave.

There are few, if any, complete measurement systems (cryostat and all) available off the shelf suited to these properties at low temperatures; a possible exception is C_P, for which two firms have recently introduced 'mini' calorimeter modules to insert in their multi-purpose measurement systems (Quantum Design and Oxford Scientific). Therefore we devote the following three sections to some details of methods and cryostats, hopefully sufficient to enable the reader to judge the accuracy that can be achieved and to locate references giving more details of the 'art' of such measurements, including those of thermal anchoring, heat switches, vacuum seals, and suitable glues or cements. We also list the most important reference materials that are often used to check or calibrate a measurement system.

Some useful books on cryogenic techniques are the following:

- *Matter and Methods at Low Temperatures* [Pob96], which emphasizes properties and refrigeration methods below 4 K;
- *Experimental Principles and Methods Below 1 K* [Lou74];
- *Experimental Techniques in Condensed Matter Physics at Low Temperatures* [Ric88], which is based on a series of lectures to graduate students at Cornell and includes many useful technical details needed for successful experiments, again with emphasis on very low temperatures;
- *An Introduction to Millikelvin Technology* [Bet89];
- *Low Temperature Laboratory Techniques* [Ros73] and *Experimental Techniques in Low Temperature Physics* [Whi79], which are each concerned with the whole temperature range below 273 K.

3.1.2. Temperature Measurement

A vital ingredient in all thermophysical property measurements is accurate knowledge of temperature. In practice we measure temperature and temperature change using various instruments and properties that happen to be suited to the particular range, accuracy and conditions, e.g., resistance thermometers, thermocouples, magnetic susceptibility, thermal expansion, etc. However, the foundations of our measurement depend on the concept of thermodynamic temperature T, which can be determined by various methods:

1. The primary method is ideal gas thermometry, which uses the equation of state $PV = RT = N_A kT$ for a perfect gas; in practice helium at sufficiently low pressure approaches 'perfection.'
2. Acoustic gas thermometry, which depends on measurement of sound velocity and requires corrections for an imperfect gas as in 1.; also dielectric constant gas thermometry.
3. Electrical noise in a resistor of Ω ohms, which gives a mean square voltage $\bar{V^2} = 4kT\Omega\Delta f$, where Δf is the bandwidth.
4. Total black-body radiation.

These are painstaking measurements to perform at high accuracy, and generally not suited to everyday measurements of physical properties. Practical temperature scales have been adopted which relate as closely as possible to the thermodynamic scale and can be realized with resistance thermometers, thermocouples etc. The development of these scales culminating in ITS-90 (the International Temperature Scale of 1990) is a long and fascinating story, well told by Quinn in *Temperature* [Qui90].

The unit of Temperature is the Kelvin, defined as 1/273.16 of the interval from 0 K to the triple point of water (0.01 °C). An outline of the text of ITS-90 which has cryogenic relevance in [Qui90, p. 59] (see also [Pre90]) says:

> '... Between 0.65 K and 5.0 K, T_{90} is defined in terms of the vapor-pressure temperature relations of ^{3}He and ^{4}He.
>
> Between 3.0 K and the triple point of neon (24.5561 K) T_{90} is defined by means of a helium gas thermometer calibrated at three experimentally realizable temperatures having assigned numerical values (defined fixed points) and using specified interpolation procedures.
>
> Between the triple point of equilibrium hydrogen (13.8033 K) and the freezing point of silver (961.78°C) T_{90} is defined by means of platinum resistance thermometers calibrated at specified sets of defining fixed points and using specified interpolation procedures ... '

Some of the defining fixed points and the uncertainty ΔT of their thermodynamic temperatures are in Table 3.1, together with some secondary points (superconducting transitions of the Standard Reference Materials (SRMs) produced by the National Bureau of Standards or NBS (now the National Institute of Standards and Technology or NIST) which are of cryogenic interest.

The ITS-90 equations for the vapor pressures of ^{3}He and ^{4}He are given in [Qui90, Pre90]. Note that the vapor pressures of the helium isotopes published in earlier cryogenic texts under the headings T_{58} and T_{62} may be in error by several millikelvins; for less precise needs there are useful tables of vapor pressures of helium, hydrogen, nitrogen and oxygen in such texts [Ros73, Whi79].

For the two ranges covered by the platinum resistance (13.8033 K to 273.16 K and 0°C to 961.78° C), there are polynomial reference functions linking the resistance ratio, $W(T_{90}) = R(T_{90})/R_{273.16}$, for a particular thermometer, see [Qui90, p. 454]. Other thermometers made of suitable pure strain-free platinum (PTRs) can be calibrated at the fixed points and deviation functions from these reference equations can be produced. For most practical purposes, we prefer the so-called Z-function, $Z(T) = (R_T - R_{4.2})/(R_{273} - R_{4.2})$, which is tabulated for a group of high-quality platinum thermometers in [Whi79, p. 310] and should be valid for others of similar quality within deviation limits of about 25 mK above 20 K [Bes78]. Resistance thermometers are available from commercial sources with calibrations (at a price) which are traceable to the ITS-90 scale through the national standards laboratories such as the National Institute of Standards and Technology (NIST, formerly NBS at Gaithersburg) and National Physical Laboratory (NPL, Teddington).

Details of the construction and performance of the commonly used thermometers are given in cryogenic texts [Pob96, Qui90, Whi79]. Quoting from a Summary in [Whi79, p. 123], we list the following:

Table 3.1. Defining fixed points of ITS–90 with estimates of their uncertainty [Qui90, Pre90]. Lower section shows some superconducting transition temperatures, T_c, of metals encapsulated in SRM 767 and SRM 768, see [Qui90, p. 183]

Fixed points	T_{90}/K	ΔT/mK
^{4}He b.p.	4.2221	0.3
(at pressure of 101325 Pa)		
e-H_2 t.p.	13.8033	0.5
Ne t.p.	24.5561	0.5
O_2 t.p.	54.3584	1
Ar t.p.	83.8058	1.5
Hg t.p	234.3156	1.5
Water t.p.	273.16	0
Ga m.p.	302.9146	1
In f.p.	429.7485	3
Also Sn, Zn, Al etc. ...		
Superconductor	T_c/K	Width/mK
W	0.016	0.7
Be	0.023	0.2
Ir	0.099	0.8
$AuAl_2$	0.1605	0.3
$AuIn_2$	0.2065	0.4
Cd	0.5190	0.5–0.8
Zn	0.8510	2.5–10
Al	1.1796	1.5–4
In	3.4145	0.5–2.5
Pb	7.1996	0.6–2

1. Those with sensitivity and stability of 1 mK:

   (a) platinum thermometers encapsulated in sheath in strain-free mount for range $T > 10K$,

   (b) RhFe resistance thermometers for range 0.5–300 K,

   (c) Ge (encapsulated) thermometers for 0.5–50 K. Some show a 'jump' (equivalent to a few mK) after cycling.

   All above are affected by magnetic fields.

2. With sensitivity of 1–10 mK and stability of < 100 mK:

   (a) platinum as thin film or in unencapsulated coil for $T > 10K$,

   (b) carbon resistors encapsulated or potted (sealed) for 0.5–100 K,

   (c) carbon-in-glass for 1–300 K, relatively insensitive to magnetic field,

   (d) capacitance (e.g., $SrTiO_3$) for range 0.5–60 K, not affected by field but calibration may be affected by cooling cycle.

3. With sensitivity of 10 mK and stability of 100 mK:

    (a) *p-n*-junction diodes for 1–300 K,

    (b) thermocouples of AuFe for 2–300 K,

    (c) CLTS (manganin + nickel) for 2–300 K.

More recently, other 'thermistor' materials with negative dR/dT characteristics have been developed for low temperature use and might be included in categories 2. or 3. above. Two which are commercially available and useful down to below 1 K are a thin-film based on RuO_2 and a thick-film (chip) using zirconium oxynitride. Commercial versions from Lake Shore Cryotronics are called Rox and Cernox respectively. They are generally less sensitive to magnetic fields than most other resistance sensors. References to these and other semiconducting materials are given in a review of progress in cryogenic thermometry between 1982 and 1996 [Rub97].

3.1.3. Temperature Control

At low temperatures both C and α vary rapidly with T and generally involve measurement of a small temperature interval, requiring temperature control at the millikelvin level. In some instances this can be achieved by controlling the vapor pressure above the liquid refrigerant by a manostat or controlling the flow rate of a cooling gas stream. More often the temperature of sample, chamber or adiabatic shield is held steady by electrical heating in response to the signal from a suitable temperature sensor selected from the groups 1., 2. or 3. listed above. For example: a carbon or small platinum resistor (group 2.) is attached to the chamber or shield and forms one arm of a phase selective ac bridge; the out-of-balance signal is amplified and fed back into a small resistance heater attached to the chamber, shield, etc. Such electronic controllers are commercially available or can be made from an ac bridge and phase sensitive detector. With thermocouple sensors, dc amplifiers can be used [Ros73, Whi79].

3.2. HEAT CAPACITY ... BY S. J. COLLOCOTT

3.2.1. Introduction

On cooling from room temperature to liquid helium temperatures the specific heat of a typical solid decreases by three to four orders of magnitude, and becomes vanishingly small at absolute zero. The small heat capacity of solids at liquid helium, and at lower, temperatures creates difficulties for the experimentalist, because small heat influxes from the surroundings, for example vibration, can lead to significant errors in the determination of the heat capacity of a solid. Heat capacity measurements become even more challenging if the specimen has mass of a few tens of milligrams (frequently samples of larger mass are not available), as there is increased difficulty in achieving adequate thermal isolation.

Heat capacity measurements reveal much information about the electronic properties of a solid, for example the density of states at the Fermi level; about the lattice or vibronic properties of a solid, in particular the low-frequency phonon density of states, acoustic, and optic modes; about phase transitions, be they magnetic, superconducting, or structural; and about a range of other low temperature heat capacity effects, for example Schottky anomalies, magnetic spin-wave contributions, and two-level systems. A consequence of this abundance of effects is that the heat capacity can vary enormously, being several orders of magnitude larger or smaller in a given specimen as a function of temperature, as well as obviously from one solid to another. For example, the heat capacity of the rare-earth metal holmium at 0.6 K, where the nuclear hyperfine heat capacity dominates, is $\approx 5.6\ \mathrm{J \cdot mol^{-1} \cdot K^{-1}}$, decreasing to less than $0.5\ \mathrm{J \cdot mol^{-1} \cdot K^{-1}}$ at 4 K, and then increasing with increasing temperature as the lattice heat capacity begins to dominate (see Section 6.4.3); and at 19.46 K there is a peak of width 0.03 K, which attains a maximum heat capacity of $145\ \mathrm{J \cdot mol^{-1} \cdot K^{-1}}$, due to a magnetic transition from a helical to a conical spiral state [Col88, Ste89]. A great strength of heat capacity measurements is that they give information on the bulk behavior of a solid, and as such are useful in determining whether an effect observed by some other technique, for example resistivity measurements, is a feature of the bulk material or due to some other minority phase. A wide variety of low temperature heat capacity effects can be investigated using a simple pumped ^{3}He cryostat, operating over the temperature range 0.3 K to about 30 K.

There are a number of excellent review articles on low-temperature calorimetric techniques [Wes68, Sto68, Hil68, Ste83, Gme87, And88, Mar88, Wes88], and these are complemented by the more general discussion of calorimetry by Hemminger and Höhne [Hem84]. This discussion draws heavily on these reviews, and the reader is referred to them for greater detail. Recently, there have been a number of new experimental developments. These have been driven by the availability of improved instrumentation, and as a result there have been advances in small sample (<200 mg) calorimetry [DeP86, Dut88], the measurement of adsorbed gases on substrates using ac calorimetric methods [Cha89, Ken90], measurements in large magnetic fields [Kla97], and increased automation of calorimeters [Pec97]. The trend to increased automation has been accelerated with the advent of a number of manufacturers offering commercial 'turnkey' systems. These new developments will also be addressed in the context of the broader discussion of low-temperature calorimetric techniques.

3.2.2. Adiabatic Calorimetry

A convenient starting point for the measurement of low temperature specific heat is the classical definition of the specific heat (per unit mass), c_P,

$$c_P(T) = \lim_{\Delta T \to 0} (\Delta Q/\Delta T)_P / M \tag{3.1}$$

where ΔQ is a heat energy input (pulse) that causes a small temperature rise ΔT in a specimen of mass M. This 'step' or 'pulse' heating technique can be traced back to Nernst (see [Gme87]), and it remains today one of the most accurate methods for obtaining specific heat data. In practice the specimen is contained in, or thermally connected to, an addenda which consists of the specimen support system or container, thermometer, resistive heater, and any other wiring — the addenda is the calorimeter — and the addenda/sample assembly is thermally insulated from the surroundings (adiabatic conditions). Thermal equilibrium with the surroundings is established before and after the heat pulse ΔQ. The temperature, T, is monitored as a function of time, and the temperatures T_i and T_f at the beginning and end of the heat pulse are corrected for any heat exchange with the environment by extrapolating T before and after the heat pulse to the time that corresponds to the midpoint of the pulse. The temperature increment is then $\Delta T = T_f - T_i$, from which C_P is obtained at the temperature $T_m = (T_i + T_f)/2$. This technique is shown schematically in Fig. 3.1. Strictly, adiabatic conditions occur only when there is no heat transfer between the calorimeter and surrounding shield. After the heat pulse the calorimeter will be at a temperature slightly above that of the shield, and there will be a downward temperature drift; and so the experimental conditions are more appropriately described as being 'quasi-adiabatic' or 'slightly isoperibol.'

In a typical experiment from 0.3 to 20 K the calorimeter is heated by series of heat pulses and the drift rates monitored before and after each heat pulse. Ideally ΔT is kept small, so that linear extrapolation of the drift rate is sufficient to determine either T_i or T_f. The shield temperature is kept constant both before and after the heat pulse, and obviously during the pulse. It is common to adjust the shield before each data point to the temperature of the calorimeter, which minimizes the drift corrections. If addenda corrections are small, the heat capacity of a specimen may be determined with an inaccuracy of order 0.2%. In this experimental technique the specimen is at thermal equilibrium with its surroundings before and after each heat pulse. This is not so in continuous heating calorimeters, where heat is added to the specimen at a constant rate and the resulting rate of increase of temperature is measured [Coc66]. In the continuous heating calorimeter, the specimen may never be in thermal equilibrium with its surroundings.

Some comments and clarification are in order on the terms adiabatic, isoperibol and isothermal (see Fig. 3.2), which are used in the literature, frequently in an imprecise manner, to describe the modes of operation of a calorimeter. The term adiabatic refers to a calorimeter where there is no heat transfer between it and its surroundings (usually a thermal shield that is part of the cryostat). In practical terms no calorimeter is truly adiabatic, as there will always be some heat input from the surroundings, though this heat leak can be minimized by ensuring the shield and calorimeter are at the same temperature and the thermal resistance between the calorimeter and the shield is very large, i.e., the best possible thermal insulation. In an isoperibol* calorimeter the surrounding shield is maintained at a constant but

*The term 'isoperibol' (uniform surroundings) was introduced by Kubaschewski and Hultgren [Kub62].

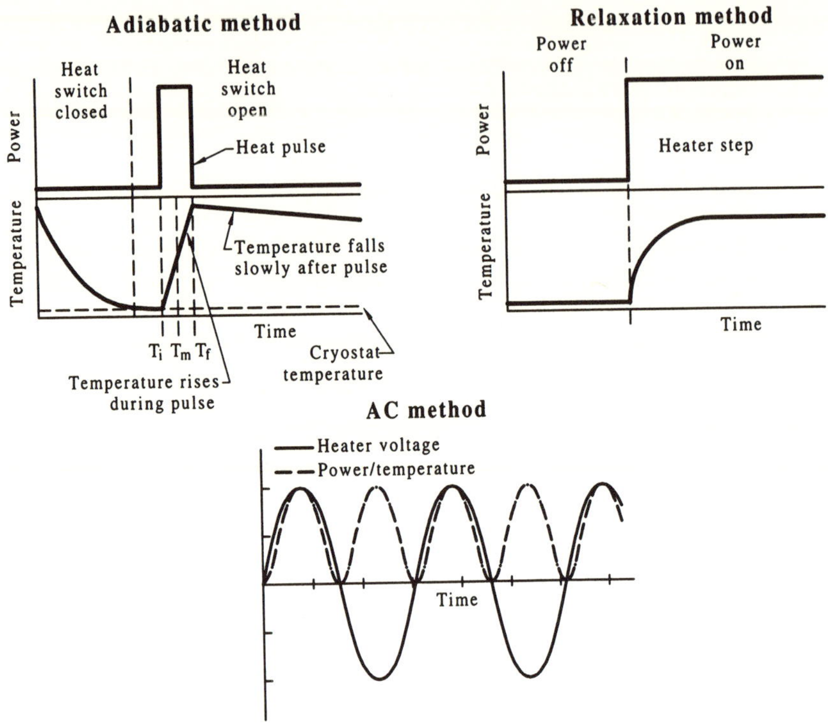

Fig. 3.1. Low temperature calorimetry methods.

different temperature to that of the calorimeter, and the thermal resistance between the calorimeter and the surrounding shield is large but of a finite value. In an isothermal calorimeter the calorimeter and surrounding shield are maintained at the same temperature and the thermal resistance between the calorimeter and surrounding shield is very small.

3.2.3. Ac-Temperature Calorimetry

The need for excellent thermal isolation and the minimization of stray heat leaks places a lower limit of about 200 mg on specimen mass for adiabatic calorimetry. The requirement for heat capacity measurements on smaller specimens has led to the development of a number of techniques, and, in 1968, Sullivan and Seidel [Sul68] introduced a technique where the specimen is heated by an ac current of angular frequency $\omega/2$ passing through a resistance heater (see Fig. 3.1). Measurement of the peak-to-peak ac temperature response, T_{ac}, by synchronously detecting the voltage across a resistance thermometer at frequency ω, using a lock-in amplifier,

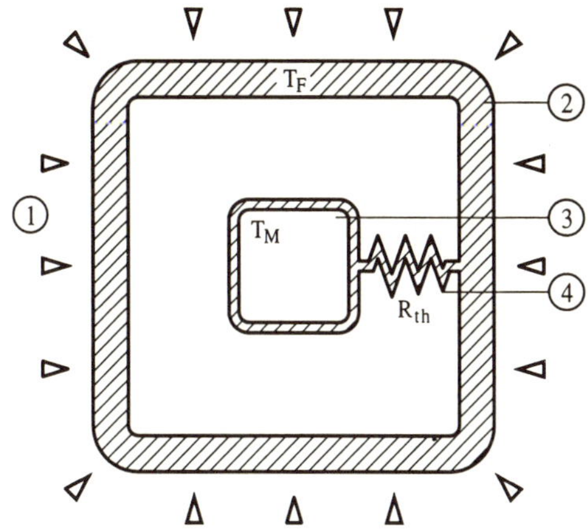

Fig. 3.2. Schematic representation of a calorimeter to highlight the various types [(1) Environment, (2) Surrounding shield, (3) Measuring system and (4) Thermal resistance, T_F Temperature of surrounding shield, T_M Temperature of measuring system and R_{th} thermal resistance]: *Isothermal* R_{th} very small and $T_F = T_M =$ constant; *Adiabatic* R_{th} very large and $T_F = T_M$; and, *Isoperibol* R_{th} fixed, T_F constant and $T_M = T_M(t)$ (Adapted from [Hem84]).

enables the total heat capacity (specimen and addenda), C_P, to be calculated from

$$T_{ac} = \frac{\dot{Q}_0}{2\omega C_P}\left[1 + \frac{1}{\omega^2\tau_1^2} + \omega^2\tau_2^2 + \frac{2K_b}{3K_s}\right]^{-1/2} \tag{3.2}$$

where $\dot{Q}_0$ is the amplitude of the sinusoidal heat flux, τ_1 is the specimen to bath relaxation time, τ_2 is the response time of the specimen, heater and thermometer to the heat input, K_b the thermal conductance of the specimen to the bath, and K_s the thermal conductance of the specimen. Equation (3.2) can be simplified through judicious choice of the experimental conditions [Ste83, Cha89, Kra84], namely $\tau_2 \ll 1/\omega$, $\tau_1 \gg 1/\omega$ and $K_s \gg K_b$, giving a simple expression for C_P,

$$C_P \simeq \frac{\dot{Q}_0}{2\omega T_{ac}} \tag{3.3}$$

Sullivan and Seidel [Sul68] demonstrated the ac-method with measurements on a 9 g specimen of indium, using an ac temperature modulation of 10 Hz with a peak-to-peak value of 4 mK. Relaxation time corrections were small and could be neglected ($\tau_1 = 2.5 \pm 0.1$ sec, $\tau_2 = (0.7 \pm 0.3) \times 10^{-3}$ sec), and they estimate an error in c_P of 1%. They report further measurements on a 82 mg single crystal of beryllium, and were able to observe relative changes in c_P of 0.04%, with an absolute accuracy of 8%. As with the adiabatic method it is necessary to correct the measured heat capacity values for the addenda contribution.

The ability of the ac-method to detect changes in heat capacity as small as 10^{-8} to 10^{-12} J·K^{-1} [Min94, Cha89, Fom97] has made it one of the favored methods for small sample calorimetry, or what has become known as microcalorimetry or nanocalorimetry. The ac-method has been used in a number of elegant experiments studying the adsorption of a range of gases on various substrates, namely ^4He on sapphire [Ken90] and on single crystal graphite [Cha89], and H_2 on gold [Bir96]. It has also been used for measurements on specimens of less than 100 mg in magnetic fields up to 20 T [Sch87a]. A variation of the ac-method is the injection of heat into the specimen by irradiation with chopped light from a tungsten lamp [Tas90] or from an electronically modulated diode laser [Mar97], instead of a resistance heater. In this case the amount of energy input into the specimen is not usually known, and a relative measure of the heat capacity is obtained, unless the calorimeter has been calibrated previously with a known, or reference, material. It should be noted that use of the ac-method is not restricted to low temperatures, and it can be used for measurements up to the melting point of refractory metals. In this broader context it is frequently referred to as 'Modulation Calorimetry,' and the reader is referred to the review of Kraftmakher [Kra84].

3.2.4. Relaxation Calorimetry

In recent times thermal relaxation calorimetry has become particularly popular as it is suitable for small samples, can be used over a wide temperature range (from below 1 K to 300 K), cryostat design and specimen mounting are simple, and signal-to-noise can be improved using signal averaging as part of a computer controlled system. Indeed, a number of the commercially available computer automated systems use the relaxation method.*

In the thermal relaxation method [Bac72, Sch74, For80, Reg86, Dut88] the specimen is connected by a weak thermal link to a constant temperature bath, at temperature T_0. The temperature of the sample is raised by a small amount, ΔT (typically $\Delta T/T \approx 1\%$), and then allowed to decay exponentially down to the bath temperature. The temperature of the specimen, T_s, is described by

$$T_s = T_0 + \Delta T \exp(-t/\tau_1) \tag{3.4}$$

where t is time and τ_1 is the specimen to bath time constant. The heat capacity, C_P, is determined from the measurement of τ_1 and the thermal conductance of the weak thermal link, K, where

$$C_P = \tau_1 K \tag{3.5}$$

Bachmann *et al.* [Bac72] used four wires of pure gold or gold alloyed with 7 at.% copper, each 25 mm long and 0.076 mm in diameter, for the thermal link. At 4.2 K the thermal conductance was 5×10^{-4} W·K^{-1} and 5×10^{-6} W·K^{-1}, for pure gold and the gold–copper alloy, respectively. Alternatively, the thermal conductivity of the

*Quantum Design, San Diego, California, USA, and Oxford Instruments, Tubney Woods, Abingdon, Oxfordshire, U. K.

link can be determined at each data point using the heater power, P, where $K = P/\Delta T$ [Sch74, Ban92]. Choice of the thermal link material gives the experimentalist control of K, and hence τ_1, which may vary typically from <50 ms to 100 secs. For the limit where K is very small, the relaxation method becomes the adiabatic method, discussed earlier. On occasions it may also be necessary to consider the behavior of τ_2, the response time of the specimen, heater and thermometer (as was defined in the ac-method). So called 'τ_2 effects' occur when τ_1 and τ_2 are comparable (for 1% accuracy it is necessary for $\tau_1 \gg \tau_2$), occurring when the specimen and thermometer are not at the same temperature, i.e., for a specimen of very low thermal conductivity. The signature of this behavior is an overshoot in the thermometer and a non-exponential shape in the decay curve, and is discussed in detail by [Bac72]. As with the adiabatic and ac methods, it is necessary to correct the measured heat capacity values for any addenda contributions. The experimental configuration can be arranged for an upward step of the constant temperature bath, and τ_1 is determined from the exponential growth towards the new higher temperature, as is shown in Fig. 3.1 [DeP86, Ban92].

The relaxation method is suitable for specimens from 1 to 100 mg. Measurement of the heat capacity of a 90 mg copper specimen by [For80] shows that the method is accurate to 1%, and resolutions in the heat capacity of 0.1 to 5 $\mu J \cdot K^{-1}$ are achievable [DeP86]. The availability of powerful personal computers combined with fast data acquisition cards has resulted in the relaxation method's becoming very popular, as data acquisition and determination of the heat capacity are done in real-time.

3.2.5. Diffusive Heat Pulse Calorimetry

There is one technique, the diffusive heat pulse method [Ber70, Fil75, Mad88], which cannot be easily categorized with the above methods. It differs in that it is not necessary for the specimen to be at a homogeneous temperature. Its advantage is that it allows simultaneous measurement of both the heat capacity and thermal conductivity, and it is suitable for specimens of small mass, such as thin films. In this method a long sample, thermally isolated from its surroundings by a vacuum, is thermally fixed at one end to a constant temperature bath. At the isolated end of the sample there is a heater, and arranged along the length of the specimen, between the heater and constant temperature bath, there may be one or more thermometers. Pulses of heat, typically of duration 10^{-4} seconds, are injected into the specimen. Bertman *et al.* [Ber70] have solved the diffusion equation for this configuration, and analysis of the pulse height and arrival time are used to determine the heat capacity and thermal conductivity, with a typical inaccuracy of ±5% [Fil75]. A number of more sophisticated numerical techniques for fitting the data have been proposed to overcome problems associated with the boundary resistance between the specimen and the attached thermometer [McM94, McM96]. In principle this method has the potential for high accuracy, as the diffusivity depends only on time measurements, and signal averaging can be used.

3.2.6. Advantages and Disadvantages of the Various Methods

It is important to tailor the heat capacity measurement method to the particular effect or anomaly being investigated. The adiabatic method works well for specimens with a mass greater than about 0.5 g and up to several tens of grams, in the temperature range from 0.3 to 30 K, and it has a high accuracy. Generally, large sample adiabatic methods are not suitable for temperatures below about 0.3 K, as it is common to use a mechanical heat switch [Hil68, Mar88] to cool the calorimeter, and operation of the heat switch at the lowest temperature injects a significant amount of heat energy into the calorimeter. (The use of helium exchange gas to cool the calorimeter is to be avoided, because heats of desorption or evaporation may be comparable with the heat capacity of the calorimeter, and it may introduce an unexpected heat leak to the surrounding temperature shield [Mar88].) Operation to lower temperatures is possible if a superconducting heat switch is used [Ste83]. Thermal isolation of the specimen may become difficult for a specimen mass below 0.5 g, as heat leaks may become large when compared to the experimental heat input. Also, for small specimens the heat capacity of the addenda may become the largest portion, 80% or 90%, of the measured heat capacity, necessitating large addenda corrections which limit the accuracy of the heat capacity determination. There are also limitations when the adiabatic method is used to examine phase transitions, as finite values of ΔT must be used. It may not be possible to resolve the fine detail of a phase transition if the values of ΔT are too coarse and the phase transition occurs over a very small temperature range.

Both the ac and relaxation methods are suitable for specimens with a mass in the range 1 to 100 mg and in the temperature range 1 to 30 K. The upper temperature limit is set by the increasing addenda contribution, but on occasions they may be used satisfactorily to higher temperatures for specimens that have a large heat capacity. Commercial instruments (see footnote[2]) use the relaxation method and operate up to 300 K. They make use of powerful personal computers, sophisticated measuring instruments, and microlithography techniques, to fabricate an addenda with a very small heat capacity. The ac-method is the best method for resolving small changes in heat capacity. At temperatures below 1 K, for specimens that have a poor thermal conductivity (e.g., pressed powders or amorphous materials), and when the condition $\tau_1^{-1} < \omega < \tau_2^{-1}$ is not satisfied, the ac-method breaks down [Ste83]. Below 1 K the relaxation method is preferred, and it can be used with specimens of poor thermal conductivity. The availability of low cost personal computers and data acquisition systems for signal averaging, with a fast response time, has simplified data manipulation when using the relaxation method. The decay or growth profile can be measured, a number of profiles averaged, fitted using a non-linear least-squares routine to obtain τ_1, the total heat capacity determined, and a correction made for the addenda to give the specimen heat capacity, all in real time.

3.2.7. Experimental Issues and Calorimeter Design

Experimental calorimetry has benefited greatly from the advances made in electronic instrumentation, personal computers, and the use of microlithography techniques to form very small addenda, but one should not lose sight of the importance of good calorimeter and cryostat design. This includes thermometry and the choice of appropriate materials for cryostat construction. A schematic diagram of a cryostat is shown in Fig. 3.3. Whilst it is beyond the scope of this section to discuss all aspects, and the subtleties, of cryostat design and construction (it is well covered in the literature [Man59, Scu65, Hil68, Joh73, Whi79, Mar88, Pob96]), it is worthwhile to review a number of aspects that are particularly relevant to low temperature calorimetry.

A number of principles of calorimeter design, with reference to [Mar88, Pec97], can be stated.

1. The accuracy of a heat capacity measurement is critically dependent on the accuracy of the energy, time, and temperature measurements.

2. There must be good thermal equilibrium within the calorimeter. The thermometer must correctly indicate the specimen temperature.

3. For adiabatic calorimeters, any heat exchange between the specimen and its surroundings must be minimized, to minimize drift rate corrections. Below about 10 K heat conduction is via electrical connections and is proportional to ΔT (attention must be paid to good thermal anchoring of all wires). Above about 50 K heat transfer by radiation begins to dominate, and is proportional to $T^3 \Delta T$. Eliminate stray heat inputs from unexpected sources, such as those from vibration and radio frequency interference.

4. Maximize the ratio of specimen heat capacity to addenda heat capacity, to minimize any addenda corrections, and thus ensure the best possible accuracy in the determination of the heat capacity of the specimen.

5. Use a simple calorimeter and cryostat design for easy loading and unloading of the specimen. Also, the calorimeter should be able to measure a large variety of different specimens.

6. Minimize the number of 'corrections' that need to be made for unmeasured material on the calorimeter (such as varnish, solder, vacuum grease, etc.). It is preferable to mount, connect or contain the specimen to the addenda, whose heat capacity has been determined in a separate experiment, rather than gluing a heater and thermometer on to the specimen.

The germanium resistance thermometer (GRT) is the thermometer of choice for heat capacity measurements, in zero applied magnetic field, between 0.3 and 30 K. GRTs are not suitable for measurements in magnetic fields, due to their large

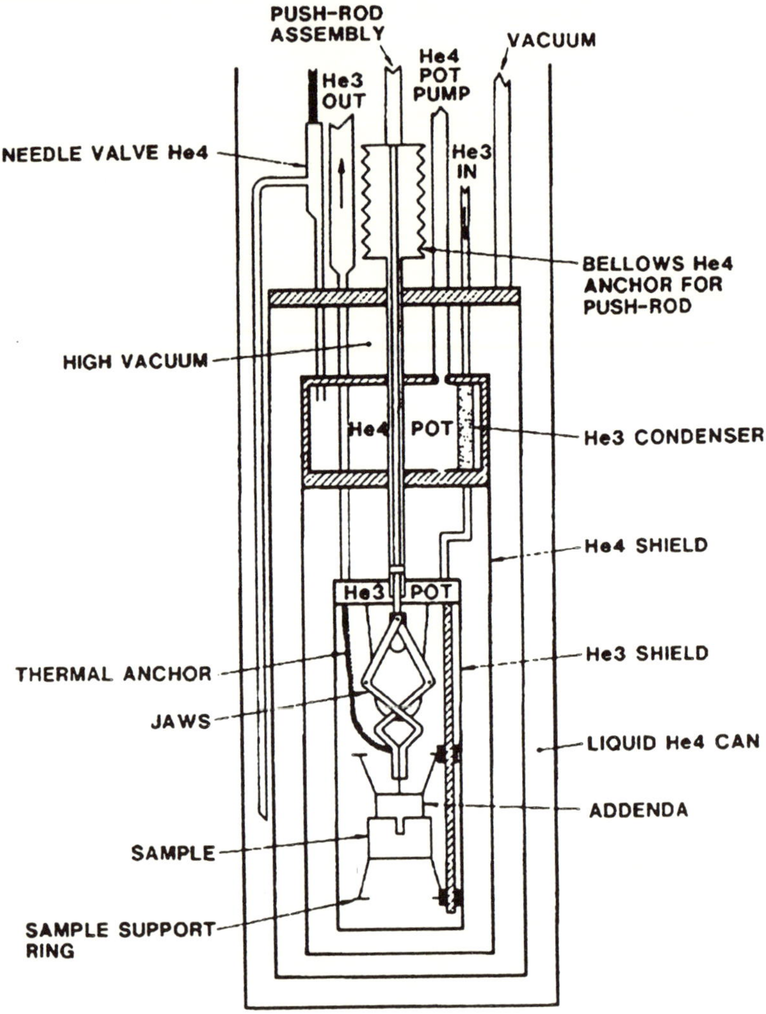

Fig. 3.3. Schematic diagram of a cryostat [Col83a].

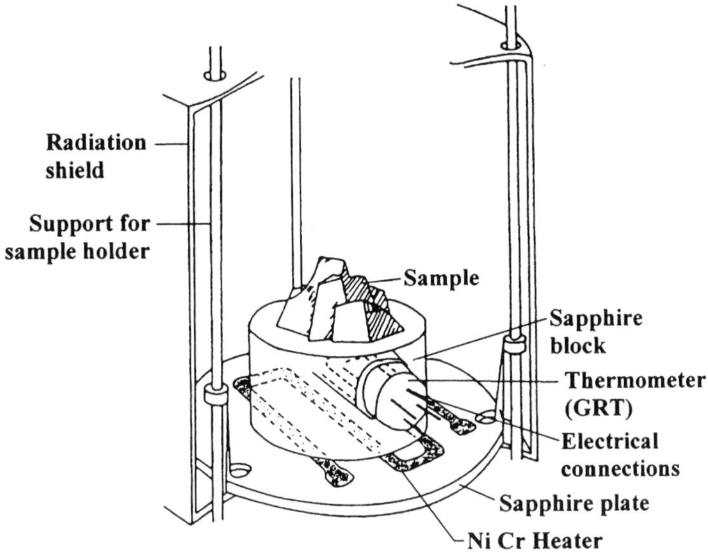

Fig. 3.4. Example of an addenda and sample holder (from [Gme81a, Var96]).

magneto-resistance; thermometers that are less sensitive to magnetic fields include carbon glass, RuO_2-based sensors and Cernox (a registered trade mark of Lake Shore Cryotronics, Inc.) [Pec97]. Measurements of heat capacity should be made on the thermodynamic temperature scale, preferably the International Temperature Scale of 1990 (ITS-90), with a thermometer calibrated appropriately [Mar75, Rub97]. A discussion of the differences between ITS-90 and earlier temperature scales (IPTS-68, EPT-76, IPTS-48 and ITS-27) is presented in [Gol92]. The effect of temperature scale differences on the analysis of heat capacity data has been addressed by [Hol72].

For best accuracy it is important to keep the addenda contribution to the total measured heat capacity as small as possible. This requires that the addenda be constructed from materials that have a low heat capacity and high thermal conductivity (to minimize thermal relaxation times), and can be well characterized, so there should be no phase transitions and amorphous materials should be avoided. In large sample adiabatic calorimeters the addenda is most commonly constructed from high-purity copper, e.g., 99.999+% ASARCO, a heater, which is bifilar wound from resistance wire, e.g., Evanohm, and a germanium resistance thermometer [Col83a]. For small specimen calorimeters, be they of the adiabatic or relaxation type, investigators have displayed considerable creativity and innovation in constructing addenda that have a very low heat capacity. Addenda described in the literature include: a holder constructed from Al_2O_3 discs with a sputtered NiCr heater, and GRT thermometer (see Fig. 3.4) [Gme81a]; thinned copper support, strain gauge heater, and unencapsulated GRT [DeP86]; Al_2O_3 holder, a heater of 160 nm thick copper sputtered onto the holder, and a thermometer made from a 0.018 cm slice cut from a carbon resistor

[Sch74]; sapphire disk, with a thin film nichrome heater and GRT bare element [Dut88]; a thin silicon membrane (2–10 μm thick), a deposited thin film CuNi heater, and a deposited thin film NbN thermometers [Fom97]; a silicon on sapphire thermometer/heater assembly [Ear81]; copper sample holder, strain gauge heater, and Cernox thermometer [Pec97]; and a sapphire disc with a AuGe thermometer and heat input by absorption of optical light pulses from a light emitting diode [Gut91].

Reference Materials. It is important to check the operation of the calorimeter by measurements on a known or reference material. This will give insight into the behavior of the calorimeter and alert the experimentalist to any odd behavior, which for example could be due to stray heat leaks. If behavior out of the ordinary is found, its source should be pin-pointed and fixed, rather than using a 'universal factor' to correct all measurements. What is thought to be a 'universal factor' may turn out not to be so, due to variations in internal thermal time constants of different materials. The reference material most often used for checking the operation of low temperature calorimeters has been the 1965 Calorimetry Conference Copper Standard, which was vacuum cast at the Argonne National Laboratory from 99.999+% high-purity copper (ASARCO Grade A-58 Copper) [Osb67]. Osborne *et al.* [Osb67] have produced the copper reference equation, a polynomial expression which gives the heat capacity between 1 and 25 K for the 1965 Calorimetry Conference Copper Standard. It has been shown by Ahlers [Ahl66] that the specific heat of vacuum annealed 99.999+% ASARCO copper differs by less than 0.1% from the 1965 Calorimetry Conference Copper Standard above 1.3 K. If a 1965 Calorimetry Conference Copper Standard is unavailable, vacuum annealed 99.999+% ASARCO copper is a perfectly acceptable substitute, for use as a low temperature calorimetric standard [Ahl66].

The temperature scale, particularly at low temperatures, has changed appreciably since the work of Osborne *et al.* [Osb67], and it is now recommended by CODATA [Whi97] and IUPAC [Mar87a] that for the range 1–30 K, the polynomial given by Holste *et al.* [Hol72] be preferred because the temperature scale (Iowa State University) on which it was based agrees most closely with ITS-90. This latter polynomial for the C_P of copper in the range 1 to 30 K is [Hol72]

$$C_P = A_1T + A_3T^3 + A_5T^5 + A_7T^7 + A_9T^9 + A_{11}T^{11} + A_{13}T^{13} \qquad (\mathrm{mJ\cdot mol^{-1}\cdot K^{-1}})$$

where the polynomial coefficients A_n (in mJ·mol^{-1}·K$^{-(n+1)}$) are as follows:

$A_1 = 0.69260$ $\quad A_7 = 1.0869\times10^{-7}$ $\quad A_{11} = 1.3343\times10^{-13}$

$A_3 = 0.047369$ $\quad A_9 = -1.9745\times10^{-10}$ $\quad A_{13} = -3.2196\times10^{-17}$

$A_5 = 1.9537\times10^{-6}$

For the range 30–300 K, IUPAC [Mar87a] produced a more lengthy (14 term) polynomial fit to C_P for Cu which agrees to better than 0.3% with other data evaluations including [Whi97, CODATA] and [Rob76, Whi84b, Mar87b]; it is the source for the selected values given in Table C.3 of Appendix C. The 14 coefficients, A_0, A_1 etc. can be found in [Mar87a, Whi97]. Measurements on the 1965 Calorimetry Conference Reference Standard and vacuum annealed 99.999+% ASARCO copper are a convenient way for investigators to assess the accuracy of their calorimetric technique, perform intercomparisons with other researchers, and to produce tables of recommended values [Ahl66, Mar67, Blo70, Hol72, Mar73, Hur74, Rob76, Ara77, Col83a, Whi84b, Mar87b, Whi97].

Another reference material recommended for temperatures above 20 or 30 K is α-alumina (sapphire), as discussed in [Mar87a, Whi97, Cas84].

Automation. A calorimetry experiment involves a series of repetitive measurements, and adjustments of shield temperatures, often over a long period of time. The experimental burden can be eased greatly by automation. Approaches to automation adopted fall roughly into two categories; data acquisition with off-line processing or on-line processing. In the latter the heat capacity is determined in real time, whilst in the former it is calculated from the acquired data on another computer at another time. The advantage of off-line processing is that all the experimental data is retained, for example temperature, time and heat input, for further reference should any unexpected behavior be observed. A disadvantage with on-line systems is that many different specimens of widely varying mass will be studied in the calorimeter, ranging perhaps from glasses to metals and superconductors with phase transitions, and it is very difficult, if not impossible, to write a software program that will adjust all the parameters given the variation in thermal time constants. (This may be a brave statement given the rapid advances being made in instrumentation and powerful personal computers.) The automation of calorimeters has tracked the improvements in measuring instruments and computers. Early systems used paper-tape, automatic ac bridges and minicomputers [Mar73, Sch75, Mos77, Mar79, Gme81b], which were soon displaced by microprocessor based systems, some of which use a high-quality digital voltmeter for thermometry [Che82, Lan81, Col83a]; these have in turn been followed by systems using the latest personal computers, and require no human intervention [Pec97]. Examples of automatic heat-pulse adiabatic calorimeters include [Mar73, Mos77, Mar79, Gme81b, Lan81, Che82, Col83a, San95, Pec97], ac calorimeters [Cha96], relaxation calorimeters [Sch75, Dut88, Ban92, Hwa97], and diffusive heat pulse calorimeters [Kwo90]. As was noted earlier, fully automatic relaxation calorimeters are available commercially.

3.3. THERMAL EXPANSION

3.3.1. Introduction

The linear coefficient of expansion α is normally measured as an average

$$\bar{\alpha} = \frac{(l_{T_2} - l_{T_1})/l_T}{(T_2 - T_1)} = \frac{1}{l_T}\left(\frac{\Delta l}{\Delta T}\right) \tag{3.6}$$

obtained from the length change (or lattice spacing change) over an interval $\Delta T = T_2 - T_2$. If l_T is taken to be $(l_{T_1} + l_{T_2})/2$ and T to be $(T_1 + T_2)/2$, the limit of $\bar{\alpha}$ as $\Delta T \to 0$ is identical to α as defined thermodynamically in Eq. (2.5):

$$\alpha = \left(\frac{\partial \ln l}{\partial T}\right)_P = \lim_{\Delta T \to 0} \frac{1}{l_T}\left(\frac{\Delta l}{\Delta T}\right) \tag{3.7}$$

In practice, the value of l_T is often replaced by the length measured at room temperature, l_{RT} (sometimes denoted as l_0), so that the experimental results usually reported are strictly for

$$\alpha^* = \frac{1}{l_{RT}}\left(\frac{\partial l}{\partial T}\right)_P = \alpha\frac{l_T}{l_{RT}} = \alpha\left(1 + \frac{l_T - l_{RT}}{l_{RT}}\right) \tag{3.8}$$

Detailed analysis of errors resulting from the finite size of intervals ΔT is discussed in [Bar98].

Alternatively values of α can be obtained by differentiating an algebraic fit to a number of readings of length (or lattice spacing) at various temperatures. In each case the temperature interval ΔT should be much smaller than T (generally $\Delta T/T \leq 0.1$) unless α is sensibly constant over a wide range, which is unlikely at low temperatures.

At ambient temperatures, where $\alpha \sim 10^{-5}$ K^{-1} for many solids, we can measure the change in length (or lattice spacing) over an interval $\Delta T \sim 10$ K. In this case, a method having sensitivity $\Delta l/l \sim 10^{-6}$ should give an inaccuracy of 1% or less from a pair of readings. This sensitivity can be achieved easily by many dilatometers but not by X-ray or neutron measurement of lattice spacing.

At low temperatures, α becomes much smaller and necessitates more sensitive methods. For example: for Cu, $\alpha = 1.0 \times 10^{-6}$ at 30 K, 0.1×10^{-6} at 15 K and 0.005×10^{-6}K^{-1} at 5 K. Therefore to measure α to 1% over a temperature interval of 1 K demands a resolution of $\Delta l/l \sim 10^{-9}$ near 15 K and very much finer at 5 K. Indeed at 5 K, even if the specimen $l \sim$ 100mm, the expansion $\Delta l \sim 0.5$ nm over a 1 K interval necessitating a sensitivity of ~ 0.005 nm (0.05 Å).

Such detection levels of 0.1 Å or less are much smaller than the average interatomic spacing in a solid and much smaller than the scale of roughness on a polished surface. Any form of dilatometry involving contacting surfaces has to take this into account: thermal cycling will often reveal hysteresis effects arising from the relative movements of the surfaces.

In the following sections we discuss the various methods in order of increasing sensitivity with comments on ease of operation and reproducibility of data. Some methods which are of use mainly at high temperatures, including telemicroscopy and γ-ray density, will not be included.

A valuable reference to methods of measurement of thermal expansion is the handbook on *Thermal Expansion of Solids* [Ho98], in which various authors give details of X-ray diffraction (H. A. McKinstry *et al.*), optical interferometry (T. A. Hahn) and high resolution techniques (C. A. Swenson).

3.3.2. X-Ray Diffraction

The normal resolution in determining changes in lattice parameter is $\Delta a/a \sim 10^{-5}$, so that with a temperature interval of 50 K an expansivity of $10^{-5}\mathrm{K}^{-1}$ could be determined with an error of 2% or less. Clearly this is a convenient method at normal temperatures for most solids (for which $\alpha \geq 10^{-5}\mathrm{K}^{-1}$), provided the temperature interval is not so large as to 'smear out' important physical features. The Debye–Scherrer (powder pattern) method is particularly convenient in not requiring large samples or single crystals; for anisotropic solids it can establish the differences in principal expansivities without the need for large single crystals.

Single crystal methods such as Bragg's, the Bond diffractometer and rotating crystal are discussed by Krishnan *et al.* [Kri79] and [Ho98, Ch. 7]. Generally resolution limits their value below 100 K or so for most solids, except for dislocation-free single crystals such as Si where triple-axis spectrometry can be used with much higher resolution.

Good examples of the use of X-ray diffraction at higher temperatures are the measurements at the University of Illinois on Cu, Ag, Al etc., where the changes in lattice spacing were compared with the macroscopic length determined from a cathetometer at 25 or 50 K intervals [Sim60]. In Cu, for example, the two methods agreed to within experimental error up to 1100 K. For higher temperatures, closer to melting, differences between $\Delta a/a$ and $\Delta l/l$ become significant enough to allow estimates of vacancy concentrations. In copper near 1300 K, they reveal vacancy concentrations of $\sim 10^{-4}$ [Sim63].

Clearly X-ray methods are generally inadequate for determination of coefficients of expansion below 100 K. Some authors have measured the lattice parameters at say 4 K, 50 K and 100 K and fitted them to a simple polynomial such as $a = a_0 + bT^4$ which may be misleading, particularly if differentiated to give 'values' of α. X-ray measurements can be useful at temperatures well below 100 K for those materials which have high expansion coefficients in this range, such as the rare gas solids.

Diffraction is the only method available for measuring the internal expansion, that is, the change with temperature of the positions of atoms within the crystal unit cell. Because the information comes from the analysis of intensities, and not from simple Bragg reflection, it is less precise than the measurement of lattice parameters. Low temperature data have large uncertainties and virtually no results are obtainable below 100 K.

3.3.3. Optical Interferometers

Interferometric measurements began with the classic experiments of Fizeau in the 1860s on mineral crystals, and are still used over wide temperature ranges as they are absolute and can now achieve resolutions of a few Å (less than a nanometer) with the aid of laser light sources.

Fizeau Technique. The usual form uses either a cylindrical hollow sample (tube) with parallel end-plates of polished silica or three rods of equal length, placed so that the separation of two etalon plates is changed as the specimen expands or contracts. Changes in the fringe pattern produced by a monochromatic light source can be measured to about 1/100 of a fringe, ≤ 10 nm. Most experimenters have quoted deviations in their data of not less than $10^{-7}K^{-1}$, a frequent source of error being tilting effects. Examples of this technique used at low temperatures include work on Cu SRM 736 [Hah70] and alkali halides [Jam65].

Polarization interferometer. Based on the work of Dyson at the National Physical Laboratory (NPL), these depend on measuring the angular rotation of the plane of polarization of a stabilized laser beam. A single beam is split and interference occurs between waves reflected from 'top' and 'bottom' of a sample.

Fig. 3.5 illustrates the arrangement of Roberts [Rob81] for measurements on an ultra-low expansion glass. Two orthogonally polarized beams from a stabilized laser pass through a Polaroid filter, a silica parallel plate beam splitter and a polarizing beam splitter (PBS). Then one beam A traverses the path D, is reflected on the shoulder of the sample back to mirror (M) and then via path G to be reflected again on the other shoulder. The double path compensates for effect of tilt. The other beam B travels via E to be reflected successively at the bottom of the sample, then by mirror M and again by the bottom plate. Finally with suitable use of 1/4 and 1/2 wave plates and superposition of beam C, an output beam reaches J and the automatic polarimeter. This output beam is linearly polarized at an angle which changes by 360° for every 1/2 wavelength change in length of the sample allowing resolution of 1/1000 of a fringe. The observation point at S is for monitoring tilt. Note that the sample (hollow cylinder in this example but solid block in others) is supported on a base of similar material (to avoid distortion during cooling) to which it is optically contacted. The length changes were measured on ULE, Zerodur, silicon to 1/1000 fringe giving an absolute precision in α of $10^{-8}K^{-1}$.

Heterodyne interferometer. Nanometer resolution is also achieved with an optical heterodyne method in which two beams of slightly different frequency are produced by acoustic-optic modulation of the beam from a stabilized laser. Length changes are measured from the phase change of the beat frequency using a frequency counter. Examples are the systems used by Drotning [Dro88] and developments by Okaji and collaborators at the National Research Laboratory for Metrology (NRLM) in Tsukuba. The latter include the following:

1. [Oka91] describes an intercomparison of results obtained for Si and silica with differing interferometers used at NRLM and at NPL by Birch for the range from 250 to 700 K.

2. [Oka95b] describes a helium flow cryostat to measure fused silica SRM 739 from 6 to 273 K with uncertainties of $\leq 2 \times 10^{-8}K^{-1}$. Temperatures were

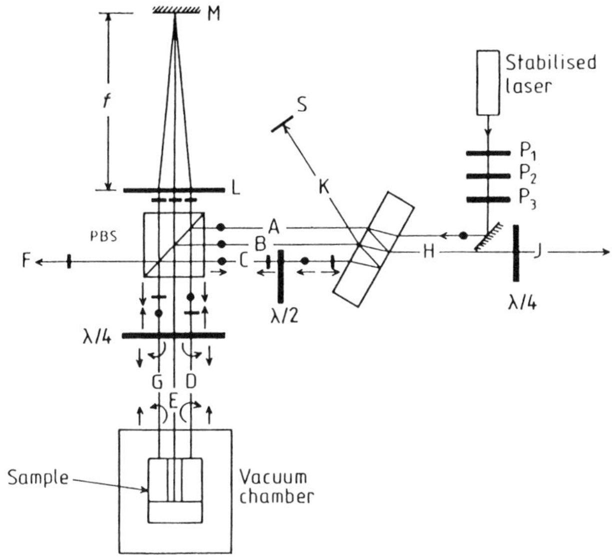

Fig. 3.5. Optical paths in a polarization interferometer [Rob81].

measured with a RhFe thermometer and controlled via a silicon diode activated system.

3. [Oka97a] covers another cell design used to intercompare various copper samples from 20 to 300 K; they confirmed that α values for high purity Cu, OFHC Cu and tough-pitch Cu differ by less than 10^{-7}K^{-1} above 20 K.

4. [Oka97b] describes the interferometer used for room temperature measurements on some standard reference materials (silica, W, and Cu).

In each of these, the optical paths are not unlike those in Fig. 3.5: tilt effects are removed by doubling the path, as for the polarization interferometer, but detection of the path change (on changing T) uses a frequency counter.

Fabry–Perot multiple-beam. Perhaps the highest precision among interferometric methods is that developed at the Optical Sciences Center in Tucson by Jacobs and colleagues, which uses the dependence of a Fabry–Perot etalon's resonant frequency on mirror separation. The sample forms a cylindrical spacer separating two mirrors (endplates) whose expansion coefficients should match that of the cylinder to avoid distortion. Shifts in the etalon resonant frequency are measured by comparing the tunable laser frequency to that of a stable reference laser. One such system was mounted in a cryostat for measuring uniformity of thermal expansion coefficient (at the 10^{-9}K^{-1} level) among samples of glasses used in large telescope mirrors [Jac84].

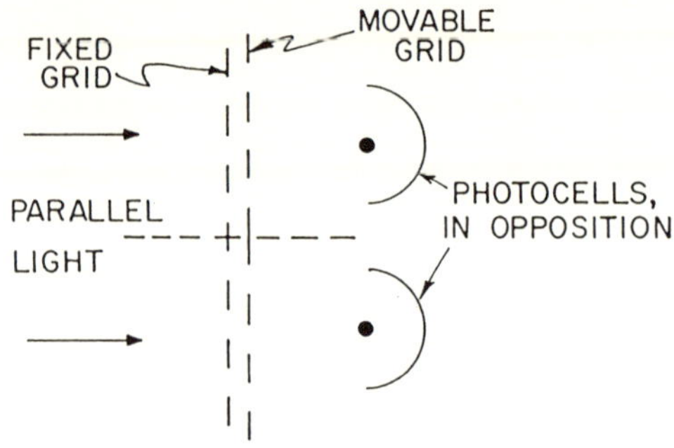

Fig. 3.6. Principles of the optical-grid system [Swe98].

3.3.4. Optical Amplifiers

Prior to the development of the three-terminal capacitance system (Section 3.3.6) some very sensitive dilatometers were developed based on optical levers and optical amplifiers, many achieving detection limits of less than an Å. Jones has reviewed [Jon61] those which he and colleagues developed for measuring angular movements as small as 10^{-10} radians and displacements of < 1 pm. Unfortunately, when applied to thermal expansion determinations, their resolution is limited by hysteresis and drift effects associated with thermal cycling and the mechanical linkage from specimen to optics.

The type of amplifier most commonly used has been the optical-grid illustrated schematically in Fig. 3.6 [Swe98]. It was used by Andres [And64] to measure the expansion between 1.5 and 12 K of a number of metals (Al, Pb, Pt, Mo, Ta etc.) with resolutions of better than 0.1 nm (1 Å) corresponding to $\sim 10^{-9}\mathrm{K}^{-1}$ in α.

3.3.5. Electrical Inductance

In an electrical inductance dilatometer, the length change of a rod is transmitted to the inner coil (secondary) of a mutual inductance and the inductance varies linearly with the displacement. Commercial push-rod dilatometers using linear variable differential transformers (LVDT) are widely used at normal and high temperatures and can have sensitivities of a few nanometers. Accuracy is usually limited by thermal problems — temperature gradients along the push-rod or sheath. These can be partially overcome by careful calibration using a reference specimen of roughly similar length and expansion to the unknown. Such reference materials include copper, silica, stainless steel, silicon, tungsten, sapphire.

Below about 100 K, the sensitivity of the commercial LVDT devices is usually insufficient. More sensitive laboratory instruments have been made using coils held at cryogenic temperatures: Carr and Swenson [Car64a] (see also [McL72]) successfully measured length changes in non-magnetic solids at liquid helium temperatures with sensitivity of 0.01 Å (1pm). Their dilatometer was absolute, requiring no calibration, because the sample was mechanically linked to the secondary coil by a sapphire–sapphire contact but thermally isolated by the high thermal resistance of this contact. However, inductive systems are generally less convenient to use than the capacitance dilatometers (Section 3.3.6 below) because of their sensitivity to magnetic effects including magnetic impurities in the specimens.

SQUID dilatometer. An ultra-sensitive dilatometer was used to measure the expansion of glasses in the 0.1 to 10 K range [Ack82]. With the use of a SQUID as a null detector, a resolution of 2×10^{-4} Å was achieved. The flux changes caused by expansion and consequent movement of a coil in a steady field were counterbalanced by a piezoelectric quartz transducer. Here again magnetic impurities and interference limit their performance.

3.3.6. Electrical Capacitance

Two different capacitance methods have been widely used for high sensitivity measurement of thermal expansion. One of these is based on measuring frequency change of a tuned oscillator circuit, but this has been now largely superseded by the other method in which a ratio transformer bridge operating at low frequency compares two three-terminal capacitors — one capacitor involving the specimen and the other being a fixed reference capacitor.

At first sight, the tuned oscillator LC circuit appears the easier of the two to use because of ease of measuring frequency and frequency change. However. in cryogenic practice, problems arise from geometry, calibration, drifts in lead capacitance, etc. In the 3-terminal method the capacitance can be well defined geometrically and 'parasitic' capacitances of leads (to earth) do not affect the bridge balance. Well-shielded ratio-transformer bridges are available which are capable of resolving capacitance changes of 1 in 10^7 and even 1 in 10^8. This translates to a length resolution of $\geq 10^{-12}$m or 0.01 Å. Some examples of both systems are discussed briefly below, and more details are given in Swenson's review of 'high sensitivity techniques,' see [Swe98].

Resonant Oscillator. Some interesting examples of high resolution measurements which have used resonant oscillators are the following:

- Tolkachev *et al.* [Tol75] for measuring the expansion of solidified gases with a tunnel-diode oscillator operating at 15 MHz;
- Van Degrift [Van74] at NBS (now NIST) with tunnel diode oscillators which could resolve movements of ~ 0.1 pm (0.001 Å);
- Kos and Lamarche [Kos69] measured expansion of Cu, Ag and Au below 15 K with sensitivity of $\Delta l/l \sim 10^{-11}$.

Three-Terminal Capacitance Method. This method stems largely from the work of Thompson [Tho58], who developed ratio transformer bridges capable of comparing capacitances with resolution of better than 1 in 10^8. The features of such bridges (Fig. 3.7) include well-shielded and tightly coupled transformer arms, shielded non-microphonic leads, stable reference capacitor C_s (e.g., made of Invar) and a detector-amplifier tuned to the bridge frequency. Not shown in this schematic diagram are the RC quadrature circuit for separating the capacitative from the conductive component and the LC tuning circuit to counteract the reduction in detector sensitivity from the effect of lead capacitance. More details of bridges and amplifiers are given in [Tho58] and of their application to thermal expansion by White [Whi61, Whi72a] (also [Car64b]). Figure 3.8 shows a dilatometer in which the capacitance between surfaces of plate (1) and the sample (2) constitutes the unknown, C_x (Fig. 3.7); the other capacitances (to the earthed shields) C_{13} and C_{23} do not affect the balance condition in a low frequency bridge.

Commercial versions of the Thompson bridge have been produced by General Radio Corp. (now QuadTech of Marlborough, Mass.) as model numbers 1615 and 1616.

A self-balancing direct read-out bridge was developed by Andeen and Hagerling (Cleveland, Ohio) with resolution of about 1 in 10^7. Such a resolution of $\Delta C/C \sim 10^{-7}$ for a parallel plate capacitor (typically $C \sim 10$ pF) having a gap of 0.1 mm corresponds to a detection limit in terms of length change of the sample (2) of 10^{-8} mm (i.e., 0.01 nm or 0.1 Å).

In calculating the thermal expansion from the change in gap g of the parallel plate sketched in Fig. 3.8, the familiar equation $C = \pi r^2/g$ must include a small correction term for the distortion caused by the guard ring. If the separation between the central electrode and the guard ring is $2w$, then from Maxwell [Whi61, Swe98]

$$C = \frac{\epsilon \pi r^2}{g} + \left(\frac{\epsilon \pi r w}{g + 0.22w}\right)(1 + w/2r) \approx \frac{\epsilon \pi r^2}{g} + \frac{\epsilon \pi r w}{g + 0.22w} \tag{3.9}$$

since $w/2r \ll 1$; here ϵ is the permittivity $\epsilon_r \epsilon_0$, and $\epsilon_r \approx 1$.

The dilatometer shown in Fig. 3.8 is a 'differential' (as opposed to absolute) cell; the length changes in the sample (2) are measured relative to the framework which is here made of OFHC copper. The copper 'plugs' (1 at the top) in the top and bottom plates are tapered and held in position with epoxy resin and mylar spacers about 0.2 mm thick so that $w = 0.1$ mm and $w/2r \sim 0.1/10 = 0.01$ in equation above. The plugs and guard-rings are lapped flat and attached to the copper cylinder with brass screws and spring washers. Parts can be gold plated to prevent tarnishing and to assist heat transfer between contacting surfaces. Usually the screened top plug (1) is connected to the 'low' voltage or detector side of the bridge and the sample (2) to the 'high' side. Within the cryostat the leads are low thermal-conductivity coaxial and outside the cryostat are non-microphonic shielded cables.

Calibration runs are usually done with Si and/or Cu samples, and are necessary at low temperatures to take account of small spurious 'expansions' arising from

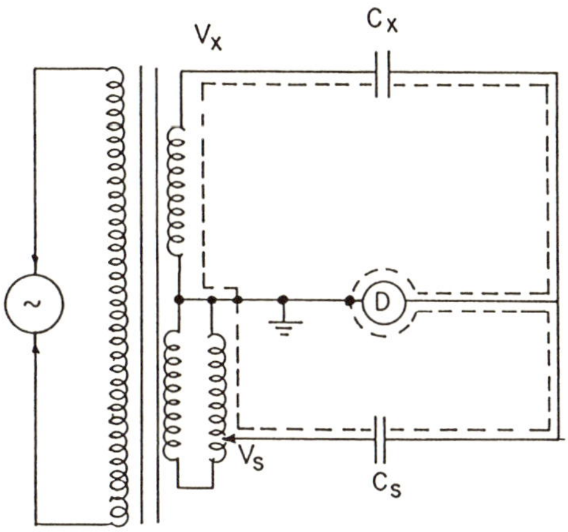

Fig. 3.7. Principal components of a three-terminal capacitance bridge: - - - denotes shielding. At balance $C_x/C_s = V_s/V_x$ [Swe98].

the epoxy joints in the end plates. For measurement of non-conducting samples, an evaporated film of silver is deposited on the surfaces. With smaller samples, copper discs can be used to make up the desired length and achieve a capacitance gap (between parts 1 and 2) of 0.1 to 0.2 mm; phosphor bronze springs hold the 'composite' sample together (more details in [Car64b, Whi72a]). When operating at a length resolution of 0.1 Å or less it is not surprising that any relative movement of contacting surfaces will produce significant hysteresis, considering that even good optical polishing leaves surface asperities $\sim$ 100 Å high. Hysteresis effects of 10-100 Å can show up after thermal cycling over say, 20 K, particularly if the materials in contact are of very different expansion coefficient. However it is remarkable how reproducible are data if 2 or 3 preliminary thermal cycles are made from say, 4.2 K to 15 and back to 4.2, then from 4.2 to 25 to 4.2 K.

This type of cell can be made 'absolute' rather than 'differential' by isolating the sample thermally from the base using sapphire spacers [Swe98].

Shown in Fig. 3.9 is a convenient copper holder (for insertion in a dilatometer such as illustrated in Fig. 3.8) which can be adjusted to take small samples of any length; on the top of the sample is a copper plate held by flexible wires [Swe98].

Pott and Schefzyk [Pot83] describe a copper cell of rather more complicated construction than that in Fig. 3.8, but which allows easier specimen changing and preparation. Their cell is open on one side for changing the sample; the sample forms a pushrod to move the capacitance plate. Parallel movement of the plate is achieved via a beryllium–copper strip acting as a spring. The gap is adjusted by a control rod from the top of the cryostat.

Many other variants of the three-terminal capacitance technique have been described, for example: for measurements on samples of large expansion coefficient,

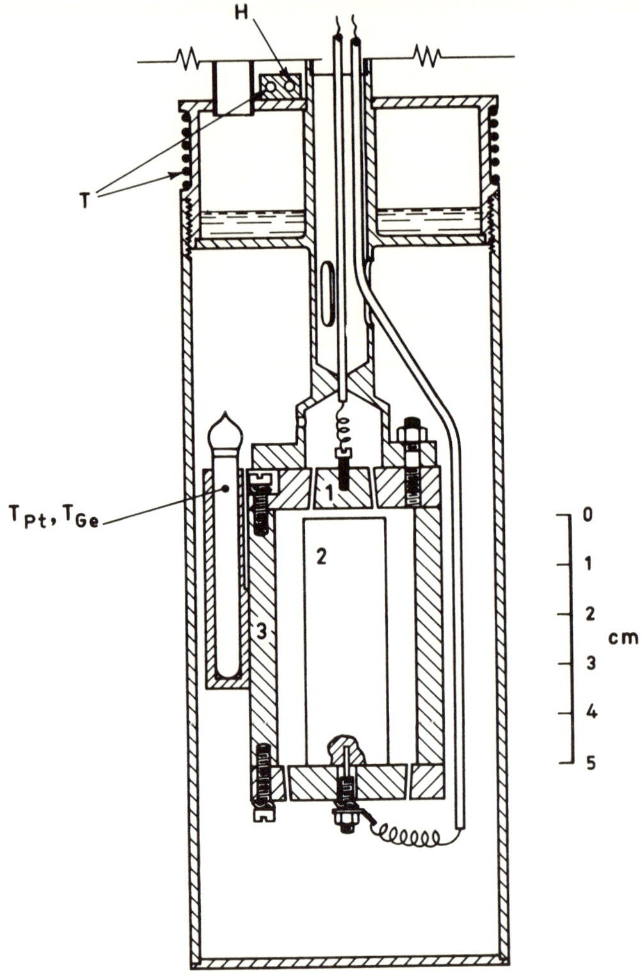

Fig. 3.8. Inner vacuum chamber (copper) of cryostat used for thermal expansion with three-terminal bridge. No. 2 denotes the sample and C_{12} is the capacitance to be measured [Whi72a].

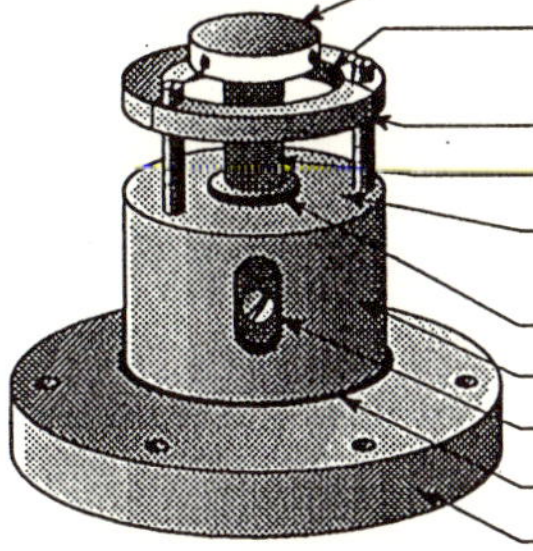

Fig. 3.9. An insert for use with small samples of different lengths to replace the base plate and sample shown in Fig. 3.8. All components are of copper excepting the manganin suspension wires and brass clamping screw [Swe98].

soft single crystals, application of a magnetic field and measurement along different axes; also with irregularly shaped samples, using tilted plates and the sample sandwiched in between them (refs. and details in [Swe98]).

There have also been cells made of low expansion materials such as Si to achieve better absolute accuracy when measuring other low expansion solids at low temperatures. One successful example [Vil80], which is based on the design in Fig. 3.8, replaces the copper cylinder (3) by three silicon posts.

A related geometry for measuring materials of large expansion coefficient (e.g., potassium) inverts the 'assembly' by replacing the cylinder with three posts made from the sample and making the central cylinder (no. 2) out of copper. During cooling, the 'soft' sample posts contract more than the central cylinder [Swe98]; therefore the capacitance gap decreases and the sensitivity of measurement is increased rather than decreased during cooling.

Another test cell, similar to that in Fig. 3.8 was made from single crystal sapphire plates and cylinders with evaporated aluminium electrodes, but was disappointing in its performance [And90b]. Hysteresis effects on thermal cycling were relatively large compared with the very low expansivity of the sapphire. The reason may have been the surface asperities which would allow less elastic deformation in sapphire than in copper [And90b].

3.3.7. Measurement of Grüneisen γ by Thermoelastic Methods

Two *direct* methods of determining the Grüneisen parameter γ avoid the necessity of determining very small expansivities (and heat capacities) which occur as $T \to 0$. One measures the change in temperature induced by adiabatic strain or volume change, and the other measures the change in stress induced by change in internal energy at constant volume. The thermodynamic relations involved are [see Eq. (2.12)]

$$\gamma = -\left(\frac{\partial \ln T}{\partial \ln V}\right)_S \tag{3.10}$$

$$\gamma = \left(\frac{\partial P}{\partial (U/V)}\right)_V \tag{3.11}$$

Boehler *et al.* [Boe77] describe the use of small adiabatic pressure changes on NaCl at various pressures to determine the Grüneisen parameter and its variation with volume. This was done at room temperature and pressures up to 3.3 GPa with an end-loaded piston-cylinder apparatus, the main aim being to determine $\gamma(V)$ and $q = (\partial \ln \gamma / \partial \ln V)_T$.

Another example of using Eq. (3.10) is to apply a single 'pulse' of strain and/or a sinusoidally varying strain to measure γ [Wri84]. Rods of silica and a polymer were strained by means of an Instron tester at temperatures down to 0.5 K. Small strains of $\sim 10^{-4}$ were determined to 2% and temperature changes of 10 mK or so were registered on a small carbon thermometer attached to the center of the rod (Fig. 3.10).

An example of the second method Eq. (3.11) is the determination of γ_0 for Si and Al [Gau71]. A thermoelastic stress is produced by a pulse of 1.5 MeV electrons, lasting about 40 nanoseconds; this stress is recorded with a quartz gauge bonded to the back face of the sample. The experiments were done over a range from 5 to 290 K. The application to Si is particularly useful, because the thermal expansivity is very small and difficult to determine below 20 or 30 K with sufficient accuracy to obtain $\gamma^{th} = \beta B_S V / C_P$ as $T \to 0$.

Volume dependence of the heat capacity. A different Grüneisen parameter γ_{SH} has been obtained from the volume dependence of the heat capacity, defined by

$$\gamma_{SH} = -\left(\frac{\partial \ln T}{\partial \ln V}\right)_{C_V} = \frac{(\partial C_V / \partial \ln V)_T}{(\partial C_V / \partial \ln T)_V}$$

It has been used particularly for compressible systems such as the rare gas solids (see [Bar80, pp. 652-6]). Its relation to other Grüneisen functions is discussed in Section 2.2.3.

3.3.8. Reference Materials

Many dilatometers give relative length changes rather than absolute values and even absolute instruments are prone to systematic errors arising from heat leaks, thermometer errors etc. With such instruments, calibration checks with samples of reproducible and well characterized materials are a virtual necessity. Fortunately there are now many suitable reference materials, some of which have been certified by national laboratories and are therefore legally traceable. Many materials such as copper, aluminium and silicon are readily available in a state of high purity so that

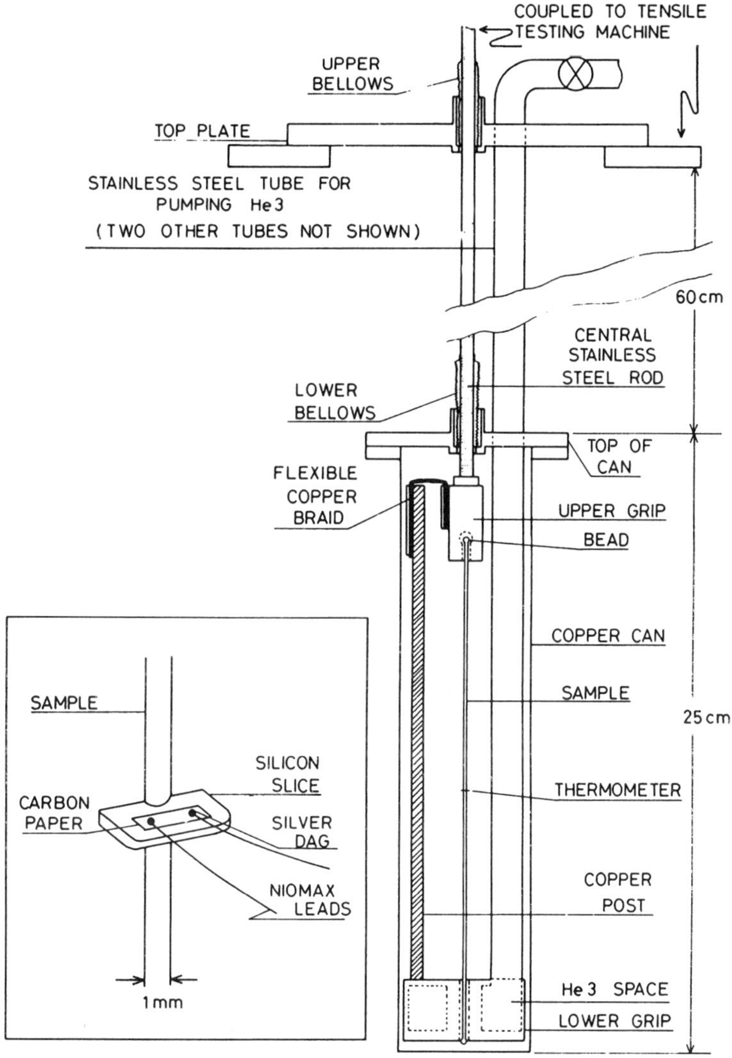

Fig. 3.10. Cryostat with thermoelastic strain technique for measuring $\gamma(T)$ from [Wri84, Fig. 1].

off-the-shelf samples will be adequate unless there is a legal necessity for 'certified' samples such as NIST Standard Reference Materials (SRMs).

Those which have been certified in the past by NBS (now NIST) include:

1. Copper SRM 736, certificated from 20 to 800 K. This is not now available but pure 99.99+% Cu is satisfactory down to 2 or 3 K. Magnetic impurities can alter α significantly at temperatures below 20 K.

2. Sapphire crystal SRM 732 (measured in direction 59° from c-axis) certificated from 293 to 2000 K, is also out of stock but similar material can be obtained commercially. It has been measured at low temperatures where the expansion is relatively small.

3. Tungsten SRM 737 measured from 80 to 1800 K at NBS is also out of stock but suitably pure rod is readily available.

4. Borosilicate glass, SRM 731, is certificated from 80 to 680 K. Being a mixed glass, off-the-shelf material cannot be relied upon.

5. Vitreous silica SRM 739, certificated from 80 to 1000 K and measured below 80 K [Oka97b] is available but is affected by exposure to elevated temperature, i.e., is sensitive to thermal history; it has a relatively large negative coefficient at low temperatures.

6. Although not certified by NIST, silicon has been well studied, is isotropic, is readily available in pure state, and has a relatively small coefficient (negative between 18 and 120 K).

For some of these materials, expansion data are given in the Appendix and further details in [Whi93b, Whi97], and [Ho98, Ch. 11].

3.4. ELASTIC MODULI

3.4.1. Introduction

This monograph includes elastic moduli because they form an important 'link' between the thermal expansion and the heat capacity; that is, they determine the changes in dimensions which result from changes in internal energy and pressure. They also furnish a means of calculating the lattice contribution to C_V in the low temperature (long wave) limit from the Debye temperature Θ_0^{el} which is calculated as an average over the $c_{\lambda\mu}$ (see Section 2.9). Likewise the pressure derivatives of the elastic moduli give values of the elastic mode gammas

$$\gamma_j = -\frac{1}{6} + \frac{1}{2}B\left(\frac{d\ln c_j}{dP}\right) \tag{3.12}$$

from which can be calculated the weighted average γ_0^{el} (see Section 2.9) and hence the lattice contribution to α or β in the low temperature limit.

This short section on techniques for measuring elastic moduli is intended to show some of the problems and levels of accuracy in determining B, χ and individual moduli $c_{\lambda\mu}$ (or $s_{\lambda\mu}$). The elastic moduli are measured in a variety of ways, both static and dynamic. The static methods include pressure–volume measurements which yield the isothermal compressibility (or bulk modulus), and linear stress–strain measurements which can give Young's modulus E, Poisson's ratio σ and shear modulus G. For isotropic materials two of these three parameters determine χ_T or B_T. For single crystals stress–strain data can also give the principal axial values of χ.

The dynamic (adiabatic) methods include measurement of low frequency torsional or flexural vibrations, ultrasonic velocities, ultrasonic resonant spectroscopy, inelastic neutron scattering, and Brillouin scattering. Ultrasonic velocity measurement is the most common as it is applicable at all temperatures to crystals of different symmetries and to polycrystals and glasses. For small ($\sim$ 1 mm) crystals, resonant ultrasonic spectroscopy has developed as a valuable tool to determine all the principal moduli from a single set of measurements and become practicable by the advent of high speed computers. The opto-acoustic techniques (Brillouin scattering and laser-induced phonon spectroscopy) are especially well suited to study of microcrystals ($\sim$ 0.1 mm).

As illustrated in Fig. 1.3 for KCl, the principal moduli vary rather slowly with temperature (except near a phase transition) and approach a constant value as $T \to 0$. At higher temperatures, $T \geq \Theta$, isothermal and adiabatic values of $c_{\lambda\mu}$ fall nearly linearly for most solids. The 'linear' slope of $B_S(T)$ is related to the volume expansion coefficient by the Anderson–Grüneisen parameter $\delta_S = -(\partial \ln B_S/\partial \ln V)_P$ (see Section 2.2.3):

$$\left(\frac{\partial \ln B_S}{\partial T}\right)_P = -\delta_S \beta \simeq -4\beta \tag{3.13}$$

e.g., for tungsten at 1000 K, $\beta = 15 \times 10^{-6}\mathrm{K}^{-1}$, so that $(\partial \ln B_S/\partial T)_P \simeq -6 \times 10^{-5}\mathrm{K}^{-1}$ and B_S falls by about 0.6% per 100 K rise in temperature.

The relatively slow change with temperature means that for most measurements of elastic moduli, temperature control is not a prime requirement. Major errors in ultrasonic data arise from other factors; they are of the order of tenths of a percent so that small drifts in T are not as serious as they are when measuring C_P or α.

3.4.2. Equation of State Methods

Dilatometric Measurements of $V(P,T)$. PVT measurements do not give individual moduli but are a useful source of compressibility (and bulk modulus) values over a wide range of temperature and pressure, particularly suited to solidified gases or very soft materials which are not easy to obtain or handle in single crystal form. Examples are the measurements on the alkaline earths [And90a] and rare gas solids.

Packard and Swenson [Pac63] describe a piston-cylinder method of measuring the change in volume of solid Xe from 20–160 K at pressures up to 2 GPa. Such soft materials have a very low shear strength so that the pressure applied via the piston is effectively hydrostatic. Values of bulk modulus resulting for the alkali metals and alkaline earths have error bars of ±0.5% [And83b].

X-Ray and Neutron Lattice Spacing. Measurements of lattice spacing by X-ray or neutron diffraction as a function of pressure (and volume) do not give individual moduli but can provide linear compressibilities as well as volume compressibility for polycrystals and single crystals. This is very useful for obtaining values of $\chi_{a,b,c}$ on sintered compacts as well as on polymers for which ultrasonic methods can be difficult; for example, the measurements of B_T for $YBa_2Cu_3O_{6+s}$ [Jor90] and polyethylene [Sak66].

3.4.3. Young's Modulus, Poisson's Ratio and Shear Modulus

By applying a tensile stress to a solid rod and measuring the change in length and diameter, values are obtained for Young's modulus E and Poisson's ratio σ from which the isothermal bulk modulus can be calculated:

$$B = E/3(1-2\sigma); \qquad B = EG/(9G-3E) \tag{3.14}$$

The vibrating reed method uses the flexural (or torsional) motion of a thin bar, clamped at one end. The free end oscillates at a natural frequency depending on length and thickness of the bar and is proportional to v_E, the Young's modulus sound velocity, i.e., from $E = \rho v_E^2$.

3.4.4. Ultrasonic Methods

The velocity of a sound wave in a uniform solid in general depends on the directions of propagation and polarization, as described in Section 2.8.7. But in a bulk sample of an isotropic material (polycrystal with randomly oriented crystallites or glassy) there are only two sound velocities, v_l (longitudinal) and v_t (transverse or shear), which together yield values of the adiabatic bulk and rigidity moduli:

$$B_S = \rho\left(v_l^2 - \frac{4}{3}v_t^2\right), \qquad G_S = \rho v_t^2 \tag{3.15}$$

These are related to other elastic coefficients such as E_S and σ_S (Section 2.8.3). E_S is also given directly by ρv_E^2 for longitudinal waves in a thin rod.

For cubic crystals, three independent velocities suffice to specify the individual moduli

- $c_{11}^S = \rho v_l^2$ from longitudinal wave in [100] direction

- $c_{44}^S = \rho v_t^2$ from transverse wave in [110] direction with [100] polarization
- $c'^S = (c_{11}^S - c_{12}^S)/2$ from transverse wave in [110] direction with $[1\bar{1}0]$ polarization.

Additional or alternative values are provided by:

- $c_l^S = (c_{11}^S + c_{12}^S + 2c_{44}^S)/2$ from longitudinal wave in [110] direction
- c_{44}^S from transverse wave in [100] direction with arbitrary polarization.

For crystals with axial symmetry, there are five or six independent moduli to be determined, and for orthorhombic crystals there are nine.

The 'standard' method of determining the velocity of a compressive sound wave or shear wave, typically of frequency 10–30 MHz, is by timing the passage of a pulse through a crystal of about 5–10 mm thickness. The sample is cut and lapped to have two parallel faces, to one (or both) of which is bonded a piezoelectric transducer. This is usually a quartz crystal, X-cut for longitudinal waves, AC- or Y-cut for shear wave; $LiNbO_3$ is also used. The bonding agents are carefully chosen to minimize attenuation and avoid cracking due to expansion mismatch. Depending on temperature range and the sample, they include Dow Corning 200 silicone, Nonaq grease, Salol, ethylene glycol, epoxy and organic liquid mixtures, e.g., 4-methyl,1-pentene [McS64, Bat67]. For higher frequencies thin films of CdS, ZnS, ZnO have been used as transducers. The path of the ultrasonic wave (or pulse) involves reflection from a specimen boundary and occupies a time interval of $\sim 10\mu$s for waves of velocity $\sim 10^5$ cm·s^{-1}. Figure 3.11 shows the measuring circuit described in McSkimin's review [McS64].

Various techniques for determining the transit times (and velocity) have been reviewed in volumes of *Physical Acoustics*, edited originally by Warren Mason and published by Academic Press from 1964 onwards. They include:

- Pulse echo technique in which the time of passage is measured directly on an oscilloscope. A single transducer or two separate transducers can be used for transmitting and receiving. A single transducer is preferred (Fig. 3.11).
- Pulse superposition method which measures the time between any given wave crest in one echo and a crest in a later echo. This avoids errors due to the time delay in the bond and which can amount to a few parts in 10^4 [McS64]. Later developments of this 'pulse echo overlap' (PEO) technique are described by Papakadis [Pap90]. Note that corrections for the transducer-bond phase shift can be made by varying the sample length or transducer configuration as shown by Jackson *et al.* [Jac81].
- Continuous wave resonance method has proved useful with thin samples and for measuring small changes in velocity [Bol63].

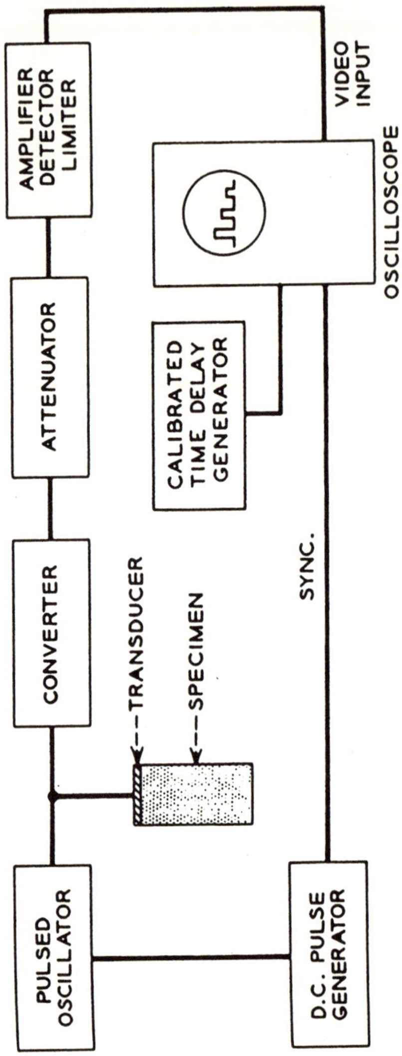

Fig. 3.11. Measuring circuit for high-frequency pulse technique [McS64].

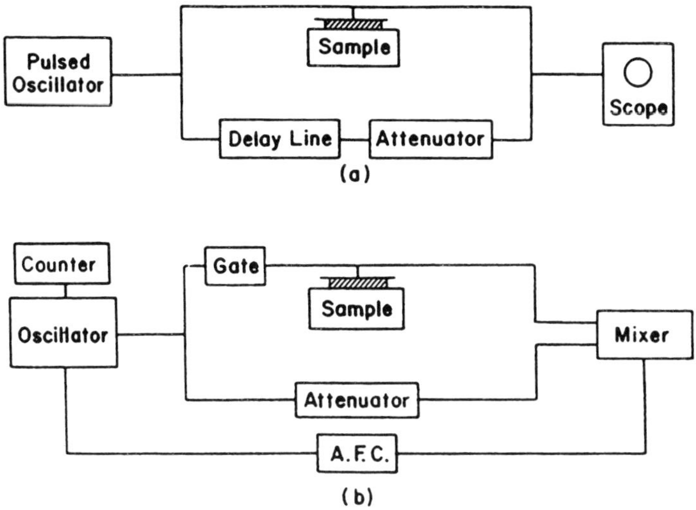

Fig. 3.12. Block diagram of the phase comparison methods in [Ale66, p. 278].

- Phase comparison methods which compare the phase of an rf signal which has traversed the specimen with a reference signal which has traversed another path. These are capable of high sensitivity and have been important in measuring the small velocity changes (≤ 1 in 10^6) which occur during a normal–superconducting transition or with the application of a strong magnetic field to a metal. Alers [Ale66] reviews variants of these methods of which two are illustrated by the block diagram in Fig. 3.12. Spetzler and colleagues have described the use of GHz sound waves for interferometry, applicable to small samples and very high pressures (see for example [Spe96]).

- The sing-around system which employs two transducers, one as transmitter and one as receiver, and can also detect very small changes in velocity. In the version developed by Forgacs (see [Ale66]), a received pulse arriving after a transit time τ is used to retrigger the transmitter, so that the circuit becomes an oscillator whose frequency is about $1/\tau$ Hz. Frequency shifts of a few parts in 10^7 can be detected.

Techniques for measuring ultrasonic velocities at very high pressures have been reviewed by Heydemann [Hey71], Jackson and Niesler [Jac82] (see also more recent measurements up to 10 GPa [Nie89, Li96]). Most involve piston or anvil methods which are not particularly suited to the cryogenic range.

Measuring systems for ultrasonic velocities have been available from various companies including Anutech Pty. (ANU, Canberra, Australia), Krautkramer Branson (Lewistown, Penn. USA), Karl Deutsch (Wuppertal, Germany), Matec Instruments (Northborough, Mass.), Parametrics Inc. (Waltham, Mass.), Ritec Inc. (Warwick,

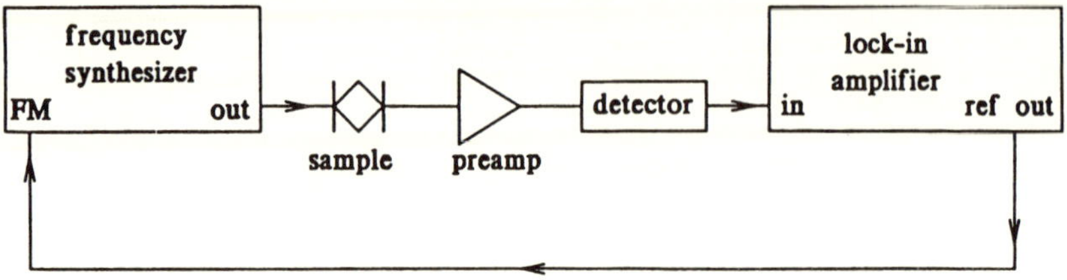

Fig. 3.13. Schematic diagram of electronics for a resonant ultrasonic system in [May92, p. 391].

Rhode Is.), Utex Inc. (Ontario, Canada).

3.4.5. Resonant Ultrasonic Spectrometry

Most of the ultrasonic methods are difficult to use with crystal dimensions less than a millimeter; for example, for a thickness of 0.5 mm pulse transit times are much less than a microsecond so that GHz transmitters are needed. The resonant spectrometry method, pioneered largely by Soga, Ohno, Kumazawa and others [And95a] and later at Los Alamos [Mig93, Mig97] is well suited to small crystals but requires sophisticated computing. This has been made easier in recent years with the development of smaller and cheaper high speed computers. The crystal is usually in the form of a polished rectangular parallelepiped to the corners of which are loosely attached two transducers. These may be small piezoelectric plastic films, one acting as driver to supply a sweep frequency to the crystal and the other monitoring the spectral response (Figs. 3.13, 3.14). Maynard *et al.* [May92] have reviewed this method and its application to small crystals of La_2CuO_4 and quartz. Values of c_{11}, c_{33} etc. for the latter differ by less than 1% from those obtained by the more usual ultrasonic pulse methods (see also book by Migliori and Sarrao [Mig97]).

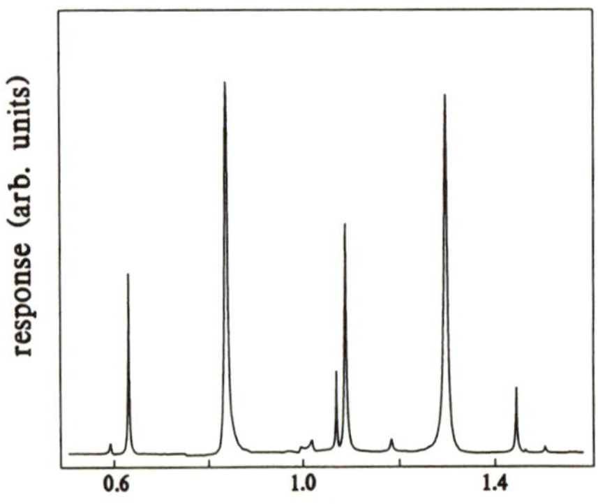

Fig. 3.14. Ultrasonic spectrum for undoped La_2CuO_4 [May92, p. 397].

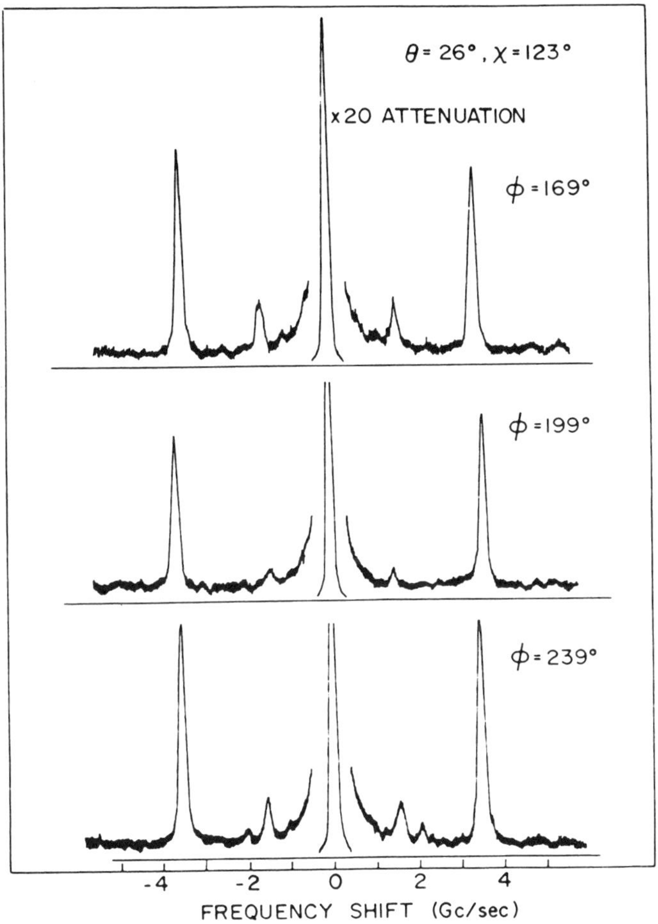

Fig. 3.15. Brillouin spectra of a xenon crystal for different orientations from Stoicheff in [Kle77, p. 1003].

3.4.6. Neutron and Opto-Acoustic Scattering

Neutron scattering. Thermal neutrons have energies and wavelengths comparable with those of the vibrating ions. Inelastic scattering of neutrons yields information about the energy–wave number relations for the phonons involved and hence the $\omega(q)$ curves for the principal modes. These can extend over the entire range of q out to the zone boundary and allow the phonon density of states to be calculated. The accuracy of determining the wave velocity $d\omega/dq$ at low frequencies is less than that achieved with ultrasonics and the method requires crystal dimensions comparable with the ultrasonic technique. For a few relatively compressible solids, measurements of $\omega(q)$ have also been done under pressure which yield volume dependences of the frequencies and hence values of γ_i up to high wave numbers. An example is RbI which was measured up to 0.3 GPa by Blaschko *et al.* [Bla75].

Brillouin scattering. Photons of visible light are also scattered by phonons and can provide information about phonon energies, particularly with the use of a laser source: there is a 'Doppler' shift of the wave velocity of the scattered light. However the photon energies are very much larger than the phonon energies and therefore energy or frequency shifts are small. The photon wave numbers are small compared with the Brillouin zone dimensions so that data are obtained only about the long wave phonons (near **q**=0). Examples of the use of Brillouin spectroscopy for determining the elastic constants of rare gas solids are given in a review by Stoicheff in [Kle77, Ch. 16]. This review shows typical cryostats and spectra for Ne, Ar, Xe etc. (Fig. 3.15). They found differences up to a few percent for these moduli compared with values obtained with ultrasonics but this may be due to imperfections in the solid gas samples.

Data on small quartz crystals (0.3 mm) have given longitudinal wave velocities with errors of less than 2%. From these measurements Weidner *et al.* [Wei75] estimate a minimum sample size of < 0.1 mm.

Laser induced phonon spectroscopy. This technique is also applicable to small crystals. Elastic waves are generated in the crystal by the interference of two laser pulses. Brown *et al.* [Bro89] describes the application to a small olivine crystal with resulting uncertainties of less than 1% in the elastic moduli.

3.4.7. Data Sources

The major compilations of elastic moduli are in:

- Volumes of *Landolt–Börnstein* III/1 [Hea66] supplemented by Vol. III/2 [Hea69]; followed by a new compilation in Vol. III/11 [Hea79] with supplement in Vol. III/18 [Hea84]; finally a replacement Vol. III/29a prepared by Every and McCurdy [Eve92].

- The handbook on *Single Crystal Elastic Constants and Calculated Aggregate Properties* by Simmons and Wang [Sim71].
- *Elasticity of Minerals, Glasses, and Melts* by Bass in thc *Handbook of Physical Constants* from the American Geophysical Union [Bas95].
- *Review of Elastic Constants of Transition Metals* by Steinemann and Fisher [Ste81].
- *Elastic Constants of Mantle Minerals at High Temperatures* by Anderson and Isaak, which includes room temperature data on MgO, CaO, NaCl, KCl, MnO, and some silicates [And95b].
- Tables by Sumino and Anderson in *Handbook of Physical Properties of Rocks*, Vol. III [Sum84].

Chapter 4

Fluids

4.1. INTRODUCTION

This monograph is primarily devoted to cryogenic solids but, inevitably, the question is asked "... What is the difference in heat capacity, thermal expansion, or bulk modulus between gas, liquid and solid?..." Gases are dilute assemblies of atoms (or molecules), so that their properties depend on the kinetic energy of translation and for molecules also on energies of rotation and other "internal energies"; at normal pressures mutual interaction is usually a small perturbation. By contrast, liquids and solids are both about 1000 times more dense, and interactions play a dominant rôle. Liquids and dense gases are more difficult to model than crystals, because they lack the long range order which allows us to apply the concept of periodicity to the vibrating atoms and the electron gas (Chs. 2 and 6); also the hindered translational and rotational motions are not approximated by harmonic vibrations. Thus the three phases require separate discussion.

The conditions under which the phases can exist are conveniently shown in a P, T phase diagram, as for the simple example of the monatomic Ar (Fig. 4.1). There are two special points labelled in this diagram: at the *triple point* (T_t, P_t) solid, liquid and gas can all coexist; and at the *critical point* (T_c, P_c) the distinction between the two fluid phases disappears. The solid–liquid coexistence (melting) curve in this range has a slope of about 4 MPa$\cdot$K^{-1} (40 bar$\cdot$K^{-1}), and is much steeper than the gas–liquid curve. Most phase diagrams are similar to this, except that often they are more complex because of the existence of different solid phases; also, for a few systems the line between solid and fluid slopes backward. The outstanding exception is helium, for which there is no triple point and the fluid phase extends to $T = 0$.

For most substances T_t is above 0 °C. The liquid does not then exist in the cryogenic region, and the vapor only at very low pressures, approximating closely to an ideal gas. But some important systems have lower triple points (Table 4.1), and so, like helium, can be used as cryogenic fluids. We therefore discuss in order dilute classical gases in Section 4.2, liquids in Section 4.3, and quantum fluids in Section

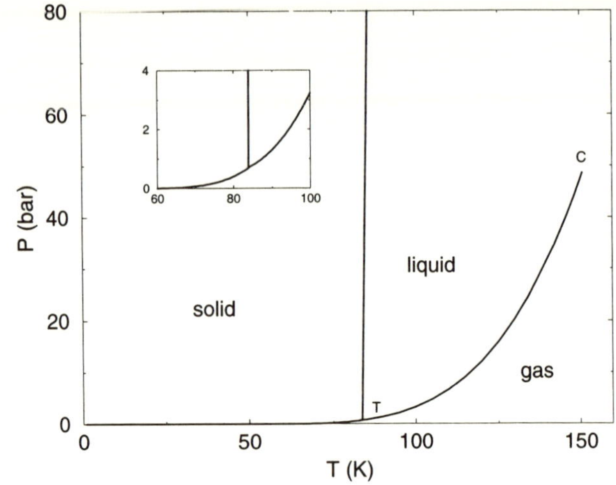

Fig. 4.1. Phase diagram of argon. T is the triple point, C is the critical point. Inset shows region near T with pressure scale enlarged.

4.4, including the unique properties of the helium isotopes.

4.2. GASES

4.2.1. Introduction

Under typical working conditions, the thermodynamic properties of most gases approximate quite closely to those of a classical ideal gas, and the ideal equation of state [Eq. (4.1)] is then adequate for most engineering purposes. However, the heat

Table 4.1. Boiling points (T_b), triple points (T_t), and critical points (T_c), for some liquids of cryogenic interest; density is at boiling point under 100 kPa (1 bar) [Jac97, Qui90]

Liquid	T_b (K)	T_c (K)	P_c (kPa)	T_t (K)	P_t (kPa)	ρ (g/cm^3)
^{3}He	3.19	3.32	115	—	—	0.059
^{4}He	4.207	5.1953	227.5	2.1768 (λ)	4.8565(λ)	0.125
n-H_2	20.345	33.19	1315	13.95	7.20	0.0707
p-H_2	20.233	32.94	1284	13.8033	7.034	0.0708
Ne	27.061	44.492	2679	24.5561	43.4	1.208
N_2	77.237	126.19	3398	63.1504	12.52	0.807
Ar	87.17	150.663	4860	83.8058	68.89	1.397
O_2	90.062	154.58	5043	54.3584	0.1463	1.141
CO_2	194.6	304.14	7375	216.589	518	1.18

capacity of an ideal gas depends on its molecular composition. In most cryogenic gases electronic and vibrational degrees of freedom are not excited, while rotational contributions to C_V have reached their full classical value. Under these conditions, the heat capacity depends only on whether the gas is composed of isolated atoms (e.g., Ar), linear molecules (e.g., N_2, CO_2), or non-linear molecules (e.g., CH_4). The values of C_V for these three groups are then respectively $\frac{3}{2}R$, $\frac{5}{2}R$, and $3R$, or numerically 12.47, 20.79, and 24.94 $\mathrm{J{\cdot}mol^{-1}K^{-1}}$; the corresponding values of C_P are 20.79, 29.10, and 33.25 $\mathrm{J{\cdot}mol^{-1}K^{-1}}$. The data for C_V in Table 4.2 for some real gases at atmospheric pressure shows that these values are a good approximation over wide ranges of temperature, although there are large deviations for hydrogen below about 200 K (Section 4.2.3), and for methane at higher temperatures due to vibrational excitation. Small deviations also occur as the temperature is lowered towards the boiling point, due to intermolecular interaction.

We treat first the ideal monatomic gas; next the ideal molecular gas, and the rotational quantum effects seen in hydrogen; and then departures from ideal behavior as the pressure is increased.

Classical Ideal Monatomic Gases. The ideal gas limit has already been discussed briefly in Sections 1.3.5 and 2.5.1. The equation of state is the same for all gases, giving

$$PV = nRT, \qquad B_T = P, \qquad \beta = 1/T, \qquad C_P - C_V = nR \tag{4.1}$$

where R is the gas constant (8.314 $\mathrm{J{\cdot}mol^{-1}{\cdot}K^{-1}}$) and n is the number of moles of the gas. The ratio of heat capacities (traditionally called γ but here denoted by γ^*) is simply related to the Grüneisen function; by Eq. (2.8)

$$\gamma^* \equiv \frac{C_P}{C_V} = \frac{B_S}{B_T} = 1 + \beta\gamma T = 1 + \gamma \tag{4.2}$$

For a monatomic gas the partition function depends only on translational degrees of freedom, and is given by Eq. (2.60) when the potential energy function Φ is put to zero:

$$Z = Z_{trans} = \frac{V^N}{N!}\left(\frac{2\pi mkT}{h^2}\right)^{3N/2} \tag{4.3}$$

From this the thermodynamic properties follow. In particular

$$C_V = \frac{3}{2}Nk = \frac{3}{2}nR \tag{4.4}$$

and the entropy at pressure P can be evaluated from the *Sackur–Tetrode* formula:

$$S = S_{trans} = nR\left\{\frac{5}{2}\ln T - \ln P + \ln\left[\left(\frac{2\pi m}{h^2}\right)^{\frac{3}{2}} k^{\frac{5}{2}}\right] + \frac{5}{2}\right\} \tag{4.5}$$

For mixtures of gases the partition function becomes

$$Z = Z_{trans} = \frac{V^N}{N_A!N_B!\cdots}\left(\frac{2\pi m_A kT}{h^2}\right)^{3N_A/2}\left(\frac{2\pi m_B kT}{h^2}\right)^{3N_B/2}\cdots \tag{4.6}$$

The equation of state and the heat capacity are unchanged, but the entropy changes both because of the distribution of masses and because there is an additional entropy of mixing

$$S_{mix} = Nk(-x_a \ln x_A - x_b \ln x_B - \cdots) = R(-n_a \ln x_a - n_b \ln x_b - \cdots) \tag{4.7}$$

where $n_a\cdots$ and $x_a\cdots$ are respectively the number of moles and the atomic fractions of the component gases.

4.2.2. Ideal Molecular Gases

The partition function. In the ideal limit the translational kinetic energy of the center of mass of a molecule does not interact with its remaining degrees of freedom, and the partition function is given by

$$Z = Z_{trans}(z_{A,int})^{N_A}(z_{B,int})^{N_B}\cdots \tag{4.8}$$

where Z_{trans} is unaltered except that m_A is now the mass of molecule A, and $z_{A,int}$ is the *internal* partition function obtained by summing over all internal energy states of the molecule, which are independent of volume. The Helmholtz energy is therefore of the form

$$F = F_{trans}(T,V) + F_{int}(T) \tag{4.9}$$

The second term does not affect the equation of state, and so the bulk modulus, thermal expansion coefficient and $C_P - C_V$ are the same as for a monatomic gas (Eq. 4.1); but it does contribute additive terms to the entropy and heat capacity.

Calculation of absolute entropies from spectroscopic data. The additive terms in F_{int} are sums over all the non-translational energy states of an isolated molecule. For many simple gases the energies of such states have been the subject of precise study by spectroscopists, thus enabling the *internal* contribution to the entropy and other thermodynamic properties to be calculated. When this is added to the *translational* entropy given by Eq. (4.5), the absolute entropy of a dilute gas at a given temperature is obtained to a good accuracy purely from theory and spectroscopic data. Calorimetric data (latent heats and heat capacities) can then be used to find the difference in entropy between this dilute gas phase and other phases over the experimental temperature range. In this way absolute entropies of condensed phases are obtained, including the residual entropies of disordered material as $T \to 0$ (e.g., [Moo62, pp. 623-624]; see also Section 2.2.5).

Rotational contributions. Usually the internal degrees of freedom are themselves separable to a good approximation. Most molecules are in their electronic ground state at cryogenic temperatures, and sufficiently rigid for the interaction between the rotational and vibrational motions to be small and for the vibrational frequencies to be too high for appreciable excitation at cryogenic temperatures. A non-linear rigid molecule has three rotational degrees of freedom, since two parameters are needed to give the direction of some axis within the molecule, and a third for rotation about that axis; and a linear molecule clearly has only two. Like the three translational degrees of freedom, the kinetic energy for each rotational degree of freedom contributes $\frac{1}{2}k$ to the classical expression for C_V, thus giving the values of $\frac{5}{2}Nk$ (linear) and $3Nk$ quoted above. Apart from some molecules with low-lying vibrational (e.g., Cl_2) or electronic (e.g., NO) states, which start to become excited below room temperature, the only exceptions are H_2 and its isotopic modifications D_2 and HD. These are important cryogenic fluids whose rotational states we shall now discuss.

4.2.3. Ortho- and Para-Hydrogen

The rotational energy levels of a diatomic molecule are labelled by quantum number J, and given by

$$\epsilon_J = J(J+1)(\hbar^2/2I) = J(J+1)k\Theta_{rot}, \qquad (J = 0, 1, 2, \cdots) \tag{4.10}$$

where I is the moment of inertia about an axis through the center of mass perpendicular to the molecular axis. The spacing between the lower levels is thus inversely proportional to I, and of all molecules only hydrogen has a small enough moment of inertia for the discrete nature of the levels to be reflected in the heat capacity when $T > T_t$. For other molecular gases typical values of the characteristic temperature Θ_{rot} lie between 1 and 15 K, much lower than the boiling points. But for H_2, HD, and D_2 the values of Θ_{rot} are respectively 85.4, 65.7, and 43.0 K, well above the boiling points.

The degeneracy (the number of quantum states in level J) depends on the constituent atoms, owing to the Pauli principle that the total wave function of the molecule must be anti-symmetric with respect to interchange of the coordinates of two identical fermions, and symmetric with respect to interchange of two identical bosons. The translational and vibrational factors of the wave-function are always symmetric, but rotational and nuclear spin factors can each be either symmetric or anti-symmetric. Furthermore, the H nucleus (proton) is a fermion with spin $\frac{1}{2}$, and the D nucleus (deuteron) is a boson with spin 1. The degeneracies are therefore different for each of the homonuclear molecules H_2 and D_2, and different again for the heteronuclear molecule HD.

Consider first HD. The nuclei are distinct, there is no symmetry requirement, and any rotational state can be combined with any of the six spin states. The rotational degeneracy (number of rotational states in level J) is $2J+1$, and the spin-rotational

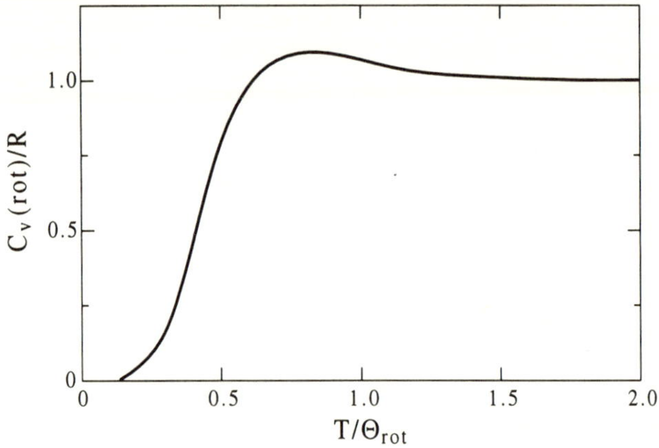

Fig. 4.2. Rotational heat capacity for a heteronuclear diatomic molecule. For HD, $\Theta_{rot} = 65.7$ K. From [Gop66, Fig. 6.1].

partition function is

$$z_{spin-rot} = 6 \sum_{J=0,1,2\cdots} (2J+1)e^{-J(J+1)(\Theta_{rot}/kT)} \tag{4.11}$$

The only thermodynamic effect of the nuclear spin is thus to contribute an additional entropy of $R\ln 6$. The rotational heat capacity derived from Eq. (4.11), which is the same for all heteronuclear diatomic molecules, is shown in Fig. 4.2. The maximum in C_{rot} is due to the low-lying triply degenerate first excited state.

For H_2 the total state must be antisymmetric with respect to exchange of the nuclei. There are three symmetric spin states ($I = 1$) and one anti-symmetric ($I = 0$). Rotational states of even J are symmetric, and to give total antisymmetry can be combined only with the single antisymmetric spin state, giving degeneracy $2J + 1$; whereas those of odd J are antisymmetric and can be combined with any of the three symmetric spin states, giving degeneracy $3(2J + 1)$. The partition function is therefore

$$z_{spin-rot} = \sum_{J=0,2\cdots} (2J+1)e^{-J(J+1)(\Theta_{rot}/kT)} + 3\sum_{J=1,3\cdots} (2J+1)e^{-J(J+1)(\Theta_{rot}/kT)} \tag{4.12}$$

From this is derived the rotational heat capacity of e-H_2 ("equilibrium hydrogen") in Fig. 4.3. The large initial Schottky-like bump is due to the now nine-fold degenerate first excited level at $2k\Theta_{rot}$; the next level is not until $6k\Theta_{rot}$, and is then only five-fold degenerate.

However, the heat capacity of e-H_2 is not that usually observed. The nuclear spins are so weakly coupled to other degrees of freedom that transitions between different

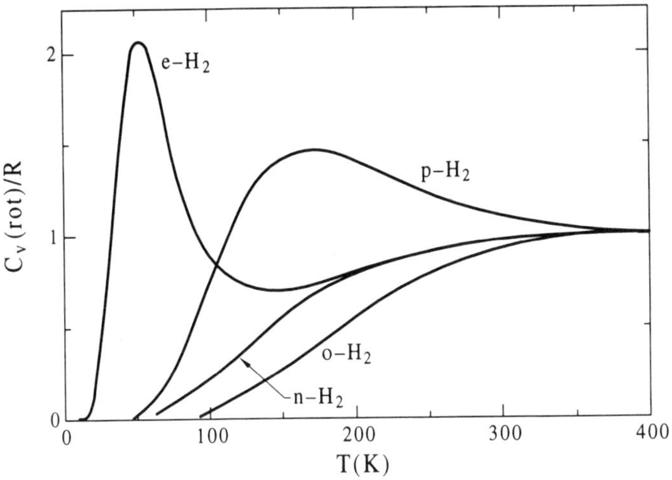

Fig. 4.3. Rotational heat capacities of e-H_2, p-H_2 o-H_2 and n-H_2. $\Theta_{rot} = 85.4$ K. From [Gop66, Fig. 6.2].

spin states seldom occur; consequently, when the temperature is altered a new thermal equilibrium is established only among the even J states and among the odd J states, and not between them. Effectively therefore we have a mixture of two different species — 'para-hydrogen' (p-H_2), whose molecules have the anti-symmetric spin state combined with one of the even J rotational states, and 'ortho-hydrogen' (o-H_2), whose molecules have one of the symmetric spin states combined with one of the odd J states. The total heat capacity is then the sum of contributions from each species. The spacing between levels for each species is greater than for e-H_2, and so their heat capacities do not approach the classical value until higher temperatures (Fig. 4.3).

The equilibrium ortho/para composition of hydrogen varies with temperature. Below about 30 K it is virtually pure para; at the boiling point of nitrogen it is about half ortho; and at room temperature and above it is three quarters ortho. The relaxation time for equilibration between the species depends upon conditions and the catalytic action of the walls of the container. Typically in the gas it is of the order of years, in the liquid of the order of days, and in the solid of the order of hours. Hydrogen is usually manufactured at room temperature or above, and then has an ortho to para ratio of 3:1. This mixture is called 'normal hydrogen' (n-H_2). Its heat capacity is shown in Fig. 4.3, and tabulated in [Jac97], as is also that of p-H_2. Other compositions varying in proportions between pure p-H_2 and n-H_2 can be obtained by passing the gas over a catalyst that facilitates ortho–para conversion at the appropriate temperature.

The conversion of o-H_2 to p-H_2 at low temperatures is exothermic, and so if normal hydrogen is rapidly liquefied and then immediately stored considerable heating and consequent evaporation occurs over a period of days. To avoid this, conversion to p-H_2 should be carried out before storage.

The same principles apply to the heat capacity of deuterium, where Θ_{rot}=43.0 K and the total wave function is symmetric with respect to deuteron exchange. Again there are two species: p-D_2 with one of three antisymmetric spin states ($I = 1$), which for bosons must be combined with rotational states of odd J; and o-D_2 with one of six symmetric spin states ($I = 0$ or $I = 2$) combined with rotational states of even J. At low temperatures the equilibrium composition is pure ortho, and for n-D_2 the ortho–para ratio is 2:1.

Further details are given in many texts, including [Rus49, Gop66, Ric67]. Similar theory is needed to account for the spin-rotation states of other molecules containing identical nuclei, although it is only for the hydrogens that the heat capacity of the dilute gas is appreciably affected. But related effects sometimes occur when molecules are able to rotate in a condensed phase, for example in some cryocrystals (Ch. 8).

4.2.4. Non-Ideal Gases: Virial Expansion

The range of validity of the ideal gas expression for the virial PV can be extended considerably by taking it only as the first term in an expansion of the inverse molar volume V_m^{-1}:

$$PV = nRT\left[1 + \frac{B(T)}{V_m} + \frac{C(T)}{V_m^2} + \cdots\right] \tag{4.13}$$

where $B(T), C(T), \cdots$ are called respectively the second, third, $\cdots$ virial coefficients. Tabulations of experimental virial coefficients for a large number of gases are given in the compilation [Dym80].

Taking the expansion up to the third term usually gives a close approximation for gases for densities less than about half the critical density. Integrating the expression for P at constant T gives the expansion for the deviation in F from ideal behavior, and subsequent differentiation with respect to T gives expressions for S and C_V. In particular

$$C_V = C_{V,ideal} - nRT\left[\frac{d^2[TB(T)]}{dT^2}\frac{1}{V_m} + \frac{1}{2}\frac{d^2[TC(T)]}{dT^2}\frac{1}{V_m^2} - \cdots\right] \tag{4.14}$$

The virial coefficients depend upon the intermolecular potential function: B depends only upon the interaction between two molecules, C also on the three-molecule potential, and so on. Since the interaction between molecules depends upon their relative orientation, the integrals for the virial coefficients involve rotational degrees of freedom (e.g., [Gra84]); they also involve other coordinates if the molecules are floppy.

The simplest application however is to a classical monatomic gas. If the pair potential is $\phi(r)$,

$$B(T) = -\frac{N_A}{6}\int\left[e^{-\phi(r)/kT} - 1\right]4\pi r^2 dr \tag{4.15}$$

At low temperatures the integral is dominated by the outer region where $\phi(r)$ is negative, and consequently B is negative. As T increases the repulsive range of the potential becomes more important, and B becomes positive at sufficiently high temperatures. The temperature at which this occurs, the Boyle point T_B, varies strongly from gas to gas. For helium it is about 22 K, for hydrogen about 110 K, for nitrogen about 330 K and for carbon dioxide over 700 K. The information contained in precise measurement of B as a function of temperature has been used in determining the shape of $\phi(r)$, particularly in the attractive range of the potential. Information about three body potentials can in principle be derived from third virial coefficients, and so on [Bar76a].

4.2.5. Numerical Data

Whereas the small departures from perfection of the equation of state of gases such as helium and hydrogen have impact on their application as thermometric standards, these departures are trivial as far as cryogenic engineering applications are concerned. Values of β are sufficiently well established for most cryogenic purposes by assuming that $PV \approx nRT$ so that $\beta = d \ln V/dT \approx 1/T$.

The heat capacity needs more discussion. The variation in C_V for a number of real gases may be judged from the data in Table 4.2. These few representative values are taken from the extensive compilation entitled *Thermodynamic Properties of Cryogenic Fluids* [Jac97]. This monograph gives both tables and computer fits for values of density, U, H, S, c_V, c_P, and velocity of sound at pressures from 0.1 MPa up to 20 MPa (or higher for some fluids) of air, Ar, CO, n-D_2, C_2H_6, F_2, n-H_2, Kr, CH_4, Ne, N_2, O_2, p-H_2, and Xe. Another useful source is the IUPAC series of *International Thermodynamic Tables of the Fluid State*, beginning in 1971 with Ar [Ang71]; subsequent volumes have been edited by S. Angus, B. Armstrong or K. M. de Reuck under the auspices of various publishers.

As discussed above, values of C_V at temperatures significantly above the boiling point are close to 3R/2 = 12.5 J·mol^{-1}·K^{-1} for the rare gases, and 5R/2 = 20.8 J·mol^{-1}·K^{-1} for oxygen and nitrogen. For n-H_2, C_V falls well below 5R/2 as T drops below about 300 K; for p-H_2, C_V has a maximum near 150 K arising from rotational energy, as shown also in Fig. (4.3). When the temperature of a dilute gas held at constant pressure is lowered towards the boiling point, significant increases can occur in C_P and C_V (see Fig. 4.4 and Table 4.2), which are accompanied by a greater increase in density than predicted for an ideal gas, giving evidence of pre-condensation clustering.

4.3. LIQUIDS AND DENSE GASES

4.3.1. Introduction

A liquid possesses some short range order in the instantaneous distribution of near neighbors about an atom or molecule, but in general the geometry of this loose packing differs radically from that in the solid phase: as well as the loss of long

Table 4.2. Values of C_V ($J{\cdot}mol^{-1}{\cdot}K^{-1}$) for some gases at pressure of 0.1 MPa (=1 bar) [Jac97] (see also ITS-90 for values of T_t [Qui90])

T (K)	Ar	^4He	n-H_2	Kr	Ne	N_2	O_2	CH_4
4.216	—	12.96	—	—	—	—	—	
10	—	12.46	—					
20	—	12.48	—	—	—			
30	—	12.48	12.57	—	13.07	—		
50	—	12.47	12.53	—	12.52			
70	—	12.47	12.87	—	12.49	—		
100	13.29	12.47	14.26	—	12.48	20.84	20.88	—
120	12.70	12.47	15.44	—	12.48	20.87	20.94	25.5
150	12.56	12.47	17.07	12.75	12.48	20.80	20.87	25.12
200	12.50	12.46	18.97	12.47	12.47	20.80	20.83	25.18
300	12.50	12.46	20.54	12.47	12.47	20.77	21.07	27.41
600	12.50	12.46	21.0	12.47	12.47	21.79	—	44.06
T_b (K)	87.17	4.207	20.345	119.62	27.061	77.237	90.062	111.51
T_c (K)	150.66	5.1953	33.19	209.43	44.492	126.19	154.58	190.56
P_c (kPa)	4860	227.5	1315.2	5510	2678.6	3397.8	5043	4599
T_t (K)	83.8058	—	13.95	115.776	24.5561	63.1504	54.3584	90.694
P_t (kPa)	68.89	—	7.2	73.0	43.4	12.52	0.1463	11.7

range order when a crystal melts, there is a discontinuous change in the nature of the short-range order. This was first illustrated by Bernal in an experimental study of the random packing of uniform hard spheres. In the close-packed crystalline form (e.g., fcc) each sphere is surrounded by twelve nearest neighbors, and 74% of the space is occupied. In various experiments Bernal found that when spheres were shaken together in a random arrangement only 64% of the space was occupied, and the coordination patterns were quite different from those in a solid (for details of this and later work see, e.g., [Cus87]).

Computer simulations (MC and MD as described in Section 2.3.3) on random assemblies of atoms, have substantially confirmed Bernal's conclusions about liquid structure, and have provided quantitative results for simple models that also enable approximate analytic theories to be tested. Such work is beyond the scope of this monograph, but some references have already been given in Section 2.3.3. To these can be added *The Physics of Structurally Disordered Matter* by Cusack [Cus87], which discusses also the electronic properties of liquids and glasses; the student text *Gases, Liquids and Solids* by Tabor [Tab91]; *Liquids and Liquid Mixtures* by the physical chemist Rowlinson [Row69]; *Theory of Simple Liquids*, 2nd. edition, by Hansen and McDonald [Han86]; and *Theory of Molecular Fluids. Vol. 1: Fundamentals* by Gray and Gubbins [Gra84].

4.3.2. Changes in Thermodynamic Properties on Melting

Except for helium at low temperatures (Section 4.4), entropy increases on melting. The molar entropy $\Delta_f S$ of fusion is of the order R: about $0.8R$ for many metals, about

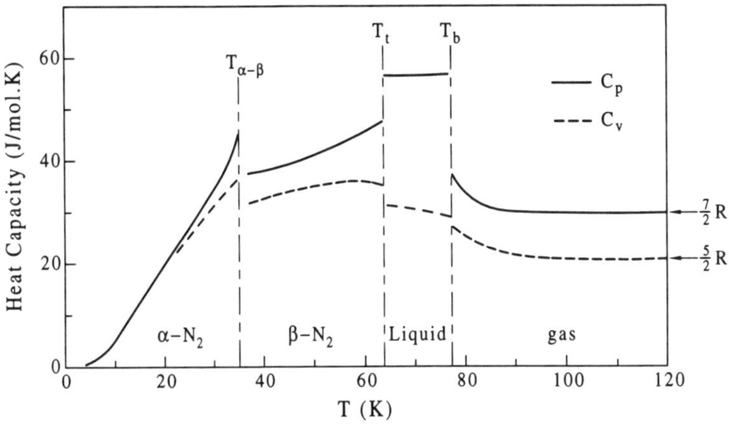

Fig. 4.4. Heat capacities C_P and C_V of solid and liquid nitrogen along the equilibrium vapor pressure curve [Man97], and of gaseous nitrogen at a pressure of 0.1 MPa (1 bar) [Jac97].

1.5*R* for the rare gas solids, hydrogen and hydrogen halides, about 2.5*R* for the semi-metals Ga, Sb and Bi, 3.8*R* for Si and Ge, and distributed in this range for various crystals of small molecules [Moe57, Wal91]. For the same groups the molar volumes increase by a few percent for most metals, by about 15% for the rare gas solids, and by 6–12% for various molecular crystals. Molar volumes *decrease* by a few per cent for Ga, Sb, and Bi, and by about 8% for Si and Ge. These values contrast with those for vaporization — typically about 12*R* for $\Delta_v S$ and 2×10^{-2} m^3 for $\Delta_v V$. Thus $\Delta_v V/\Delta_f V > 10^3$, whereas $\Delta_v S/\Delta_f S \leq 10$; by the Clapeyron Eq. (2.26) this explains why solid–liquid equilibrium lines in *P*–*T* phase diagrams are much steeper than liquid–vapor lines (Fig. 4.1). Similarly, solid–liquid lines for substances whose volumes decrease on melting have a negative slope. These include the semimetals and semi-conductors mentioned above, in which the closer packing in the liquid is associated with radical change in the electronic structure [Wal91]; and also ice, where on melting there is a partial breakdown of the tetrahedral hydrogen bonding between the oxygen atoms.

For the heat capacity, we must distinguish between C_V and C_P. Since each translational or rotational degree of freedom contributes only $\frac{1}{2}R$ to C_V when the molecule is free in an ideal gas, but double that amount for classical harmonic vibrations, we might expect C_V to decrease as the molecules become freer on melting. This is seen to happen in all three phase transitions shown for N_2 in Fig. 4.4. In the α solid phase there is long range orientational order; this is lost in the β solid phase, where however there is believed to be correlated rotation retaining considerable short range order. C_V decreases also when there is further loosening of the structure both on melting and on vaporization. In contrast, the difference in heat capacities $C_P - C_V$ increases on melting, presumably due to an increase in compressibility and hence in thermal expansion.

4.3.3. The Critical Region

Both liquid and gas have random structures. At temperatures just above the triple point the molecules of a liquid cohere because of the attractive part of the intermolecular potential. At the saturated vapor pressure the liquid has the same Gibbs free energy $H - TS$ as the vapor, whose greater enthalpy is compensated by its greater entropy. But at temperatures high enough for kT to be much greater than the depth of the potential well, only the repulsive part of the potential is important, and there is no tendency for the molecules to cohere. Then, as the pressure is increased, the random structure of the fluid changes continuously from dilute gas to dense fluid, and there is no phase change until the Gibbs free energy of the ordered crystalline structure becomes less than that of the fluid. There is therefore a critical temperature T_c above which the distinction between gas and liquid disappears. The fluid can change continuously from low temperature gas to low temperature liquid by passing above T_c (Fig. 4.1).

It was J. D. van der Waals who produced the first and simplest theoretical model which demonstrates this behavior. The ideal equation of state is modified to give

$$\left(P + \frac{a}{V_m^2}\right)(V_m - b) = RT \tag{4.16}$$

where the effective reduction in molar volume to $(V_m - b)$ simulates the effect of the hard core inner repulsion, and a/V_m^2 simulates the effect of the attractive forces. The success of this beautifully simple approximation in the interpretation of fluid behavior is described in countless books; for a full and clear discussion see [Dom96, Ch. 2]. The critical point is given by

$$RT_c = \frac{8a}{27b}, \qquad V_c = 3b, \qquad P_c = \frac{a}{27b^2} \tag{4.17}$$

The second virial coefficient is $B(T) = b - (a/RT)$, illustrating the general dependence on temperature discussed in Section 4.2.4, and the Boyle temperature is thus $T_B = a/Rb$. C_V is finite but has a discontinuity at the critical point and C_P and χ_T diverge.

Such a simple model does not of course represent quantitatively the behavior of even the simplest real fluids. For example, Eq. (4.17) gives a value of $8/3 = 2.67$ for the dimensionless ratio RT_c/P_cV_c, compared with values for Ar and n-H_2 of 3.43 and 3.28 respectively. Moreover, the nature of the singularity at the critical point (for example, how χ_T and C_P approach infinity) is different from that observed. Experimental study of the singularity demands high precision as the critical point is approached, and theoretical study demands extensive calculations using either series expansions or computer simulation. As an introduction to the vast literature on this subject, see [Sta71, Dom96]; for critical phenomena in liquid mixtures, see [Row69].

4.4. QUANTUM FLUIDS; LIQUID HELIUM

4.4.1. Quantum Ideal Gases

The ideal gas is a system for which the effects of quantum statistics can be calculated exactly. As in Section 4.2.1, we consider a monatomic gas of N non-interacting particles of mass m, occupying a volume V, so that the particle density is $\rho = N/V$. Quantum theory gives the density of states for a particle in a large box as

$$n(\epsilon) = 2\pi\sigma V(2m/h^2)^{\frac{3}{2}}\epsilon^{\frac{1}{2}} \tag{4.18}$$

where $n(\epsilon)d\epsilon$ is the number of states between ϵ and $\epsilon + d\epsilon$, and $\sigma = 2s+1$ is the spin degeneracy for a particle of spin s. This density of states is written as continuous, because when V is of macroscopic size the energy intervals between states are very small. But strictly the levels are discrete, and there is a unique lowest energy state, with an energy $\left(3h^2/2mV^{\frac{2}{3}}\right)$ that tends to zero for a large system.

At $T = 0$ the gas is in its ground state. For a Bose gas all the particles are in the lowest energy state, and there is no zero-point energy. For a Fermi gas there is one particle in each of the N lowest energy states; the highest occupied state has energy

$$\epsilon_{F0} = (h^2/2m)(3\rho/4\pi\sigma)^{\frac{2}{3}} \tag{4.19}$$

and the zero-point energy is $\frac{3}{5}N\epsilon_{F0}$.

As T increases the distribution of particles among the energy states spreads out, until at high enough temperatures the probability of occupation of any given state becomes small and both systems behave like a classical gas, with thermodynamic functions as given in Section 4.2.1. At intermediate temperatures their behavior is determined by the Fermi–Dirac or Bose–Einstein distribution functions derived in Section 2.3.1; the average number $f(\epsilon)$ of particles in a given state of energy ϵ is either

$$f_{FD}(\epsilon) = \frac{1}{e^{(\epsilon-\mu)/kT}+1} \quad \text{or} \quad f_{BE}(\epsilon) = \frac{1}{e^{(\epsilon-\mu)/kT}-1} \tag{4.20}$$

as plotted in Fig. 4.5. At each temperature the chemical potential μ has a value such that the total number of particles is N:

$$\sum_i f(\epsilon_i, \mu, T) = \int f(\epsilon)n(\epsilon)d\epsilon = N \tag{4.21}$$

At $T = 0$, $\mu_{FD} = \epsilon_{F0}$ and $\mu_{BE} = 0$.

Ideal Fermi Gas. When $kT \ll \epsilon_{F0}$ the occupation numbers are affected only for those states for which $|\epsilon - \epsilon_{F0}| \sim kT$. The number of particles that are affected is of

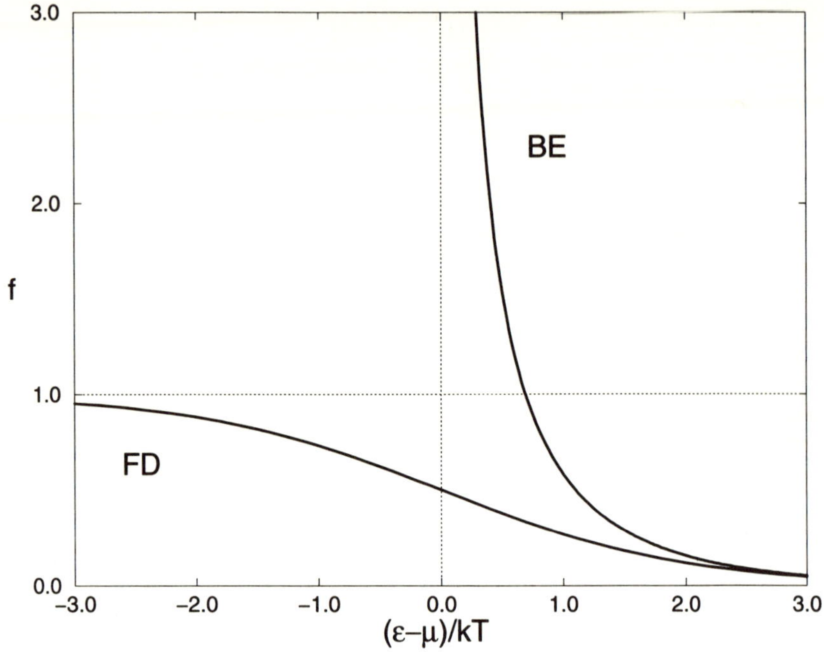

Fig. 4.5. Fermi–Dirac and Bose–Einstein distribution functions, plotted against $(\epsilon - \mu)/kT$.

order $n(\epsilon_{F0})kT$, and the mean increase of energy for such particles is of order kT. The total increase of energy above the ground state is therefore of order $n(\epsilon_{F0})(kT)^2$, and the heat capacity is of order $kn(\epsilon_{F0})(kT)$. For a general density of states detailed calculation leads to the numerical result of Eq. (1.19). For the ideal gas C_V increases as

$$C_V/R = \frac{1}{2}\pi^2\frac{T}{T_F}\left[1 - \frac{3\pi^2}{10}\left(\frac{T}{T_F}\right)^2 - \cdots\right] \tag{4.22}$$

while μ falls with temperature:

$$\mu = \epsilon_{F0}[1 - (\pi^2/12)(T/T_F)^2 + \cdots] \tag{4.23}$$

On further increase of temperature the thermodynamic functions change smoothly up to the classical limits (Fig. 4.6).

Ideal Bose Gas. The occupation of each level increases with increase of chemical potential. Since $f_{BE}(\epsilon)$ becomes infinite when $\mu = \epsilon$, μ cannot exceed the energy of the lowest state, which we have seen is effectively zero. At low temperatures μ has

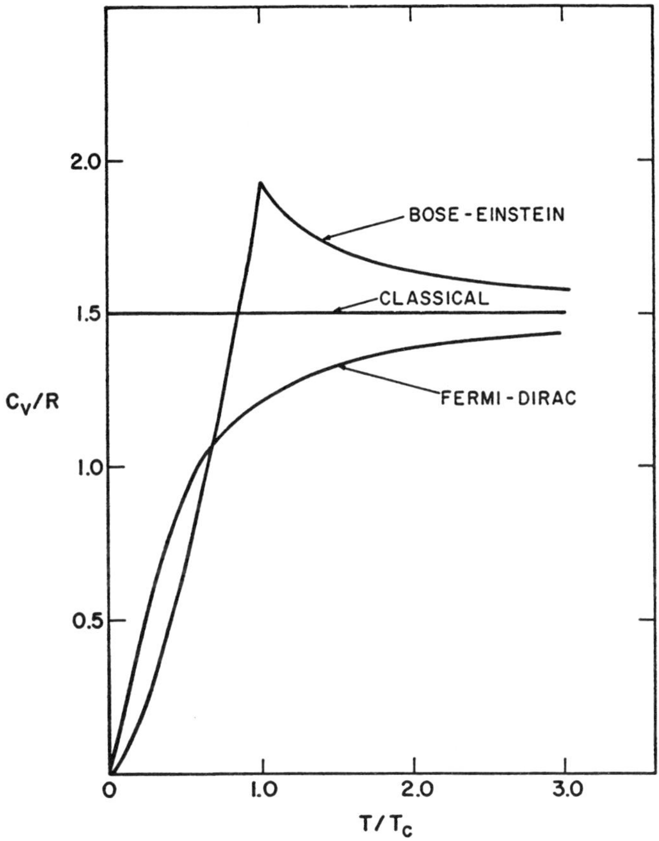

Fig. 4.6. Heat capacity as a function of T/T_c for Fermi, Bose, and classical ideal gases, all for the same density of states. T_c is the critical temperature of the Bose gas, and $T_F = 2.29T_c$. From [Dug96, Fig. 14].

this maximum value, so that the number of particles in excited states is

$$2\pi\sigma V\left(\frac{2m}{h^2}\right)^{\frac{3}{2}}\int_0^\infty \frac{\epsilon^{\frac{1}{2}}}{e^{\epsilon/kT}-1}d\epsilon \tag{4.24}$$

This is less than the total number of particles N in the system at all temperatures up to

$$T_c = \frac{h^2}{2\pi mk}\left(\frac{N}{2.612\sigma V}\right)^{\frac{2}{3}} \tag{4.25}$$

Below this critical temperature the rest of the particles are in the lowest state, with energy zero. The energy is therefore $\int f(\epsilon)\epsilon\, d\epsilon$, and C_V and S follow thermodynamically:

$$U = \frac{3}{2}PV = 0.770RT\left(\frac{T}{T_c}\right)^{\frac{3}{2}}, \quad C_V = 1.92R\left(\frac{T}{T_c}\right)^{\frac{3}{2}}, \quad S = 1.28R\left(\frac{T}{T_c}\right)^{\frac{3}{2}} \tag{4.26}$$

Above the critical temperature μ starts to fall, and there is a cusp in C_V which gradually falls to its classical value of $1.5R$ (Fig. 4.6). The nature of the transition at T_c is discussed by London in [Lon54, Section 7]. Although C_V is continuous at T_c, it is not a third order phase transition (Section 2.2.4). Below the critical temperature P depends only on T and not on V, and so the compressibility is infinite, and C_P diverges; and when P is greater than the critical pressure the volume collapses to zero. The accumulation of particles in the lowest energy state which starts to occur below T_c is called *Bose–Einstein condensation*. The condensate has zero entropy and energy, and makes no contribution to the thermoelastic properties.

Real atomic and molecular gases do not exhibit these quantum properties, because condensation to liquid or solid occurs well above T_F or T_c. But related phenomena occur in the liquid helium isotopes.

4.4.2. Helium

Near their boiling points the liquids of ^{4}He and of the rarer isotope ^{3}He resemble the other rare gas liquids, except that the helium liquids have larger molar volumes. But they differ radically at lower temperatures, where quantum effects dominate. The difference is due to the very low binding energy of the interatomic potential, combined with the low mass, which results in a ‘zero-point’ energy of motion (i.e., even in the quantum ground state) which almost cancels the binding energy of the potential. This not only “blows up” the volume of the liquid, but also prevents it from solidifying under normal pressure even at the lowest temperatures. As in solids, marked departures from classical behavior are seen as the temperature is lowered to absolute zero; and because helium is the only element which remains liquid down

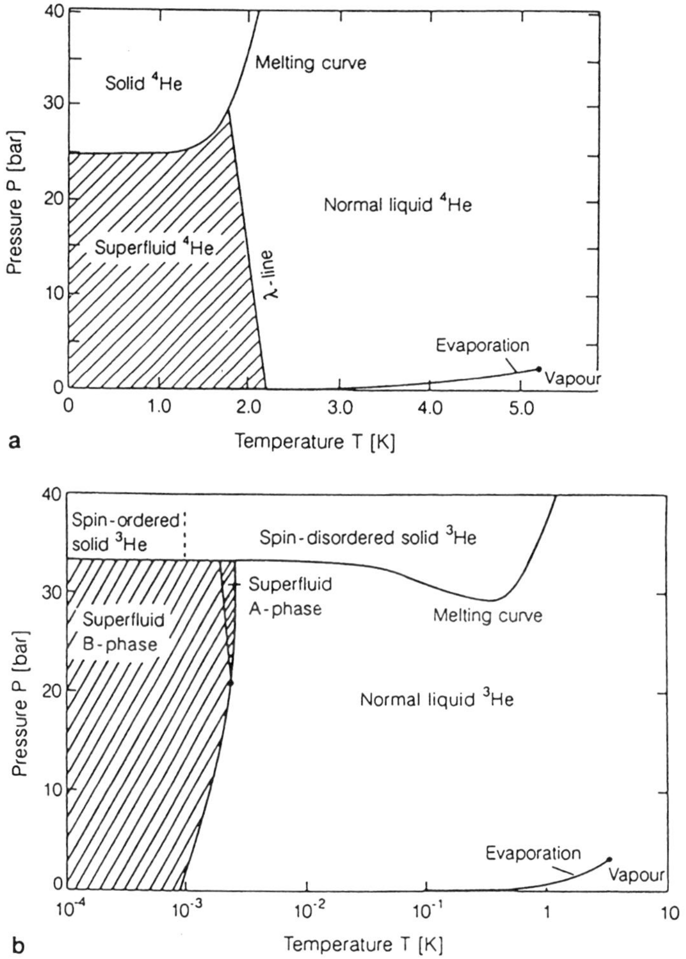

Fig. 4.7. Phase diagrams at low pressures for (a) ^{4}He, (b) ^{3}He. From [Pob96, Fig. 2.4].

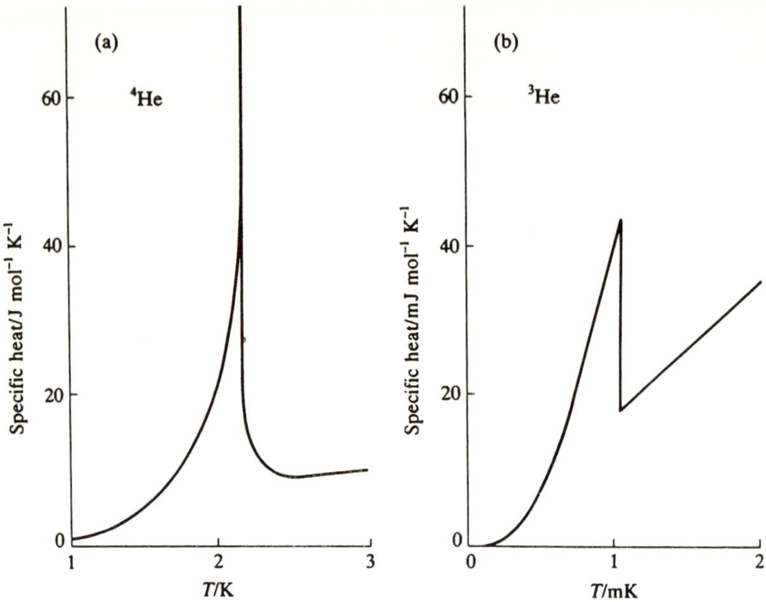

Fig. 4.8. The heat capacities of liquid ^{4}He and ^{3}He near T_λ. Note the difference in the temperature scales. From [Wil87, Fig. 1.3].

to $T = 0$, this "quantum liquid" behavior is unique. Moreover, the ^{4}He nucleus is a boson, with nuclear spin $I = 0$, while the ^{3}He nucleus is a fermion, with nuclear spin $I = \frac{1}{2}$; and so the two liquids show quite different quantum effects.

Phase diagrams for the two isotopes are shown in Fig. 4.7, where the temperature scale for ^{3}He is logarithmic. The "λ-line" for liquid ^{4}He marks the famous λ-shaped transition in the heat capacity, below which the liquid becomes *superfluid* (vanishing viscosity) as the Bose particles condense into a ground state (the phases above and below the transition are often called He I and He II). No such transition occurs in the Fermi liquid ^{3}He, which is in some ways like the electron "sea" in a metal. ^{3}He does not become superfluid until much lower temperatures are reached (about 2.6 mK), and then the peak in the heat capacity is a different shape (Fig. 4.8) more reminiscent of a superconducting transition (cf. Fig. 6.19). The requirement of different types of Cooper pairing from that in a typical superconductor, and the effect thereon of a magnetic field (including the existence of a magnetic superfluid phase), has provided a system of great complexity for the delight of theorists [Vol90]. Further discussion is however beyond the scope of this monograph.

Among the many books and reviews on this subject are *Helium-3 and Helium-4* by Keller [Kel69], *Liquid and Solid Helium* by Wilks [Wil67] and the later *An Introduction to Liquid Helium* by Wilks and Betts [Wil87], *Helium Cryogenics* by Van Sciver, with tables of thermodynamic data for ^{4}He above 1 K in an Appendix

[Van86], *The Superfluid Phases of Helium 3* by Vollhardt and Wölfle [Vol90], and the more general reference *Matter and Methods at Low Temperatures* by Pobell [Pob96].

4.4.3. ^{4}He

There are two complementary simple models or ways of thinking about liquid ^{4}He — the two fluid model of London and Tisza, based on analogy with the ideal Bose–Einstein (B–E) gas, and the theory of Landau, based on the quantization of hydrodynamics. We consider first the two fluid model.

The statistical mechanics of the ideal Bose gas suggests that we may picture the gas as consisting of two components — one the 'condensate' of atoms in the ground state, the other comprising the remaining atoms in excited states. The phenomenological two fluid model for the liquid ^{4}He below T_c follows by analogy, with a *superfluid* component carrying no entropy, and a *normal liquid* component. It has proved successful in correlating many of the transport properties of liquid ^{4}He. However, the analogy with an ideal Bose gas should not be pushed too far. In particular the behavior of the heat capacity is very different. For example, below about 0.6 K, ^{4}He has $C_V = 0.0816\ T^3\ \mathrm{J{\cdot}mol^{-1}{\cdot}K^{-1}}$, compared to the ideal Bose gas for which C_V varies as $T^{\frac{3}{2}}$.

In contrast, Landau proposed that the lowest lying excitations were quantized longitudinal sound waves (phonons), which as in a solid give C_V proportional to T^3 at low temperatures, and also that at higher values of the wave-number there was a pronounced minimum in the dispersion curve which he ascribed to some kind of localized rotational motion (rotons). Later inelastic neutron scattering measurements confirmed this general shape of dispersion curve. The coefficient of T^3 derived from the measured sound velocity is in good agreement with the calorimetric value, and the joint contribution of phonons and rotons to C_V approximates well to experiment until fairly close to the lambda transition. In the neighborhood of the transition measurements of increasing precision are consistent with a logarithmic divergence both below and above T_c (Fig. 4.9), quite different from the ideal Bose gas. Extensive theory is discussed in [Kel69, Wil87].

4.4.4. ^{3}He

Once again it was Landau who gave the seminal model for the energy states, this time for a Fermi liquid. The system is far removed from an ideal Fermi gas of non-interacting atoms, but the model is rather similar except that 'quasi-particles' of effective mass m^* take the place of the bare atoms (cf. Section 6.1). As in metals, the heat capacity is predicted to be proportional to T at low temperatures (but not of course in the superfluid region).

Interest in ^{3}He as a Fermi liquid has led to many measurements of heat capacity: for example those of Greywall in Fig. 4.10 [Gre83] showing that C_V at a fixed volume varies roughly linearly with T at temperatures above 0.3 K. Below 0.1 K data may

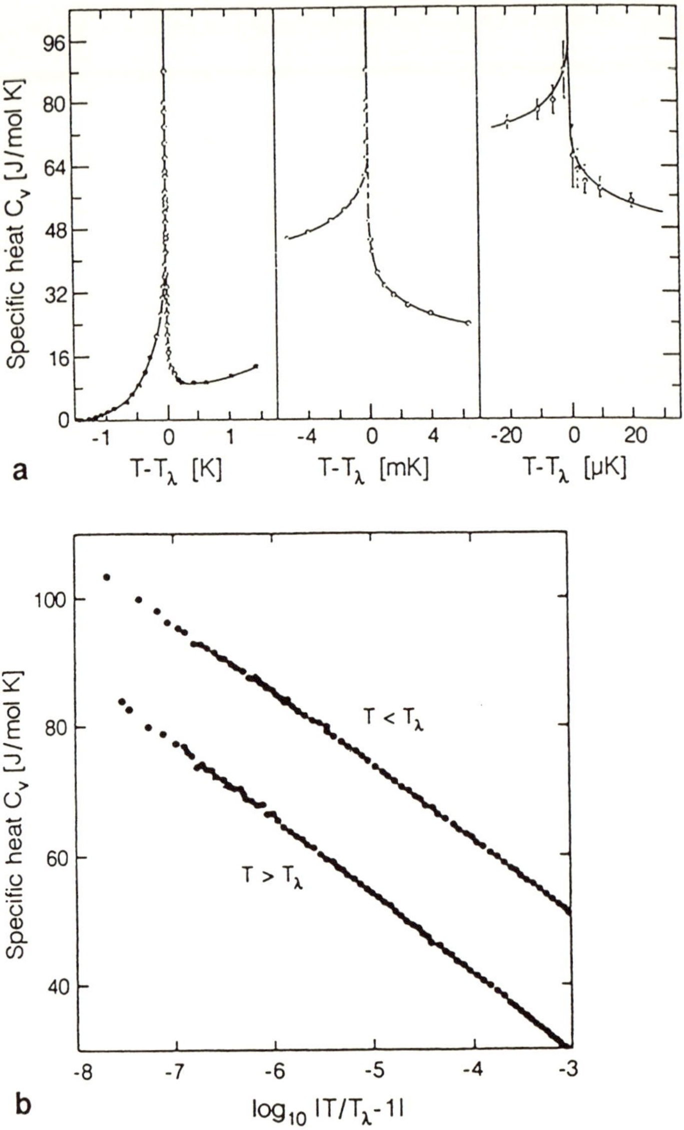

Fig. 4.9. Heat capacity of liquid ^{4}He at saturated vapor pressure near the lambda transition: (a) with increasing T-resolution on a linear temperature scale; (b) on a logarithmic temperature scale. From [Pob96, p. 23].

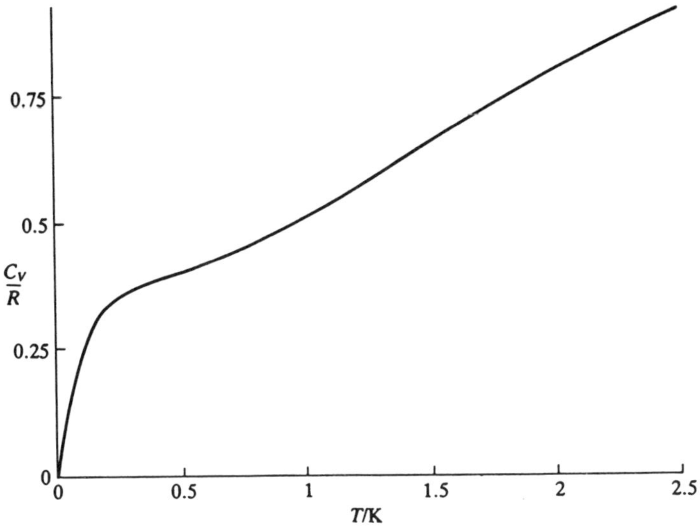

Fig. 4.10. C_V/R for liquid ^{3}He at a fixed volume of 36.82 $\text{cm}^3 \cdot \text{mol}^{-1}$, see [Wil87, p. 8] and [Gre83].

be fitted to a theoretical relation

$$C_V/R = \Gamma T + BT^3 \ln(T/\Theta) \tag{4.27}$$

where Γ, B and Θ are volume-dependent parameters; for a volume of 36.74 $\text{cm}^3 \cdot \text{mol}^{-1}$ they have respective values of 2.78 K^{-1}, 35.4 K^{-3} and 0.458 K. The second term in Eq. (4.27) is ascribed to the effect of local fluctuating ferromagnetic alignments.

Pomeranchuk refrigerator. The melting curve in Fig. 4.7(b) is determined by the properties of the solid as well as the liquid. In the solid we can regard the atoms as labelled by their lattice sites, and the Pauli principle (though still operative) can be ignored. The nuclear spins at each site can take either value, and because the interaction between them is weak they are not ordered except for temperatures of the order of mK and below; above this the spin entropy is $R\ln 2$. In consequence the entropy of the liquid is less than that of the solid up to about 0.32 K, and it follows from the Clapeyron equation Eq. (2.26) that the slope of the melting curve is negative in this region. Appreciably below this temperature the entropy of the liquid is therefore equal to that of the solid at a much lower temperature. This was predicted as early as 1950 by the theorist Pomeranchuk [Pom50], who also drew the important conclusion that freezing the liquid by adiabatic compression could be used to reach temperatures in the range corresponding to spin ordering in the solid. An extensive discussion of the practical use of Pomeranchuk cooling is given by Pobell, see [Pob96, Ch. 8].

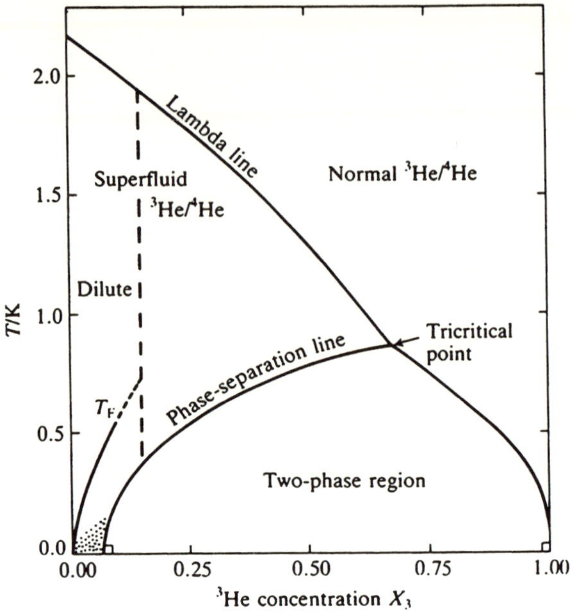

Fig. 4.11. The phase diagram of liquid ^{3}He/^{4}He mixtures at saturated vapor pressure. From [Wil87, p. 106].

4.4.5. ^{3}He/^{4}He Mixtures

The phase diagram (Fig. 4.11) shows that as the temperature is lowered below the tricritical point at 0.7 K a mixture of liquid ^{3}He and ^{4}He separates into two phases over an increasing range of concentrations. As $T \to 0$ ^{4}He becomes totally insoluble in ^{3}He, but ^{3}He remains up to 6.6% soluble in ^{4}He. In such dilute solutions, at temperatures well below the λ-line, the ^{4}He component in the solution is almost wholly superfluid, providing an inert background diluting the ^{3}He so that it becomes a Fermi gas rather than a Fermi liquid, but with an effective mass m^* of about 2.4 times the bare ^{3}He atomic mass. Heat capacity measurements for dilute solutions confirm the expected Fermi gas behavior, with T-dependence at low temperatures changing smoothly to the classical value of $3R/2$ per mole of ^{3}He at higher temperatures (e.g., [Edw65, And66b]). Figure 4.12 shows this, and also demonstrates the effectiveness of heat capacity measurements in determining points on the phase separation curve.

Dilution refrigerator. At the 1951 Low Temperature Conference in Oxford, H. London suggested that the adiabatic dilution of a solution of ^{3}He in liquid ^{4}He would prove an effective means of cooling. This is the principle of the ^{3}He/^{4}He dilution refrigerator, now widely used to achieve temperatures down to about 5 mK. Methods for the effective realization of the principle, and the construction and

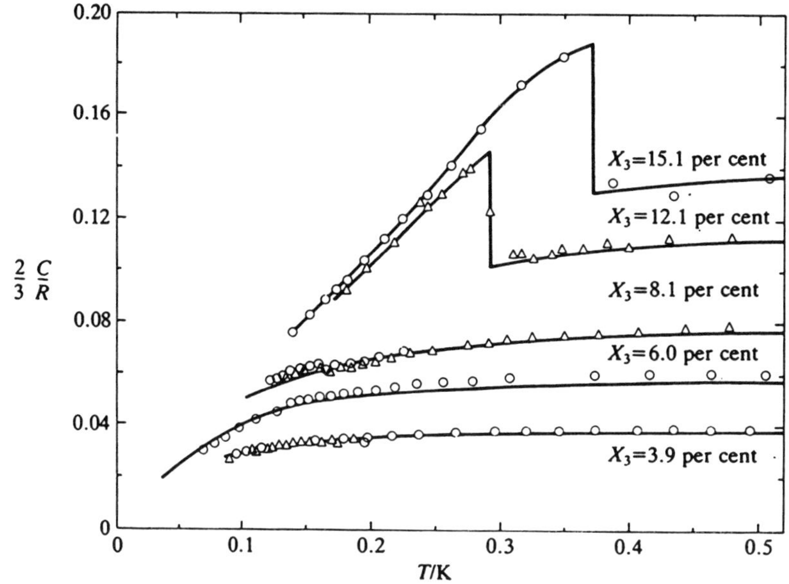

Fig. 4.12. Heat capacity per gram-atom of liquid $^3He/^4He$ mixtures at the saturated vapor pressure, plotted as $C_{sat}/\frac{3}{2}R$. From [Wil87, p. 107].

operation of different types of equipment, are discussed by Pobell [Pob96] (see also [Lou74, Bet89, Whi79]).

Chapter 5

Non-Metals

5.1. INTRODUCTION

In this chapter we give examples of the behavior of heat capacity and thermal expansion for some of the important classes of non-metallic solids, in which itinerant electrons make little or no contribution. Usually the dominant contributions to the heat capacity and thermal expansion are those of the lattice vibrations. At very low temperatures these can be related to the elasticity, and as the temperature increases to the phonon frequencies as determined by neutron diffraction (Section 2.6.2). But some of the solids also display non-vibrational effects, due for example to localized magnetic moments, or to quantum tunnelling of atoms between different sites.

We begin with the 'simplest' family — the rare gas solids — traditionally beloved by the theorist because of their electrically neutral closed shell atomic structures and close-packed crystal structures, enabling them to be approximated quite well by short-range central force models.

Then we turn to other important cubic systems: the strongly ionic alkali halides of rocksalt structure; the alkaline earth fluorides (fluorite structure); and the tetrahedrally bonded crystals of diamond and zincblende structures, together with those of the associated hexagonal wurtzite structure. These are followed by a structurally disparate group of oxides — magnesia, alumina, α-quartz, zirconia and titanium dioxide (rutile); then glasses and glass ceramics, including the ultra-low expansion materials based on aluminosilicates; and then highly anisotropic crystals such as graphite (layer structure), tellurium (chain structure), and the chain-like polymers. Although polymer crystals are chain-like, many commercial samples are in the isotropic or 'amorphous' condition and display some glass-like behavior in their thermal properties.

The chapter continues with the ceramic high-temperature superconductors, which are generally non-cubic, although their structure is related to that of perovskite. They behave thermally much like other ceramics, except in the vicinity of their superconducting transition and at liquid helium temperatures. The chapter concludes with magnetic crystals, and the effect of impurities such as magnetic ions, electric dipoles, and heavy or light ions.

5.2. RARE GAS SOLIDS

The history of the discovery of the rare gas elements, and of fifty years of progress in unravelling their physical properties, is well told by George Horton in the opening chapter of the two volumes of *Rare Gas Solids* [Kle76, Kle77]. Both theory and experiment are covered comprehensively in succeeding chapters.

All these solids are weakly bound by van der Waals interatomic forces, which can be represented fairly closely by pair potentials of similar shape [Bar76a], scaled by parameters ϵ_0 for energy and r_0 for distance. For the heavier rare gases this leads to a *law of corresponding states*, so that they have similar properties when scaled to units derived from these two parameters and the atomic mass (see, for example [Row69]). But departures from this simple law of corresponding states become larger for the lighter rare gas solids, as the ratio of the zero point vibrational energy to the binding energy of the interatomic potential increases from about 0.03 for Xe to 0.25 for Ne and nearly unity for He. This ratio depends on the de Boer dimensionless parameter

$$\Lambda^* = h/\sqrt{m\epsilon_0 r_0^2} \tag{5.1}$$

tabulated in Table 5.1 [deB49]. In helium the zero point energy is so large that even at the lowest temperatures solidification occurs only under pressure of at least 25 bar, and even then the vibrations are strongly anharmonic. For this reason we consider first the heavier rare gas solids.

5.2.1. Ne, Ar, Kr and Xe

Purely central forces would give the hcp structure, but small many-body effects cause all these crystals to be fcc [Nie76]. Because the bonding is weak they have low melting points and large expansion coefficients, making pure strain-free single crystals difficult to produce and presenting a challenge to the experimentalist. Many measurements have been made on samples which were condensed in metal or glass cylinders and thereby severely strained, and so not truly representative. Thermal and elastic properties are reviewed by P. Korpiun and E. Lüscher in the second volume of *Rare Gas Solids* [Kle77], and high pressure thermodynamic data by C. A. Swenson in the same volume. Data for Ne, Ar, Kr and Xe from these two sources are given in Table 5.1, together with values for the two helium isotopes (see Section 5.2.2 and the review by H. R. Glyde in [Kle76]).

Figure 5.1 illustrates the temperature dependence of Θ^C for Ar [Bea61]. For Kr, Xe and Ne the graphs are very similar in shape to that for Ar. The reduced values of $\Theta^C(T)/\Theta_0$ versus T/Θ_O are quantitatively similar for all four, with a shallow

Table 5.1. Data for rare gas solids. Λ^* is the de Boer parameter, T_t is the triple point and * denotes solid phase under pressure of 3.3 MPa (^{3}He) and 2.5 MPa (^{4}He) [Kle76, Kle77]

Property	^{3}He	^{4}He	Ne	Ar	Kr	Xe
Λ^*	3.08	2.67	0.593	0.186	0.103	0.063
T_t(K)	—	—	24.556	83.806	115.763	161.391
ρ_0(g/cm^3)	0.123*	0.190*	1.507	1.771	3.093	3.781
V_0 (cm^3/mol)	24.3*	21.0*	13.4	22.6	27.1	34.7
B^S_{10}(GPa)	0.019	0.027	1.12	2.9	3.5	3.7
Θ^C_0 (K)	15–18*	26*	74	92	72	64
γ_0		2.9*	2.6	2.7	2.6	2.5

minimum near $\Theta_0/10$. The rise in $\Theta^C(T, V_0)$ at high temperatures indicates that the anharmonic contribution to C_V is negative, as often seen in a close-packed crystal.

Thermal expansion measurements by X-rays and capacitance dilatometry and also equation of state data (see [Bar80, p. 654]) lead to limiting values of $\gamma_0 \simeq 2.6$ and values in the range of 2.7 to 2.9 at higher temperatures (Fig. 5.2).

The most accurate elastic data appear to be those derived from Brillouin scattering or from the limiting slopes of phonon dispersion curves obtained by inelastic neutron scattering (see [Kor77]). At low temperatures departures from the Cauchy relation $c_{12} = c_{44}$ are small, as expected for solids dominated by pair potentials (Section 2.8.6); The crystals are elastically anisotropic, with $2c_{44}/(c_{11} - c_{12}) \simeq 2.4$ as $T \to 0$ [Kor77].

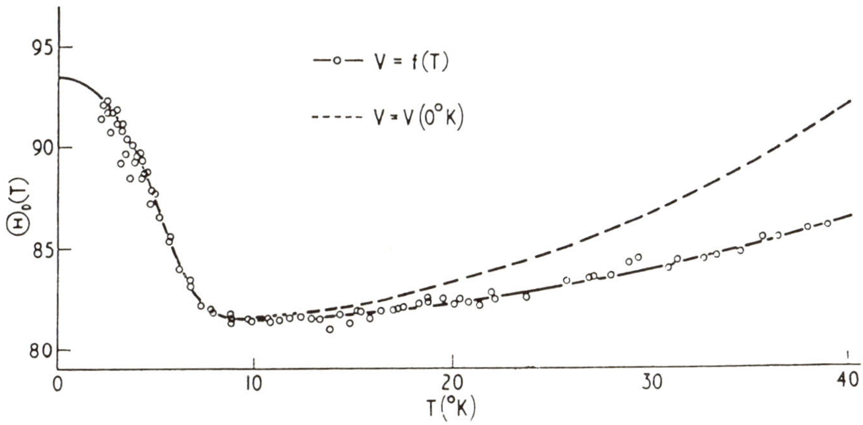

Fig. 5.1. Variation of $\Theta^C(T)$ for Ar [Bea61].

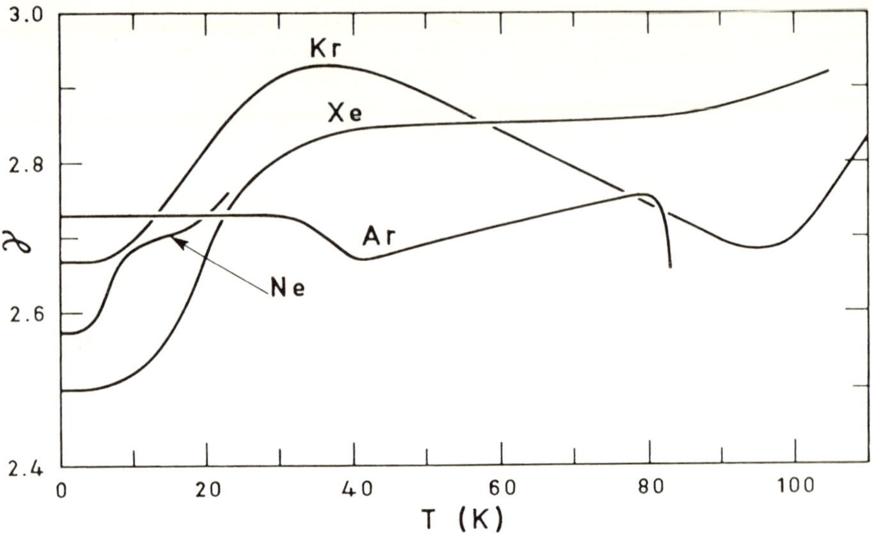

Fig. 5.2. $\gamma(T)$ for Ne, Ar, Kr and Xe [Bar80, p. 655].

5.2.2. Helium

Helium gas, at least at low pressures, may be close to 'perfection' as far as the equation of state is concerned, but the liquid and solid states present a complex picture which is still being unravelled and would take a volume to cover in detail. The weak binding forces and large zero point energy prevent solidification under normal pressure (see phase diagram in Fig. 4.7): the more common isotope ^{4}He liquefies under the standard pressure of 100 kPa (1 bar) at 4.21 K, and ^{3}He at 3.19 K. The minimum pressure to produce solid ^{4}He increases from about 2.5 MPa (25 bar) near absolute zero to about 8 MPa near 3 K. Even under these pressures the volume is considerably larger than the equilibrium volume of the static lattice, and gives a negative static lattice compressibility; it is the zero-point vibrational energy that stabilizes the structure at these volumes. But although the quasi-harmonic approximation thus breaks down completely, the heat capacity and thermal expansion behave similarly to other solids, with a renormalized effective vibrational spectrum (Section 2.6).

The thermodynamic properties above 1 K are well established. The solid phase is hcp at moderate pressures, except for a narrow range along the melting curve between 1 and 2 K where it is bcc. For ^{3}He the solidification pressures are higher (Fig. 4.7 and Table 5.1), and there is a more extended bcc phase region at lower pressures. Both isotopes transform from hcp to fcc at high T and P.

Because of the large compressibility of solid helium, there have been many measurements of both heat capacity and $(\partial P/\partial T)_V$ as functions of volume and temperature, which lead to values of $\Theta^C(V,T)$ and $\gamma(V,T)$. For ^{4}He, γ_0 increases from about 2.5 at a molar volume $V = 17.5\ \text{cm}^3$ to 2.9 at $V = 21\ \text{cm}^3$, indicating that γ/V is fairly constant and $q = d\ln\gamma/d\ln V \simeq 1$ (e.g., [Swe77]).

Dugdale and Franck [Dug64] made extensive measurements of C_V of solid (and fluid) ^{3}He and ^{4}He from 3 K up to the melting points at different densities. Curves of Θ^C versus V measured near $T = \Theta/20$ show parallel behavior of the two isotopes, each giving $\gamma = -d\ln\Theta/d\ln V = 2.4$. At much lower temperatures ($\sim$ mK) the nuclear spin of ^{3}He gives rise to magnetic properties in both the bcc and fcc phases. These, and also the properties of ^{3}He–^{4}He mixtures, are described in *Solid Helium Three* [Dob94].

Theoretical treatment of these highly anharmonic crystals is reviewed extensively in [Gly76]. More general references are the monographs by Wilks [Wil67], Keller [Kel69], Dobbs [Dob94], and Pobell [Pob96].

5.3. ROCKSALT STRUCTURE

5.3.1. Alkali Halides

Most of the alkali halides crystallize in the 6-coordinated cubic rocksalt structure; cations and anions each form a fcc lattice, and together occupy the sites of a simple cubic lattice. At high pressures the crystals undergo a first order transition to the more compact 8-coordinated cesium chloride structure, where cations and anions each form a simple cubic lattice, and together occupy the sites of a bcc lattice. Transition pressures increase from a few GPa for Rb halides to tens of GPa for Na halides; they have been predicted by *ab initio* electronic computations for several of the salts in good agreement with experiment [Sim98]. The remaining halides, CsCl, CsBr and CsI, have the cesium chloride structure at atmospheric pressure, at least up to room temperature.

Qualitative differences in the lattice dynamics and thermal properties of rocksalt halides may be expected to reflect differences in the ratios of atomic masses and of ionic radii, and also in the nature of the short-range interatomic potential. The differences in ionic radii are illustrated in Fig. 5.3, which compares LiF with RbBr and CsF using values for the Pauling radii and nearest neighbor distances taken from Born and Huang, see [Bor54, p. 18]. It is clear that the rocksalt structure can comfortably accommodate a wide range of ratios of ionic radii. The ionic masses do not directly affect the elasticity, but do affect the vibrational spectrum and hence the heat capacity and thermal expansion. In contrast, the nature of the interatomic potential affects all three properties.

Elasticity. We consider first the volume and the elasticity. Table 5.2 shows how the molar volume V increases with increasing size of each ion. At the same time the bulk modulus B decreases, so that the product VB remains roughly constant throughout the table. The shear stiffness for transverse waves polarized perpendicular to $\{100\}$ planes, c_{44}, also decreases with the size of either ion, but more drastically

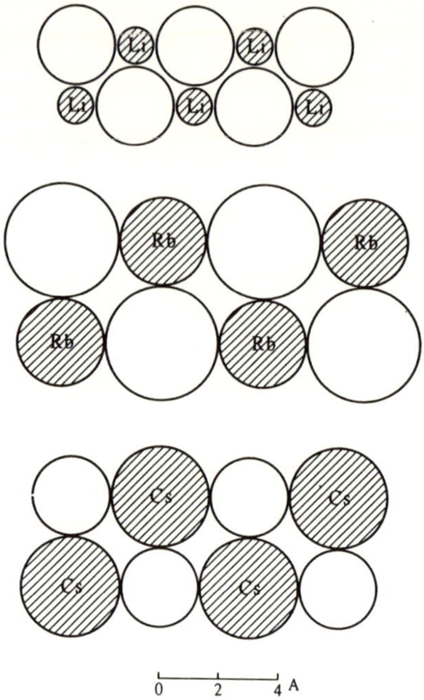

Fig. 5.3. Schematic view of a (100) plane in (a) LiF, (b) RbBr and (c) CsF.

when the cation is changed. The ratio c_{44}/B thus depends mainly on the type of alkali ion; and so also does its pressure derivative, which becomes negative for the heavier cations. This has important consequences at low temperatures, affecting the heat capacity and particularly the thermal expansion. Our main source of elastic data for this table are from [Lew67, Har79] and compilations by Hearmon [Hea66, Hea69].

Heat Capacity. The first definitive measurements of C_P by Clusius *et al.* [Clu49] were followed by extensive work at the National Research Council (Ottawa) on K and Na halides [Bar57b]. The latter achieved sufficiently high accuracy (error bars of $\pm 0.1\%$) for the Taylor expansion in Eq. (2.87) to be fitted to the data below $\Theta/25$, which gave the higher order T^5 and T^7 terms as well as the T^3 term and hence the corresponding terms in ω^4 and ω^6 for the frequency distribution. The T^3 term gave values for Θ_0^C in good agreement with best elastic values.

At higher temperatures the accuracy was sufficient to allow extrapolation of $(\Theta^C)^2$ versus $1/T^2$ to give the harmonic high temperature limit Θ_∞^C for several salts (see Table 5.2 and Section 2.6.2). For other salts in Table 5.2, Eq. (2.76) was used to derive Θ_∞^C from values of $< \omega^2 >$ computed from the DD3N models of Hardy and Karo [Har79].

Plots of $\Theta^C(T)$ show a characteristic pattern, as in Fig. 5.4 (see [Bar57b, p. 485]), with a minimum near $\Theta_D/15$ and an increase towards Θ_∞ at higher temperatures; the

Table 5.2. Data for alkali halides of NaCl structure. Most are from [Bar80] with elastic data from [Lew67]. Asterisk* indicates use of room temperature data, ^{el} use of elastic data, and ^{m} use of $< \omega^2 >$ [Har79]

	M.W.	V_0 (cm^3)	B_0 (GPa)	c_{44} (GPa)	$\frac{dc_{44}}{dP}$	Θ_0 (K)	$\Theta_\infty(V_0)$ (K)	γ_0	α_{293} (10^{-6}/K)
LiF	25.9	9.81	69.8	65	1.38	735	662^m	1.7	33.9
LiCl	42.4	20.0	35.4	26.9	1.70	429	419^m	2.0^{el}	44
LiBr	86.8	$24._0$	26.3	20.5	1.80	274	338^m	1.9^{el}	49.5
LiI	133.8	$33._0$	19.2*	14.0*	1.96*	178^{el}	281^m	—	—
NaF	42.0	14.7	51.4	28.8	1.21	488	438^m	0.92	32.5
NaCl	58.45	26.4	27.0	13.2	0.37	321	290	1.06	39.5
NaBr	102.9	31.2	22.6	10.6	0.42	225	236^m	1.0	42.0
NaI	149.9	39.9	17.9	7.9	0.59	164	195	1.03	45.5
KF	58.1	23.0	34.2	12.3	−0.45	333	334^m	0.49^{el}	31.2
KCl	74.55	36.1	19.7	6.7	−0.39	235	235	0.35	37.0
KBr	119.0	42.1	17.7	5.1	−0.33	174	188	0.29	39.0
KI	165.0	53.2	12.8	3.7	−0.29	132	163	0.28	40.5
RbF	104.5	$36._0$	30.1	9.6	−0.70	221	273^m	0.06^{el}	—
RbCl	121.0	42.9	18.7	4.7	−0.61	168	198^m	0.0	35.5
RbBr	165.4	48.2	16.0	4.1	−0.59	134	146^m	−0.03	37.5
RbI	212.4	57.9	13.3	2.9	−0.51	107	122^m	−0.11	39.0
CsF	151.9	32.3	26.5*	8.0*	—	174^{el}	254^m	0.3	33.9

fall-off in Θ^C at still higher temperatures for the potassium halides is an anharmonic effect, due to a small positive contribution to C_V. The shapes of these curves depend both on the mass ratio and on the type of alkali ion. Θ_0 depends through the elastic wave velocities on the density, and hence on the average mass $(m_1+m_2)/2$, whereas to a rough approximation Θ_∞^C depends on the reduced mass [Bla42]:

$$(\Theta_\infty^C)^2 \propto \frac{1}{2}\left(\frac{1}{m_1}+\frac{1}{m_2}\right)$$

For crystals with similar interionic forces, the ratio Θ_0/Θ_∞^C thus becomes proportional to $\sqrt{1-\eta^2}$, where $\eta = (m_1-m_2)/(m_1+m_2)$. To a good approximation this factor accounts for changes in Θ_0/Θ_∞^C as the halide ion is changed. But in addition, the lowering of the ratio c_{44}/B with increasing atomic number of the alkali ion affects the weighting of the low-lying transverse modes, producing a relative lowering of Θ_0, so that the mass-corrected ratio depends chiefly on the cation. For example, KCl and RbBr both have ionic mass ratios close to unity but different ratios Θ_0/Θ_∞^C.

Thermal Expansion. At room temperature linear expansion coefficients lie mainly in the narrow range 36 to 45 $\times 10^{-6}$ K^{-1}. This is a consequence of a similarly narrow range (1.4–1.7) for the Grüneisen function at $T \sim \Theta_D$ for these salts (Fig. 5.5), since VB varies little. Values for the fluorides are lower. There is a clear trend for α_{273} to increase with the size of the halide ion, and except for CsF

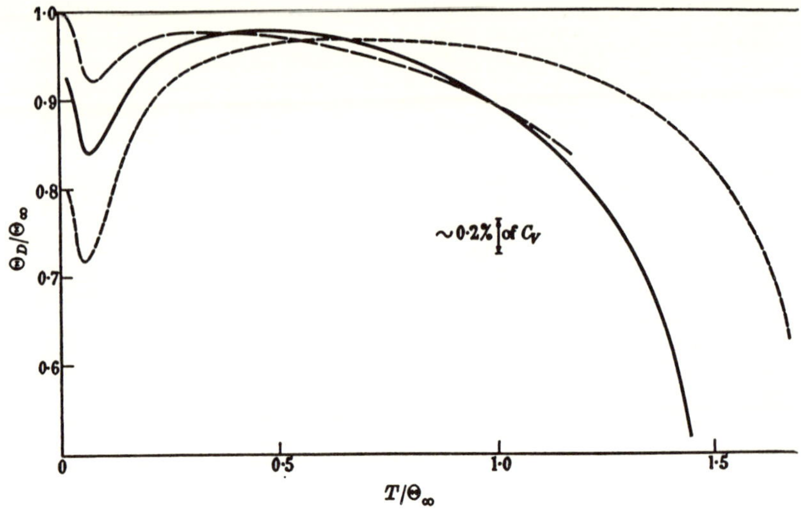

Fig. 5.4. Reduced plot of Θ^C against T for potassium halides [Bar57b].

to decrease with the size of the alkali ion. Sources of α values in Table 5.2 are the reviews [Bar80, Mer73] and the book by Krishnan *et al.* [Kri79].

At low temperatures γ decreases markedly (excepting for the lithium salts). This is another dimensionless property that depends primarily on the alkali ion (Fig. 5.5), as can be seen also from the values of γ_0 given in Table 5.2; these correlate well with the change in the alkali ion and agree within limits of measurement with values of γ^{el} calculated from the pressure dependence of the ultrasonic velocities. In particular, transverse modes propagating in $\{100\}$ planes and polarized normal to the planes do not bring into play the strong force-constant of the nearest neighbor pair potential, and so tend to have low frequencies and small or negative gammas because of the tension effect (Section 2.6.3). In the acoustic limit these are governed by c_{44}, which has a negative pressure dependence for both the potassium and rubidium halides; and this plays a major part in lowering the average γ_0. For example, RbCl has $dc_{44}/dP = -0.6$, which gives a value of about -1.3 for the associated mode γ. The weighted average over all the low frequency modes of RbBr gives $\gamma_0^{el} \simeq -0.05$, agreeing well with $\gamma_0^{th} = -0.03$.

By contrast, for the three Cs halides which have the CsCl structure $\gamma \simeq 2.0$ over the whole range from 2 to 300 K (see review [Bar80, p. 661]).

5.3.2. Other Crystals of Rocksalt Structure

Numerous other solids crystallize in the rock salt structure, including the alkaline earth oxides (MgO is discussed in Section 5.6) and many compounds of the transition and post-transition metals, including PbS, PbSe, PbTe, SnTe, etc. For many of

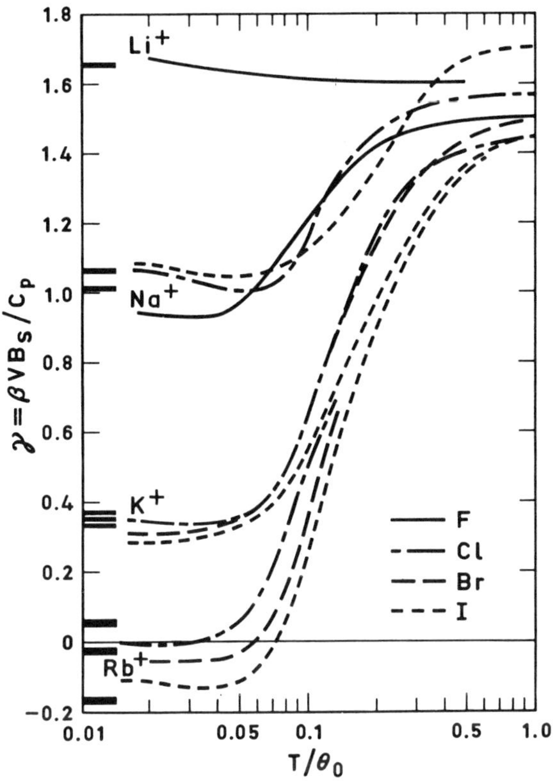

Fig. 5.5. γ for alkali halides as function of reduced temperature, T/Θ_0 [Bar80, p. 658].

these there are heat capacity data in Vol. 5 of *Thermophysical Properties of Matter* [Tou70b] and thermal expansion data in Vol. 13 of the same Series [Tou77]. They are not generally of major cryogenic interest.

5.4. FLUORITE STRUCTURE

The best known member of this family is fluorspar, CaF_2, which may be described crystallographically as a face-centered cubic array of Ca^{2+} ions with F^- ions occupying the 8 tetrahedral sites in the cube. Physically it is useful to think of the structure as a simple cubic array of F^- ions with centers of alternate cubes occupied by the smaller Ca^{2+} ions, especially when considering the anion disordering which occurs in these compounds at high temperatures ($T \sim 0.8\ T_m$).

Table 5.3 lists relevant data for three alkaline earth fluorides. Measurements of heat capacity show that Θ^C decreases from its $T = 0$ value by about 10% with a

Table 5.3. Data for some crystals of fluorite structure [Bar80, Whi80, Col83b]

	M.W.	V (cm^3)	B_0 (GPa)	c' (GPa)	$\frac{dc'}{dP}$	Θ_0 (K)	Θ_∞^C (K)	γ_0	α_{293} (10^{-6}/K)
CaF_2	78.1	24.3	88.2	63	0.8	508	500	1.0	18.5
SrF_2	125.6	29.1	74.5	40	0.4	382	420	0.64	17.9
BaF_2	175.3	35.6	62.6	25.7	−0.2	286	360	0.23	18.2
PbF_2	245.2	30.8	≈ 70	27	0.2	230	$\simeq 350$	0.5–0.9	29.0

minimum near $\Theta_0/15$. Their lattice properties display a rather similar pattern to that shown by the NaCl structure, but with the shear modulus $c' = (c_{11} - c_{12})/2$ replacing c_{44} in the dominant rôle. As the alkaline earth ion changes from Ca to Sr to Ba, the increase in ion size brings a progressive weakening of the shear stiffness c' and a decrease in its pressure derivative. Figure 5.6 shows the pattern wherein the γ falls from a high temperature value between 1.6 and 1.9 to a low temperature limit of 1.0 for the calcium fluoride, 0.64 for strontium fluoride and 0.23 for the barium fluoride.

An interesting contrast to the alkaline earth fluorides is PbF_2 (Fig. 5.6), with a large peak in $\gamma(T)$ around 25 K. The heat capacity and expansion data are each well represented by an Einstein contribution centered about $\Theta_E \approx 65$ K as well as a Debye term [Whi80], in agreement with the difference in shape of the frequency distribution $g(\omega)$ from those of the alkaline earth fluorides [Dic78, Bil79].

Two oxides of fluorite structure which are of high temperature interest are ThO_2 and UO_2. They are high melting point ceramics, and any cryogenic interest rests with a first order antiferromagnetic transition in UO_2 near 30 K (see Section 5.11).

5.5. TETRAHEDRALLY BONDED CRYSTALS

This family of open structured crystals with low coordination number ($z = 4$) includes many important minerals and semiconductors and has attracted much experimental and theoretical attention. Their vibrational and associated thermal properties show interesting features, including a low frequency transverse acoustic branch which is important at low temperatures and leads to serious departures from Debye-like behavior, and in many cases to negative values of the expansion coefficient. Such behavior is likely in open structure crystals (Section 2.6.3), being due to low frequency transverse modes (likened to the vibrations of a guitar-string) which are preferentially excited at low temperatures [Bla57, Bar57a]. This pattern, already seen to a lesser extent in the alkali halides, extends also to other systems of open structure, including silicate crystals and silica glasses.

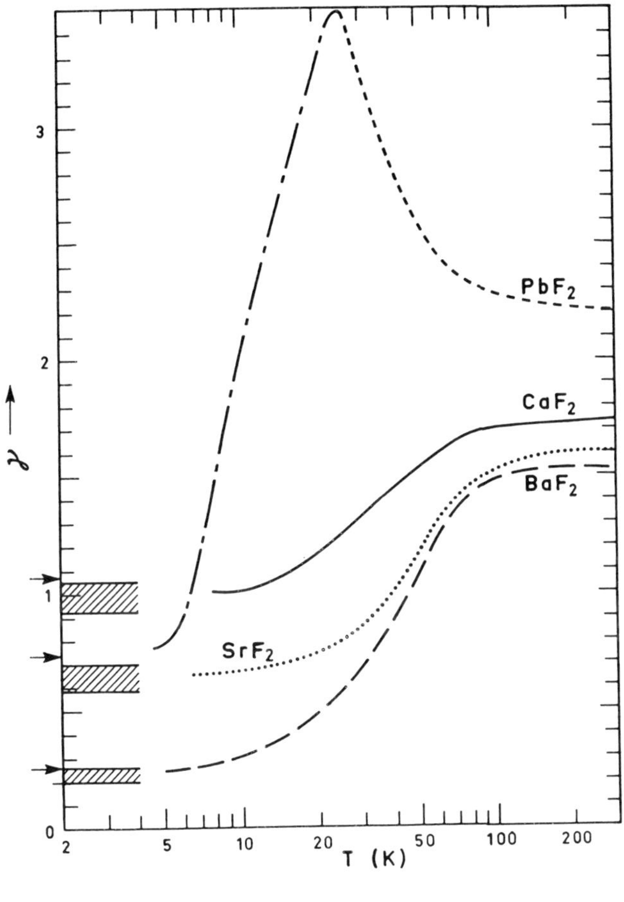

Fig. 5.6. $\gamma(T)$ for alkaline earth fluorides and PbF_2. Hatched areas denote limiting values of γ_0^{th} for the alkaline earth fluorides, and arrows show γ^{el} [Whi80].

5.5.1. Diamond and Zincblende Structure

As in the fluorite structure, the lowest acoustic velocity in these cubic crystals is for c' transverse modes propagated along the [110] direction with a $[1\bar{1}0]$ polarization; but now *all* the ions are tetrahedrally coordinated. If we look at a three-dimensional model of this rather open structure, it is easy to see how such an acoustic wave (having an open space normal to the propagation direction) propagates along 'chains' without altering distances between nearest neighbors. Unless the bonds have appreciable covalency and hence strong angular rigidity, such modes will not only be of relatively low frequency but will also soften under pressure like a relaxed guitar string (Section 2.6.3). Ultrasonic measurements confirm that the velocity of this wave decreases under pressure for many members of the family; dc'/dP is negative, and increasingly so for the more ionic (less covalent) members (Table 5.4). A measure of the ionicity is a factor, f_i, defined and tabulated by Phillips [Phi73] (see Table 5.4 below). Crystals for which f_i exceeds 0.78 are not likely to be stable in this cubic structure.

Neutron spectroscopy confirms a high degree of dispersion for these transverse acoustic modes. At temperatures in the region of $\Theta/25$ they may be fully excited, and contribute strongly to the sharp drop in $\Theta_D(T)$ (see Fig. 5.7) and the minimum in $\gamma(T)$ (see Fig. 5.8). The decrease in γ from γ_0 to γ_{min} results from two factors: dispersion, which weights these TA modes more heavily, and more importantly a decrease in the mode gammas (to become negative — or more negative) with increase in wave number, revealed both by model [Dol66] and *ab initio* [Xu91] calculations, and by neutron diffraction under stress [Pay64]. Measurements of the Raman spectra under pressure for Zn chalcogenides also confirm that γ_{TA} at the zone boundaries may be much lower than at the low-frequency zone center (e.g., [Wei77]).

Table 5.4 lists diamond or zincblende structure crystals for which there are extensive low temperature data [Bar80]. Excepting diamond itself they all appear to have low-temperature ranges over which α is negative; thus γ_{min} is negative, although γ_0 is sometimes positive (though small). Barron *et al.* [Bar77a] show a correlation between γ_0, c' and the ionicity, f_i, for a number of these crystals. Values of γ_0^{th} marked with a query (?) in Table 5.4 are less reliable than elastic values, γ_0^{el}, due to uncertainty in measuring the very small expansion below 10 K.

Most of these crystals have γ values at higher temperatures which approach 0.7 or 0.8. An exception is HgSe, for which $\alpha < 2 \times 10^{-6}\ K^{-1}$ up to 500 K [Zhd66], giving $\gamma_\infty \simeq 0.2$. Pressure derivatives of both the shear moduli (c' and c_{44}) are negative for HgSe, and the minimum value of α is comparable with that for CuCl (Fig. 5.9), namely about $-8 \times 10^{-6}\ K^{-1}$ near 30 K.

There are limited thermal and elastic data for some other III-V compounds such as the phosphides. Useful sources of available data on the Group IV and III-V compounds are the Landolt–Börnstein volumes III/17a [Mad82]. Slack and Bartram [Sla75] have reviewed thermal expansion data for a number of diamond-like crystals, with emphasis on high temperature data and the importance of matching expansion values for technical purposes; the crystals include AlN, cubic BN, BP, GaP, cubic and polycrystalline SiC, diamond, Ge, Si and polycrystalline BeO (a wurtzite structure).

Table 5.4. Data for crystals of zincblende structure (and wurtzite in brackets). Most are from review [Bar80]. Asterisk denotes room temperature data, ^{el} denotes use of elastic moduli and f_i is Phillips ionicity factor. For (?) see text

	Θ_0 (K)	B_0 (GPa)	c' (GPa)	$\frac{dc'}{dP}$	γ_0^{th}	γ_0^{el}	f_i	α_{293} (10^{-6}/K)
C	2230	442	477	1.46	—	0.84	0	1.0
Si	645	99.5	51	0.075	0.44(?)	0.24	0	2.56
Ge	374	76.5	40.9	0.35	0.5	0.48	0	5.70
GaAs	345	78.9	32.7	0.105*	0.32	0.38	0.31	5.74
GaSb	269	58.0	24.7	0.16*	0.30(?)	0.39	0.26	6.3
InAs	251	63.6	19.7	−0.20*	0.28(?)	0.01	0.36	≃ 5
InSb	206	48	15.6	−0.15*	0.21(?)	0.05	0.32	5.0
ZnS	339	77	17.9	0.0*	−0.14	−0.17	0.623	6.40
ZnSe	271	64.7	18.0	−0.25*	0.0	−0.05	0.63	6.90
ZnTe	223^{el}	52.8	15.8	−0.16	0.07	0.05	0.61	8.25
CdTe	160	45	8.6	—	−0.85	—	0.72	4.9
HgSe	151^{el}	57	9.0	−0.52	—	−1.2	0.68	1.5
HgTe	148	47.5	9.0	—	−1.2	—	0.65	4.8
CuCl	179	46	4.7	−0.50	−2.3	−2.0	0.75	14.0
(ZnO)	416	139	—	—	—	−1.1	0.62	4.65(⊥) 2.8 (∥)
(CdS)	215	62*	15.5	—	−0.8	−1.4	0.69	4.3(⊥) 2.6(∥)

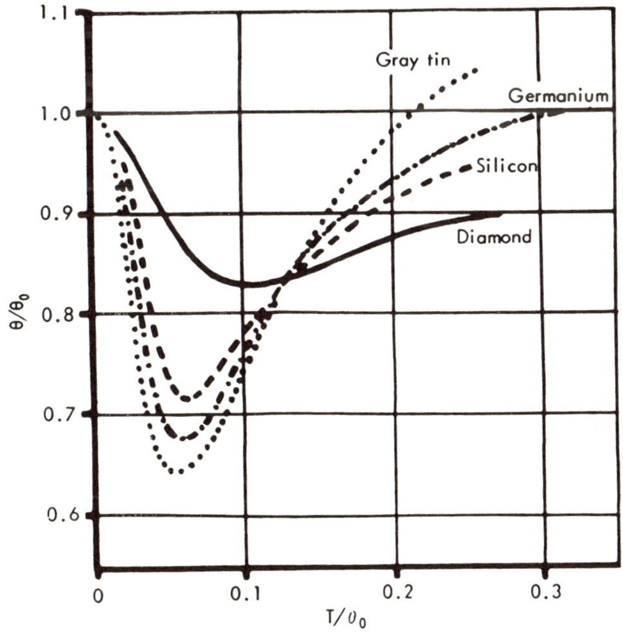

Fig. 5.7. Reduced plot of Θ^C against T for some diamond-type crystals [Gop66].

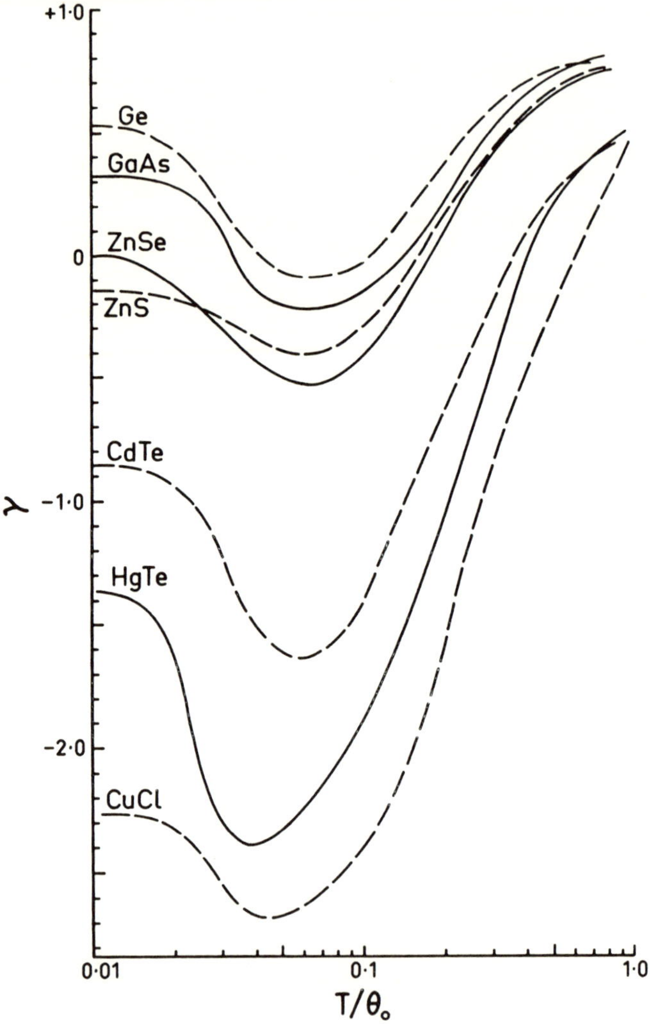

Fig. 5.8. γ for some diamond-type crystals as function of reduced temperature [Bar80, Fig. 5.10].

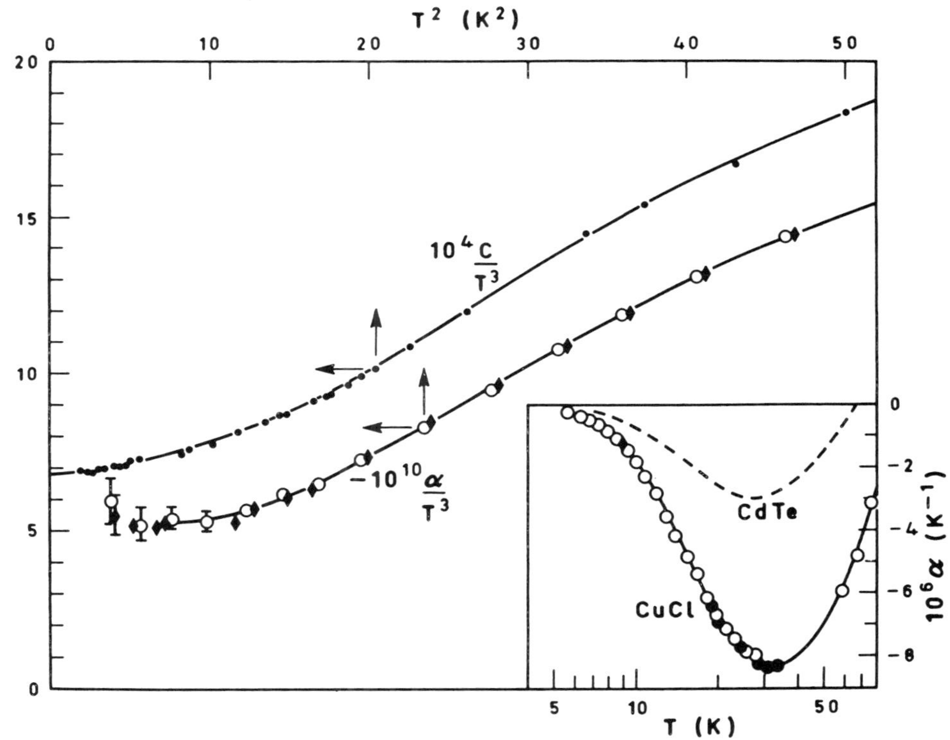

Fig. 5.9. Values of C/T^3 (J·mol^{-1}·K^{-4}) and $-\alpha/T^3$ (K^{-4}) versus T^2 for CuCl. Inset shows $\alpha(T)$ for CuCl compared with CdTe. From [Bar80, Fig. 5.9].

Silicon. This has particular significance as a reference material for thermal expansion (see Section 3.3): it is cubic, readily available in a state of ultra-high purity (and therefore very reproducible), and the expansion has been carefully measured over a wide temperature range. Levels of uncertainty in α are $\leq 10^{-8}$ K^{-1} below room temperature. Values recommended by CODATA (Committee on Data for Science and Technology) [Whi97] are given in Table C.3 of Appendix C.

5.5.2. Wurtzite Structure Including Ice

Closely related to the zincblende structure is the wurtzite structure. This has hexagonal symmetry (ABAB... packing) rather than cubic (ABCA...), and the thermal expansion can therefore be anisotropic. A few of the III–V, II–VI, and I–VII compounds may exist in either zincblende or wurtzite form. Others such as ZnO, AlN, AgI, CdS, BeO are always wurtzite. The low-lying TA modes correspond to the c_{44} and c_{66} elastic constants, which because of the different crystal axes correspond roughly to c' in zincblende.

There is not a great body of thermal data at low temperatures on this group. The Landolt–Börnstein volume on semiconductors [Mad82] gives information on band structure, optical properties etc. for many of the III–V compounds, but few thermal data at low temperatures.

CdS. Cadmium Sulphide has attracted some attention for its electroacoustic properties, but values for α at temperatures below about 30 K are confined to polycrystalline compacts. Curves of $\alpha_{av}(T)$ and $\gamma(T)$ for CdS are similar to those of CdTe (zincblende structure), after allowing for differences in Θ [Bar80, p. 669].

ZnO. Values of the expansion coefficient of zinc oxide along the principal axes are shown in Fig. 5.10, α being negative below about 100 K.

Ice. The oxygen atoms in the hexagonal (normal) form of ice also form a wurtzite structure, being linked by hydrogen bonds. The light mass of the hydrogen or deuterium atoms and consequent low moment of inertia of the water molecules causes the frequencies of the translational vibrations of the molecules to be well-separated from the higher rotational frequencies; frequencies of the intramolecular vibrations are much higher again, and their contributions to C_V can be neglected over most of the cryogenic range.

The thermodynamic properties have been discussed in a classic analysis by Leadbetter [Lea65], who exploited differences between H_2O and D_2O ice to separate the contributions of translational and librational modes to the heat capacity and thermal expansion. He thus obtained moments $< \omega^n >$ and approximate Grüneisen parameters for the different parts of the spectrum. These correlated well with spectroscopic and Debye–Waller data. Small additional effects in C_P due to incipient orientational ordering of the molecules [Hai72] are shown in Fig. 1.12.

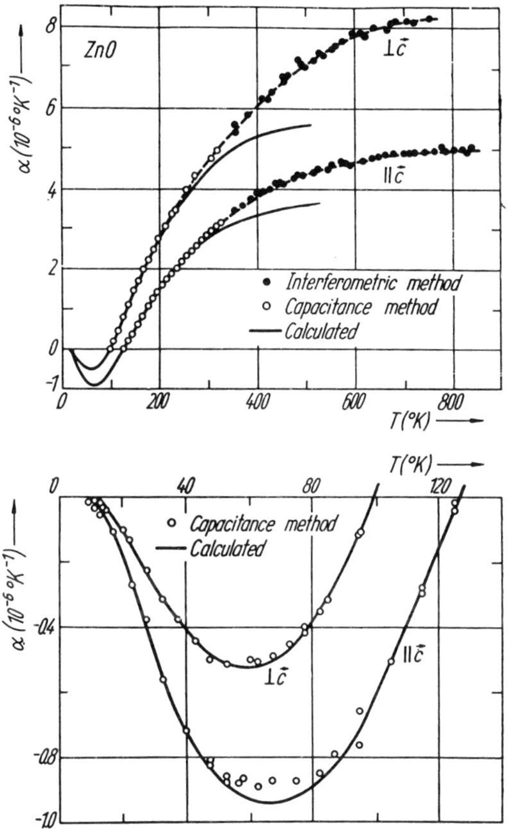

Fig. 5.10. Linear coefficients of expansion of ZnO [Iba69]. The calculated curves were obtained by fitting to a frequency distribution of two Einstein peaks ($\Theta_E = 107$ K and 590 K).

The thermal expansion is not markedly anisotropic: the principal coefficients of linear expansion differ by less than 2% near 273 K, with values of about $50 \times 10^{-6} K^{-1}$. Crystals of both H_2O and D_2O have been measured down to 20 K [Dan62b], showing similar α values for the two isotopes which become negative below 63 K. At 20 K, $\gamma(T)$ is about -0.9, rising to about 0.6 above 150 K [Lea65].

5.5.3. Phenacites, Cuprite

The mineral phenacite, Be_2SiO_4, is another open structure, with tetrahedrally coordinated cations (Be^{++}) and 3-coordinated anions. Slack and Huseby [Sla82] have reviewed the thermal behavior of a number of phenacite-type compounds because of their potential interest as low expansion materials. They are also reviewed by D. C. Palmer in the chapter on 'stuffed derivatives of the silica polymorphs' in *Silica*, see [Hea94, Ch. 3].

Many show a marked departure from Debye behavior in heat capacity with Θ^C falling to $\Theta^C_{min} \sim 0.6\ \Theta_0$ at $T \sim 0.06\ \Theta_0$. They are all non-cubic, and those with small average expansion coefficients at 300 K include:

- Be_2SiO_4 ($\alpha_{av} = 3.0 \times 10^{-6}\ K^{-1}$),
- Zn_2SiO_4 ($\alpha_{av} = 0.8 \times 10^{-6}\ K^{-1}$),
- Zn_2GeO_4 ($\alpha_{av} = 0.7 \times 10^{-6}\ K^{-1}$),
- β-Si_3N_4 ($\alpha_{av} = 1.5 \times 10^{-6}\ K^{-1}$).

For willemite (Zn_2SiO_4) there are expansion data on single crystals extending down to 2 K [Whi88]; $\alpha_\perp$ is negative below 290 K and $\alpha_\parallel$ is negative below 150 K. There are no reliable data on C_P or on B below 50 K, but extrapolation indicates that γ falls from 0.6 at high temperatures to about 0.1 at room temperature and less than -1 at low temperatures (see Section 8.4.3).

Cuprite. Another material of potential technical interest because its low coordination leads to negative expansion is the cubic oxide of copper, cuprite (Cu_2O), see [Bar80, p. 674]. Structurally it consists of two interpenetrating networks not connected to each other by any primary CuO bonds; in each network, the O atoms form a diamond structure, being linked by Cu atoms midway between them, so that each O is surrounded tetrahedrally by Cu and each Cu has two nearest neighbor O atoms linearly arranged. Both the c' and c_{44} stiffness constants have negative pressure derivatives, leading to $\gamma_0 \simeq -4$ and negative values of expansion coefficient below 280 K. Other parameters are $\Theta_0 \approx 185$ K and $B_0 = 110$ GPa.

Table 5.5. Data for some ceramic oxides used at low temperatures [Bar80, Tou77]. $\Delta l/l = (l_{293} - l_{4.2})/l_{293}$. Asterisk denotes room temperature data

	Θ_0 (K)	B_0 (GPa)	c_P^{293} (J/g·K)	α_{293} (10^{-6}/K)	$\Delta l/l)$ (10^{-6})	γ_0	γ_{293}
α-Al_2O_3	1035	251	0.765	5.80($\parallel$) 5.06($\perp$)	635(av)	1.24^{el}	1.31
MgO	953	165	0.92	10.4	1390	1.61	1.50
TiO_2	780	223	0.68	9.1($\parallel$) 7.1($\perp$)	1470(av)	0.5^{el}	1.65
α-SiO_2	558	37	0.733	7.54($\parallel$) 13.75($\perp$)	1700(av)	0.4	0.67
ThO_2	415	193*	0.27	7.2	1160	1.7	2.0
ZrO_2 (stab.)	540	192	0.495	7.8	1300	5	1.7

5.6. USEFUL OXIDES: α-Al_2O_3, MgO, α-SiO_2, TiO_2, ThO_2, ZrO_2(STAB.)

There are a number of ceramic oxides which are used over a wide temperature range for structural supports, insulators, reference standards, film substrates etc. They have little in common crystallographically but are tough and stable materials. In Table 5.5 are given technical data on Al_2O_3 (alumina or sapphire), MgO (magnesia), TiO_2 (rutile), α-SiO_2 (quartz), ThO_2 (thoria), and yttria stabilized zirconia ($ZrO_2 + 9$ mol% Y_2O_3). Some further data on length changes at intermediate temperatures are given in Table C.2 of Appendix C. Data sources are compendia of CINDAS [Tou77], American Institute of Physics Handbook [Kir72], and [Bar80, p. 674]. For the yttria stabilized zirconia, data are from Collins *et al.* [Col85a]. The data for this zirconia should be fairly representative also of calcia- and magnesia-stabilized zirconia.

Sapphire (α-alumina). α-Al_2O_3 is trigonal and so has anisotropic expansion; but it has been available in single crystal form as a Standard Reference Material for both heat capacity (SRM720) and thermal expansion (SRM732), being recommended as a reference material by both CODATA [Whi97] and IUPAC [Mar87a] (see also Ch. 3).

Magnesia. MgO has cubic rocksalt structure, is stable with a high Debye Θ, and is used as a substrate for thin films of high temperature superconductors.

Quartz. SiO_2 can exist in many forms: α-quartz, β-quartz, cristobalite, tridymite, coesite, stishovite, as well as in the vitreous state (see [Hea94, Ch. 1]). The stable form below 846 K is α-quartz, which has an open structure of trigonal symmetry composed of regular SiO_4 tetrahedra linked by shared oxygen atoms at each corner. This linkage is a feature of many forms of silica and silicates, and usually such open structure should give rise to small or negative expansion.*

*Professor Volke Heine and others (e.g., [Pry96]) have discussed to what extent the negative thermal expansion often exhibited by such systems is due to vibrational motion of effectively rigid tetrahedra, but as yet with no application to low temperatures.

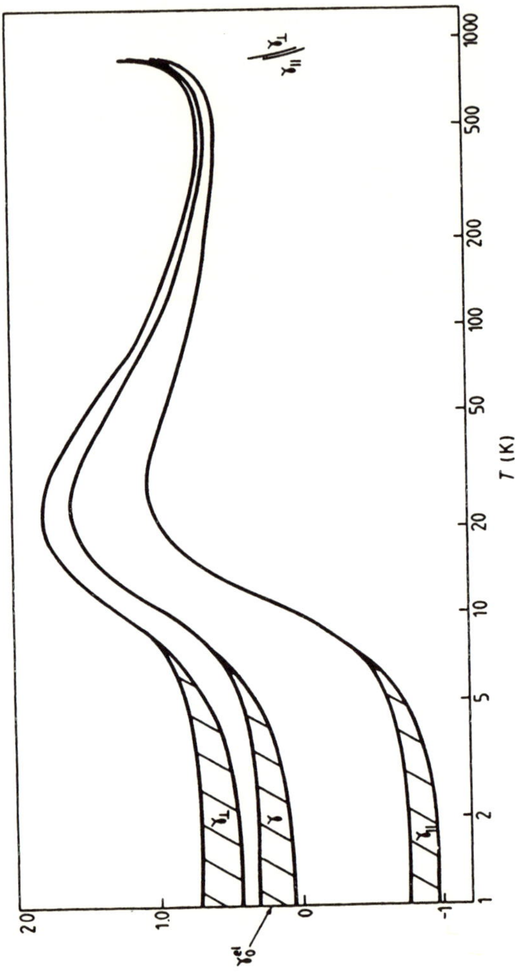

Fig. 5.11. Grüneisen functions for quartz up to 923 K [Bar82].

In α-quartz, however, helical arrangements of the tetrahedra occur parallel to the hexagonal c axis which permit the Si–O–Si angle joining the tetrahedra to change by cooperative rotation or 'tilting' of successive tetrahedra. This flexibility accounts for the relatively large compressibility (see low value of B in Table 5.5) of α-SiO_2 (and of α-GeO_2), and the large *positive* thermal expansion at room temperature compared to other forms of silica [Wri76]; the tetrahedra themselves are fairly rigid, but the change in macroscopic dimensions when tilting is induced by pressure or temperature is relatively large (see for example the neutron diffraction studies under pressure [Car81]). Barron *et al.* [Bar82] discuss the expansion, heat capacity and elastic constants over the range of stability and show how Θ^C and the principal γ's change with temperature (Fig. 5.11). Theoretical models indicate that although tetrahedral tilt is the largest effect in the change of geometry with temperature [Bar87], tetrahedral distortion makes a comparable effect on the macroscopic expansion at low temperatures.

Rutile. Many oxides have the structure of the mineral rutile (TiO_2), including GeO_2, SnO_2, MnO_2. The symmetry is tetragonal and the expansion anisotropic. There are few low temperature data excepting for TiO_2, whose properties are strongly influenced below 150 K by optic modes. These modes, in particular a soft 'ferroelectric' mode (A_{2u}), are responsible for a deep minimum in Θ^C near 30 K ($\Theta_{min} = 440$ K) and a maximum in the principal values of γ also near 30 K. Values of γ(T) are shown in Fig. 5.12 [Col84a] for TiO_2 together with those for MgF_2, which has the same structure but does not contain a transition metal atom. The latter has a less marked minimum in Θ^C and a deep *minimum* in γ, arising in part from an optic B_{1g} mode for which Raman spectroscopy gives $\gamma_j \simeq -5$.

Another tetragonal compound with the rutile structure is paratellurite (TeO_2), for which $\Theta_0^C = 265 \pm 10$ K. At room temperature $\alpha_{\parallel} = 5.6 \times 10^{-6}\text{K}^{-1}$ and $\alpha_{\perp} = 10 \times 10^{-6}\text{K}^{-1}$, giving $\gamma = 0.9$, while at liquid helium temperatures $\gamma \approx -1$ owing to the influence of a soft c' transverse mode [Whi90a].

Thoria. ThO_2 is another stable high melting point ceramic, having the cubic fluorite structure. It is discussed briefly in Section 5.11.3, in association with its isomorph UO_2 which becomes antiferromagnetic below 30 K.

Zirconia. ZrO_2 also has the cubic fluorite structure at high temperatures. The structure can be stabilized at lower temperatures by addition of a few per cent of yttria, calcia or magnesia. It is then important as a high strength ceramic, an oxygen sensor and as a substrate for film deposition. The low temperature heat capacity has a linear (T-) term, indicative of 'disordered' tunnelling states, discussed in more detail under Section 5.7 on glasses and glass-ceramics.

Low temperature values of c_P/T^3 and α/T^3 for $ZrO_2 + 9$ mol% Y_2O_3 are shown in Fig. 5.13 [Col85a] with data below 1 K from [Ack84]. The measured heat capacity includes a Schottky contribution from magnetic impurities, evident below 5 K and centered about 1.5 K. Subtracting this term leaves:

$$c_P \text{ (in } \mu\text{J} \cdot \text{g}^{-1} \cdot \text{K}^{-1}) = 1.9T + \text{Debye } T^3 \text{ term } (\Theta_0 = 540 \text{ K})$$

Values for γ are about 4 near 10 K and 1.7 at 293 K (see Table 5.5).

5.7. GLASSES AND GLASS CERAMICS

5.7.1. Introduction

Glasses have some features in common with liquids. They have no long range order, but some degree of order at short range. The scale of the short range order can be estimated from the phonon scattering which determines the thermal conductivity, λ. At normal temperatures the thermal conductivity and heat capacity of glasses are fairly temperature independent. The values indicate that mean free paths for the higher frequency lattice waves are less than or equal to 10Å (1 nm).

Early recognition of unusual thermal properties for a glass at low temperatures came from heat capacity measurements on silica [Flu59]: at the low-temperature limits of measurement near 2 K, C_P showed increasing departures from the expected T^3 phonon contribution (Fig. 5.14). There were also observations of a 'plateau' around 10 K in the thermal conductivity, $\lambda(T)$, both for SiO_2 and Se. Later measurements on many other glasses showed that the magnitude of λ in the region near 10 K was surprisingly similar, differing from 0.1 $\text{W}\cdot\text{m}^{-1}\cdot\text{K}^{-1}$ by factors of 2 or less. Another feature observed in glasses of high silica content or with other types of tetrahedral bonding was that the thermal expansion at lower temperatures is negative and relatively large (e.g., review by Leadbetter [Lea68]).

It was then recognized [Zel71] that the departure of C_P from a T^3 dependence at the lowest temperatures for silica and many glasses was due to the presence of a dominant 'linear' T-term, and that λ varies approximately as T^2 in this same region below the 'plateau' range. The magnitude of the 'linear' terms in the heat capacity ranges from 1 T to 6 $T\,\mu\text{J}\cdot\text{g}^{-1}\cdot\text{K}^{-1}$ (Fig. 5.15).

A model was proposed by Phillips [Phi72] and Anderson *et al.* [And72] to explain the origin of this linear term in C: in a disordered solid, some atoms might have available two or more potential wells and could tunnel between them. At low temperatures (over a limited range of temperature or energy) the random nature of the disorder could lead to a constant density of energy-states and hence to a linear contribution to the heat capacity like that produced by electrons in a metal. This is often referred to as the TLS or two-level-system model; it also leads to a T^2 dependence for the lattice heat conductivity at the lowest temperatures. The tunnelling concept also offers an explanation of the large values of Grüneisen parameter observed in many disordered solids at the lowest temperatures, since tunnelling will usually be sensitive to strain; for example, large expansion anomalies occur in alkali

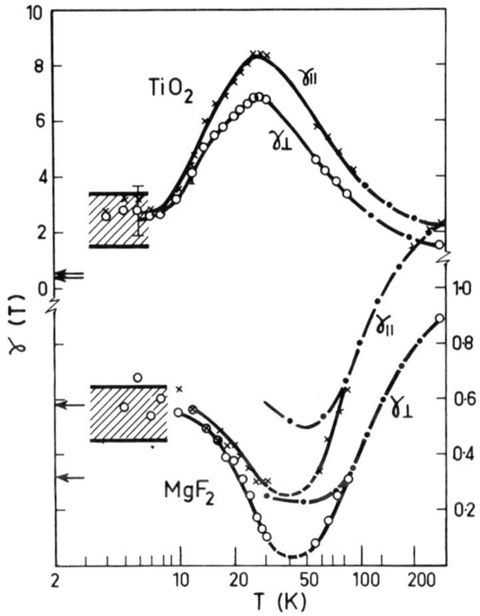

Fig. 5.12. Principal Grüneisen functions for TiO_2 and MgF_2 [Col84a]. Dot–dash curves are from [Kir67] for TiO_2 and [Bro75] for MgF_2. Arrows denote values of γ^{el}.

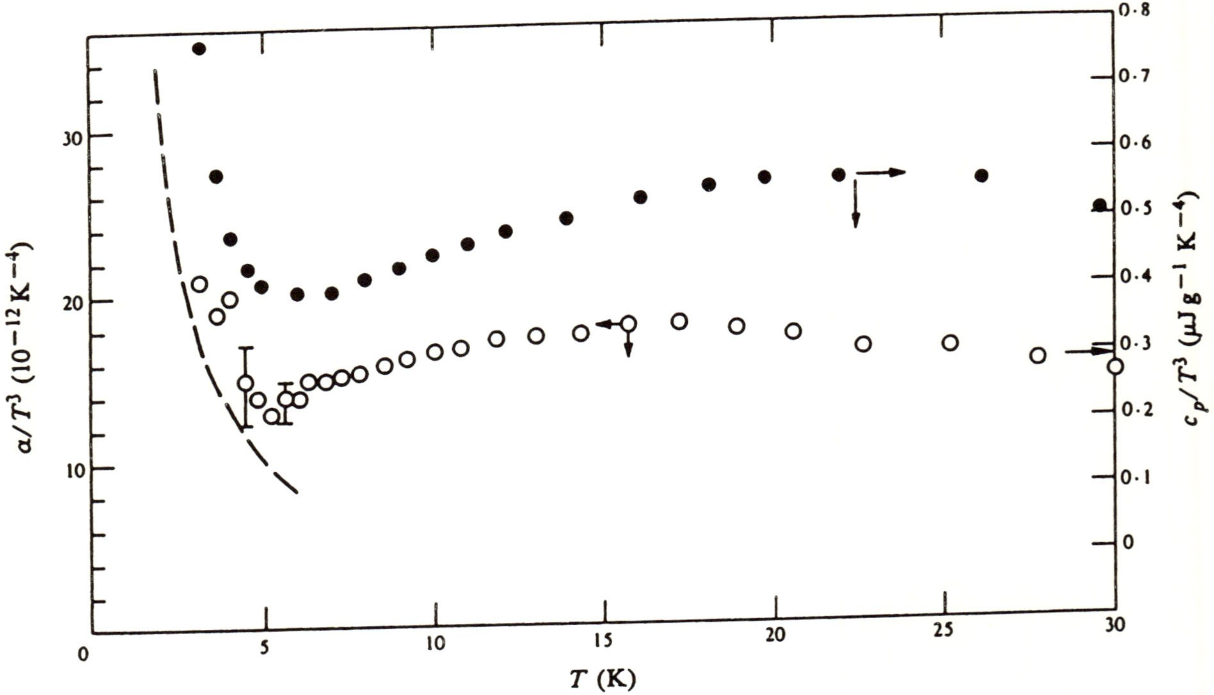

Fig. 5.13. Plots of α/T^3 and c_P/T^3 (per gram) for ZrO_2 + 9 mol% Y_2O_3 [Col85a]. Dashed curve is α/T^3 from [Ack84], and arrow at right shows Debye term with $\Theta_0 = 540$ K.

halide crystals from impurity dipoles which can tunnel between a number of equivalent orientations (Section 5.12). However the TLS model does *not* give a specific microscopic picture of what tunnelling sites actually exist in individual glasses; so that it cannot, for example, make quantitative predictions. Nor, of course, does the TLS model account for all the aspects of thermal behavior at temperatures of 10 K or so. There are considerable differences in $C_P(T)$ between various glasses in the region $T \sim 10$ K: the tetrahedrally bonded glasses generally depart from Debye-like behavior much more than the non-tetrahedral glasses.

There is such a wealth of data at low temperatures on the heat capacity, ultrasonic attenuation, conductivity and thermal expansion of some glasses that it is easy to be confused and lose track of the salient or connecting features. These topics are reviewed in more detail by a number of authors in *Amorphous Solids: Low Temperature Properties* [Phi81]. There are more recent measurements on thermal expansion below 1 K for silica, PMMA, As_2S_3, epoxy polymer and 'defect' crystals (Naβ-alumina and ZrO_2:Y_2O_3) [Ack84], giving a wide variety of values for the Grüneisen γ calculated from the T-terms.

In Section 5.7.2 we discuss separately the tetrahedral glasses, with emphasis on the silica family, and a few representative non-tetrahedral glasses, before reviewing some technically important glass ceramics. Later in Section 5.9 we shall return to glass-like behavior in some polymers, where the degree of glassiness or crystallinity varies from one extreme to the other.

5.7.2. Tetrahedral Glasses

All these amorphous solids display negative values of α at temperatures above the 'linear T' region (e.g., Fig. 5.16), and also associated negative pressure coefficients of the shear modulus G and bulk modulus B. Kurkjian *et al.* [Kur72] measured the ultrasonic velocities and thermal expansion of various glass formers — SiO_2, GeO_2, BeF_2, $Zn(PO_3)_2$ — and noted that these all had negative expansion coefficients at low temperatures, negative values of $(\partial G/\partial P)_T$ and positive values of $(\partial G/\partial T)_P$, in contrast to another glass former, B_2O_3, which is not tetrahedral.

It is not obvious that TLS tunnelling controls this behavior above 2 or 3 K. It is more likely that the tension mechanism is responsible, as with the transverse modes in crystalline Si, Ge, GaAs, etc. It is interesting to speculate whether the bridging oxygen atoms contribute to this effect as well as providing 'tunnelling sites' (see Fig. 5.17 in [Phi81, Ch. 6]).

A related puzzle concerns the nature of the departures of C_P from the Debye model in the temperature region of 10 to 20 K (near $\Theta/40$). The departures appear to be larger for many tetrahedral than for non-tetrahedral glasses, and they are also large for many diamond-like crystals and α-cristobalite (see Fig. 5.18).

Silica and Silicates. The thermal properties of vitreous silica show significant changes with change in density. Density differences amounting to parts in a thousand

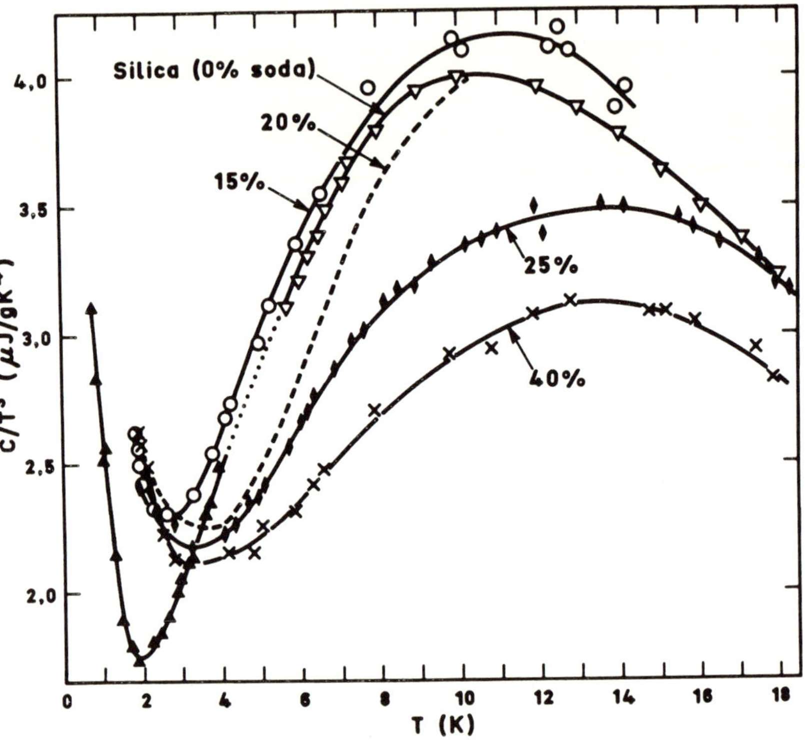

Fig. 5.14. c_P/T^3 (per gram) for two samples of pure silica (triangles) and four sodium silicate glasses [Whi77].

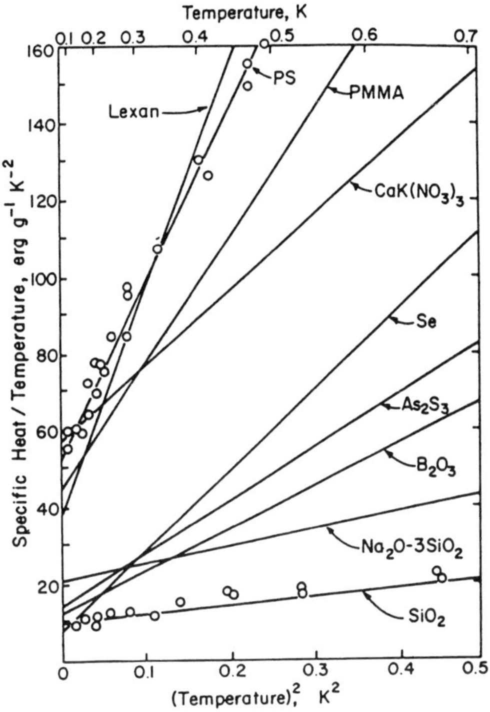

Fig. 5.15. c_P/T (per gram) for various non-crystalline solids below 1 K [Ste73].

may be produced by thermal ageing and from presence of hydroxyl ions; the latter depends on the origin of the silica samples, e.g., whether made by electrical fusion of quartz or from breakdown of tetrasilicide compounds. Neutron irradiation can produce density increases of many percent and large effects on thermal properties. Table 5.6 gives data for two samples of silica aged at 1000 and 1400 °C, SRM 739, a Spectrosil sample after neutron irradiation dose of 5×10^{19}nvt, an ultra-low-expansion glass (SiO_2 + 7.5% TiO_2) and a series of sodium silicate glasses; also included is a borosilicate similar in composition to Corning Pyrex, containing 4% Na_2O, 12–15% B_2O_3, and 2% Al_2O_3. For pure SiO_2, $\Theta_0^{el} \approx 495$ K.

Table 5.6 and other sources show that the addition of soda, a network filler, to silica does not change the heat capacity by more than a few percent at room temperature. Measurements of c_P in the range 20–100 K show changes with soda content are generally less than 10% [Kru72]. Near 10 K there is a significant decrease (see Fig. 5.14) with increasing soda content, but at lower temperatures where the tunnelling term becomes dominant the heat capacity does not change much with composition, although there is a marked difference from pure SiO_2 (see Fig. 5.19 for potash and soda glasses).

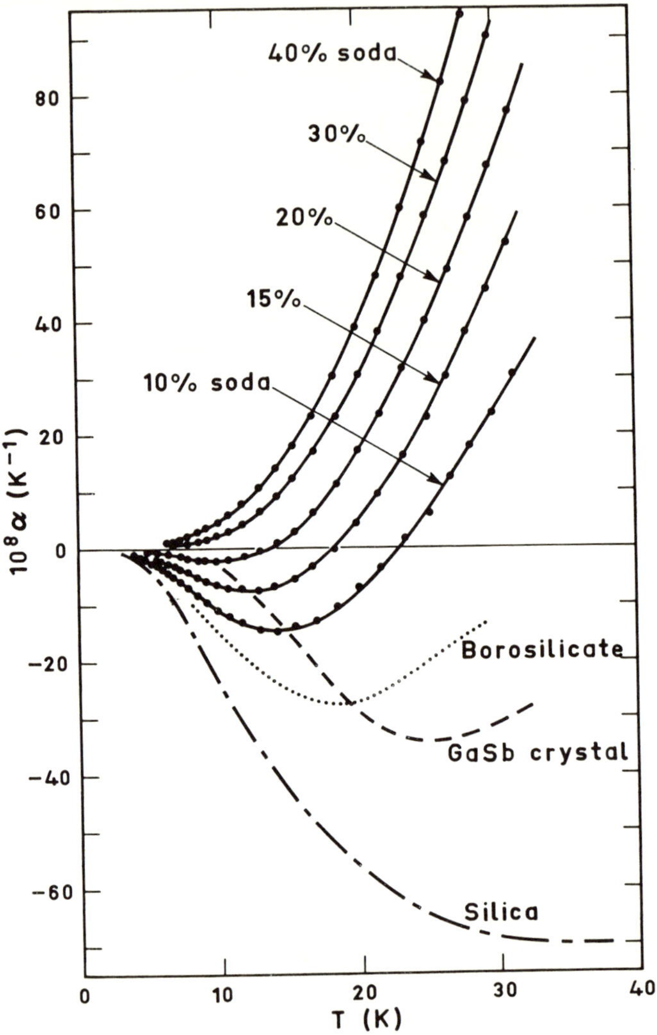

Fig. 5.16. $\alpha(T)$ for silica and sodium silicate glasses, compared with a GaSb crystal [Whi77].

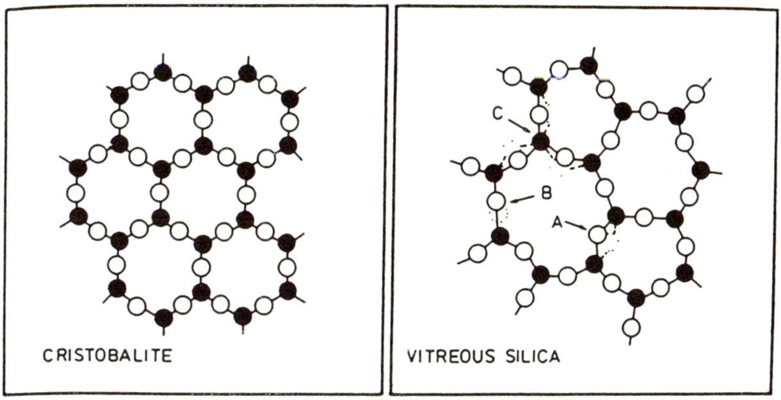

Fig. 5.17. Schematic two dimensional picture of cristobalite and vitreous silica: full circles, Si; open circles, O. Arrows indicate types of defect. From Hunklinger and Schickfus, [Phi81, p. 85].

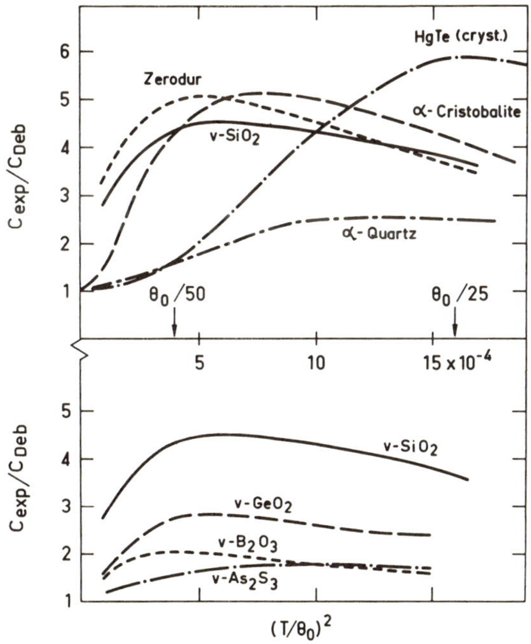

Fig. 5.18. Reduced plot of $(C/T^3)_{expt}/(C/T^3)_{Debye}$ versus $(T/\Theta_0)^2$. Values of Θ_0 are 271 K (B_2O_3), 165 K (As_2S_3), 495 K (GeO_2), 560 K (quartz crystal), 475 K (cristobalite crystal), 530 K (Zerodur glass ceramic), 148 K (HgTe crystal) [Whi84c].

Table 5.6. Data for some silica based glasses at room temperature from [Bar80, Tou77]. $\Delta l/l = (l_{293} - l_{4.2})/l_{293}$

Material	ρ (g/cm^3)	B_S (GPa)	c_P^{293} (J/g·K)	α_{293} (10^{-6}/K)	$\Delta l/l$ (10^{-6})	γ_0^{el}
SiO_2(1000°)	2.200	36	0.73	0.50	20	−2.3
SiO_2(1400°)	2.203	36	0.73	0.40	70	-
SiO_2(SRM739)	2.20	—	0.73	0.49	47	—
SiO_2(n-irrad)	2.253	—	—	—	-	-
ULE(Corning)	2.199	—	—	0	—	-
Vycor(SiO_2+ 4%B_2O_3)	—	—	0.70	0.75	—	—
Borosilicate	2.23	—	0.72	3.0	540	-
SiO_2+10%Na_2O	2.29	33.9	≈0.75	5.1	-	−1.4
SiO_2+20%Na_2O	2.38	35.4	—	9.1	≃ 1700	−0.44
SiO_2+30%Na_2O	2.47	39.0	—	12.3	—	0.26
SiO_2+40%Na_2O	2.52	43.1	—	15.0	—	0.76

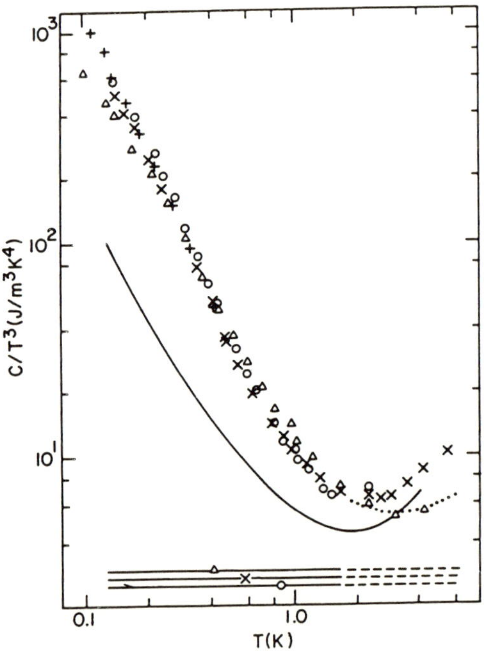

Fig. 5.19. Log–log plot of C/T^3 (per volume) for silicate glasses. Solid line indicates data for pure silica while other points include K concentrations of 0.08, 0.10 and Na concentrations of 0.25, 0.27. Horizontal lines show Debye phonon contributions (details in [Mac85]).

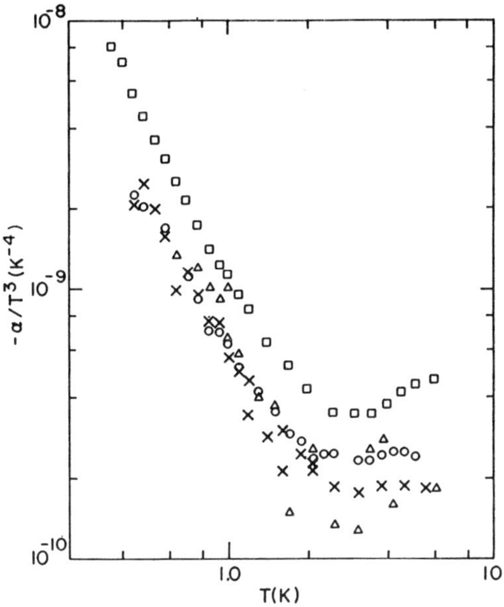

Fig. 5.20. Log–log plot of $-\alpha/T^3$ versus T for silica-potash glasses. Concentrations of K are x=0 (□), x=0.05 (×), x=0.08 (△), x=0.10 (○) [Mac85].

Table 5.6 and Fig. 5.20 show that the thermal expansion is affected much more than the heat capacity by change in composition. The addition of network fillers like soda or potash inhibits the mechanisms responsible for negative expansion in pure silica; presumably they break down the open Si–O–Si bonding. The effect is also evident from the experimental values of γth over the range down to about 3 K (Fig. 5.21). Note these become lower than ultrasonic values of $\gamma^{el} = -2.3$ for silica, -0.44 for silica with 20% soda etc., see [Bar80, p. 670] and [Whi77]. Below 1 or 2 K (Fig. 5.20) α is dominated by a negative linear term $\sim 10^{-9}T$ K^{-1} which is less sensitive to composition.

Other Tetrahedral Glasses (GeO_2, BeF_2 etc.). These are of less technical importance than the silica family, but are of basic interest in that they show similarities to the silica systems in having negative expansivities at low temperatures and negative pressure dependence of elastic moduli. The heat capacity of BeF_2 in the 5 to 15 K range is very like that of silica in so far as C/T^3 exceeds the expected Debye term by a factor of 5 or 6 near 7 K ($\Theta_0/50$) [Lea71], whereas for vitreous germania the increase is less than a factor of 3 at the maximum (also near $\Theta_0/50$) (see Fig. 5.18). Experimental data for these and some non-tetrahedral glasses (see Section 5.7.3) are given in Table 5.7 and are selected from sources given in the review [Bar80, p. 669] and [Ste73, Ack84].

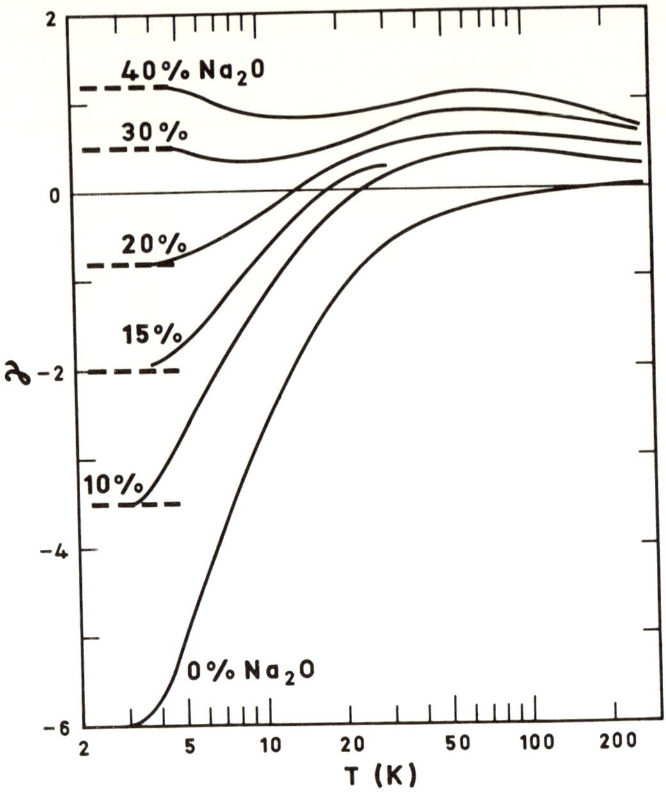

Fig. 5.21. Variation of the thermal Grüneisen parameter with temperature for sodium silicate glasses above about 3 K. Dashed lines were calculated from the effective T^3 terms in α and C_P between 2 and 4 K, which is above the region where T-terms dominate ([Whi77] and [Bar80, p. 671]).

5.7.3. Non-Tetrahedral Glasses

Apart from polymers, low temperature data for non-tetrahedrally bonded glasses are largely restricted to B_2O_3 [Ste73, Whi84a], As_2S_3 [Bar80, Mor81a], and As_2Se_3 [Hor78, Cla78], as summarized in Table 5.7. Chief features are as follows:

1. the maxima in C_P/T^3 (or the equivalent minima in Θ^C) are less pronounced than for the tetrahedral glasses or crystals (see lower part of Fig. 5.18);

2. the elastic moduli do not soften under pressure;

3. there is no indication of negative expansion, at least above 1 K. Ackerman *et al.* [Ack84] have measured one of these glasses, As_2S_3, below 1 K; their extrapolation suggests that a tunnelling (T-) term has an associated γ value of about -2.

Table 5.7. Experimental data for some non-silicate glasses from references in the text including [Bar80, p. 669] and [Ste73, Ack84]. Vitreous SiO_2 is included for comparison. γ_{10} is from thermal data at 10 K and γ_0^T is from the respective T-terms (tunnelling) below 1 K [Ack84]

Material	B_S (GPa)	Θ_0 (K)	α_{293} (10^{-6}/K)	γ_{293}	γ_{10}	γ_0^{el}	γ_0^T
Tetrahedral							
SiO_2	36	492	0.5	0.03	−3	−2.3	−60
GeO_2	24	309	6.9	0.2	−0.2	−0.9	—
GeS_2	—	170	14	—	1.2	—	—
BeF_2	19	390	7.5	—	—	−1.4	—
Non-tetrahedral							
B_2O_3	13	259 271^{el}	15.5	0.35	0.8	0.28	—
As_2S_3	13	165	22.5	0.55	1.8	1.8	−2
As_2Se_3	14	144	22	0.6	2.1	2	—
PMMA	8	~ 270	80	—	—	—	−1

5.7.4. Glass Ceramics

The term 'glass ceramics' describes those materials which can be formed in the glassy state and then heat treated to partially recrystallize with the help of a nucleating agent. The resulting mix of small crystallites (~ μm or less) in a glass matrix is achieved with very little change in volume or shape, and has zero porosity and high mechanical strength.

The history of the development and technical importance of low thermal expansion glass ceramics is discussed in a book by Hans Bach of Schott Glaswerke [Bac95]. The process of photonucleation was discovered at Corning (New York) about 1940. There followed research on the lithium-alumina-silicates such as β-eucryptite ($Li_2O{\cdot}Al_2O_3.2SiO_2$) which have negative values of volume expansion at ambient temperatures, and also the discovery of nucleation by addition of TiO_2. Another convenient nucleating agent was found to be ZrO_2. Development work at Corning, Owens-Illinois and Schott centered on the LiAlSiO family with β(hi)-quartz structure. This structure can be stabilized below the α–β quartz transition (573°C) by additions of MgO, ZnO, Al_2O_3, etc. Generally the addition of $LiAlO_2$ leads to strong negative values of α, $ZnAl_2O_4$ gives smaller negative values, $AlPO_4$ has little effect, and $MgAl_2O_4$ gives a strong positive contribution to the expansion.

These developments led to the production of glass ceramics for kitchen ware which were resistant to thermal shock. Later came their potential use for large zero-expansion blanks for telescope mirrors. This imposed additional requirements of homogeneity, suitable polishing characteristics and adhesion of Al films. One such material from Schott was called Zerodur and typically contained (in wt%) about 57% SiO_2, 25% Al_2O_3, 6.5% P_2O_5, 3.4% LiO_2, 1% each of MgO and ZnO, and 2% each of TiO_2 and ZrO_2. Later Schott developed Zerodur M with less MgO to improve stability on thermal cycling (see also [Hea94, Ch. 14]).

Partly because of their use in space telescopes, there are thermal data extending to liquid helium temperatures, particularly for thermal expansion. At temperatures below about 100 K, the expansion coefficient is negative with values comparable to those for vitreous silica.

Measurements of heat capacity have been largely confined to below 30 K, due largely to a fundamental interest in the magnitude of T- and T^3- terms in C_P. At temperatures between about 50 K and room temperature the observed values of C_P for these aluminosilicates are not very different from values for silica or many silicate glasses.

β-Eucryptite. $LiAlSiO_4$ or ($Li_2O{\cdot}Al_2O_3.2SiO_2$) is a hexagonal Li-stuffed derivative of β-quartz, and plays an important rôle in determining the expansion coefficient of ceramic glasses. There are no high resolution expansivity data, but there are lattice spacing measurements at 20, 100, 200, 300 K, and also at higher temperatures, which reveal markedly anisotropic expansion. Along the c-axis, $\alpha_{\parallel}$ is negative above 20 K, with a value of $\simeq -20 \times 10^{-6}K^{-1}$ at 300 K; while $\alpha_{\perp}$ is positive (except possibly below 50 K) with a magnitude roughly half that of $\alpha_{\parallel}$. The volume coefficient is near zero over a wide range. The relative importance of low-frequency TA modes, rigid unit modes (which may include some low frequency TA modes) and the Li^+ ion is yet unresolved; see [Pil73, Lic98] and [Hea94, Ch. 3].

Glass Ceramics. The heat capacity of most glass ceramics at room temperature is about 0.8 $J{\cdot}g^{-1}{\cdot}K^{-1}$ or 48 $J{\cdot}mol^{-1}{\cdot}K^{-1}$, compared with 44 $J{\cdot}mol^{-1}{\cdot}K^{-1}$ ($\pm 1\%$) for α-quartz, cristobalite and vitreous silica at 293 K. For all these latter three forms of SiO_2, C_P values agree within 1 or 2 per cent from 300 K down to 80 K [Wes63]. They diverge increasingly below this temperature: at 25 K, $C_P = 1.32$ $J{\cdot}mol^{-1}{\cdot}K^{-1}$ for quartz, 2.33 for cristobalite, 2.26 for silica and 1.7 $J{\cdot}mol^{-1}{\cdot}K^{-1}$ for Cer-Vit.

Some of the measurements of C_P on samples of Cer-Vit (Owens-Illinois) and Zerodur (Schott) are illustrated in Fig. 5.22 [Col85b]. The Cer-Vit samples had varying degrees of crystallinity and were stated [Lea77] to be ' ... essentially a lithium aluminium silicate containing several percent of TiO_2 and ZrO_2 as nucleating agents ... '. Below 4 K, data could be represented by $c_P \approx (5T + 0.8T^3)$ $\mu J{\cdot}g^{-1}{\cdot}K^{-1}$, the linear term being of similar magnitude for differing proportions of glass-to-crystallite.

The Zerodur samples had average crystal sizes 50 to 135 nm and had 70 to 80% crystallinity [Col85b]. Below 5 K, measurements fitted $c_P = AT + BT^3$ with values of A from 5.4 to 7.8 $\mu J{\cdot}g^{-1}{\cdot}K^{-2}$ and $B \approx 0.7$ $\mu J{\cdot}g^{-1}{\cdot}K^{-4}$; these are rather similar to the Cer-Vit data. Note that the linear (tunnelling) term is much larger per gram than the $1.2\mu J{\cdot}g^{-1}{\cdot}K^{-2}$ observed for vitreous silica (Fig. 5.15).

The thermal expansion coefficient for these aluminosilicates is much more sensitive than C_P to composition, particularly at intermediate temperatures, say 50 to

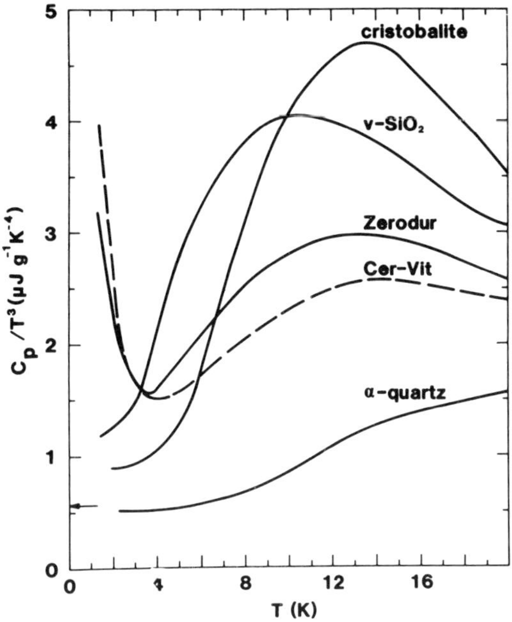

Fig. 5.22. c_P/T^3 against T for some silica-based materials [Col85b].

350 K. Berthold and Jacobs [Ber76a] (see also [Bar80, p. 673]) measured the expansion of more than 40 samples of Cer-Vit between 150 K and 540 K, finding α at 150 K varying from -0.3 to $+0.8 \times 10^{-6}\text{K}^{-1}$. Figure 5.23 shows some selected data on Zerodur samples and MGC (or MACOR, a machineable glass ceramic from Corning); and for comparison vitreous silica (aged at 1000°C) and Corning ULE [Col91, Whi76a]. The maximum in α near 100 K for many samples of Cer-Vit and Zerodur raises interesting questions. Obviously there is a balance between negative and positive contributions to α, but it is difficult to reconcile the large differences near 100 K with the rather similar values below 10 or 15 K. At the lowest temperatures transverse modes of vibration (giving negative expansion as in vitreous silica) must be dominant, and near 100 K other modes become important. Near room temperature there may be a strong negative contribution from a crystalline phase such as β-eucryptite.

The machineable glass ceramic, MGC or MACOR, contains small (5–10 μm) blocks of a fluorophlogopite mica phase crystallized in a boraluminosilicate matrix. Its machinability, shock-resistance, and non-magnetic qualities make it useful for cryogenic equipment. The heat capacity per gram measured below 20 K by Lawless [Law75] is similar to that of silica. The density is 2.52 g·cm^{-3}, and $\alpha_{293} = 8.8 \times 10^{-6}$ K^{-1}. Values of α down to 2 K are shown in Fig. 5.23 [Whi76a].

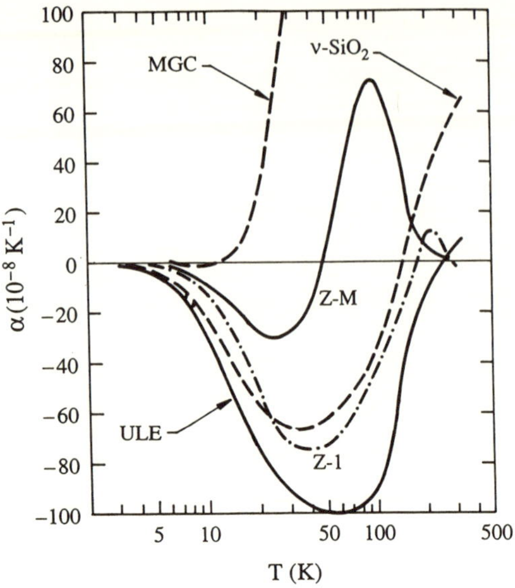

Fig. 5.23. α(T) of Zerodur samples, Z-1 and Z-M, vitreous SiO_2, ULE 7971 and MGC (Macor) [Whi76a, Col91].

5.8. HIGHLY ANISOTROPIC CRYSTALS

5.8.1. Introduction

We usually associate the most extreme cases of anisotropy with easily cleaved crystals such as mica and graphite, or with the fibrous chain-like crystals of asbestos and some polymers. Their anisotropy arises from having weak Van der Waals forces between planes (or chains) and stronger covalent bonding within planes (or chains). Generally this anisotropy shows up as high compressibility in one direction and low compressibility in the other. For example, in layered crystals of axial symmetry such as graphite $\chi_{\|} \gg \chi_{\perp}$; while for chain-like crystals such as Te $\chi_{\|} \ll \chi_{\perp}$. Indeed, for Te and Se at room temperature $\chi_{\|}$ has a small negative value; under hydrostatic pressure the spiral chains decrease in radius and expand slightly in length. Another measure is the ratio of the elastic compliances along and perpendicular to the symmetry axis, s_{33}/s_{11}, which is about 28 for graphite and 0.6 for Te.

The vibrations that depend primarily on the weak force-constants have much lower frequencies than those that are governed by the strong force constants, and at low temperatures they are excited preferentially and dominate the heat capacity and thermal expansion, with low values of Θ^C. Above the low temperature region $\Theta^C(T)$ increases with T until the high frequency vibrations become excited at much higher temperatures. Likewise changes in the Grüneisen γ occur, because the uniaxial strain dependences of low and high frequency modes are very different and may differ also with orientation.

5.8.2. Layer Structures

A simple layer structure has strong elastic anisotropy like that of graphite, with only small cross compliance between the $\|$ and $\perp$ directions. The low frequency modes are polarized roughly perpendicular to the layers and the high frequency modes roughly perpendicular to the axis. The high frequencies are therefore weakened by stretching the crystal perpendicular to the axis but scarcely affected by stretching along the axis, so that $\gamma_\perp$ is large and positive and $\gamma_\|$ is small. In contrast, the lower frequencies are weakened by stretching along the axis (which reduces the restoring forces for motion in this direction) and are strengthened by stretching perpendicular to the axis (which increases the restoring force perpendicular to the layers because of the tension between neighbors), so that for these modes $\gamma_\perp$ is negative and $\gamma_\|$ large and positive. At low temperatures when only low frequency modes are excited there is a relatively large expansion in the soft direction along the axis, and a small contraction within the layers. As T increases the excitation of the high frequency modes will have little effect on the expansion along the axis, but their large $\gamma_\perp$ will drive the expansion within the layers positive, though it remains small because of the high stiffness in the layers. A theoretical example is provided by one form of the rhombohedral model discussed in Section 2.6.3.

Graphite. In graphite, each carbon atom is bonded to three other carbons forming a network of planar hexagonal rings. The distance between carbon atoms in a plane is 0.142 nm (1.42 Å), and planes are about 0.33 nm (3.3 Å) apart. It is a good electronic conductor within the planes, giving an electronic ΓT contribution to the heat capacity at low temperatures. The weak bonding between planes results in a compressibility at room temperature of $\chi_\| = 27 \times 10^{-12}$ Pa^{-1}, compared with $\chi_\perp = 0.5 \times 10^{-12}$. This layer structure led to an early theoretical prediction that the vibrational heat capacity at low temperatures might vary as T^2 rather than T^3, and indeed the first definitive measurements on a natural crystal from 13 to 300 K [DeS53] showed $C_P \propto T^2$ from 13 to 50 K. But at low temperatures the weak interlayer forces become important, and later measurements [Van63] extending down to 0.4 K on a similar natural crystal gave $C_P = 13.8T + 27.7T^3\,\mu$J·g-at^{-1}·K^{-1} below 1.2 K, confirming that in the low temperature or long wave limit the Debye T^3 law applies with $\Theta_0 \approx 413$ K. These and other measurements of C_P on various graphites (Canadian and Madagascar natural graphites, pile graphite, pyrolitic graphite, etc.) showed that the values obtained below 20 K were highly dependent on crystallite size, stacking faults, etc., with the natural crystals giving the lowest values of heat capacity (Fig. 5.24). Data in [Tou70b] suggests that at higher temperatures, above say 50 K, differences are small. At 300 K, $C_P = 8.58$ J·g-at^{-1}·K^{-1} [DeS58].

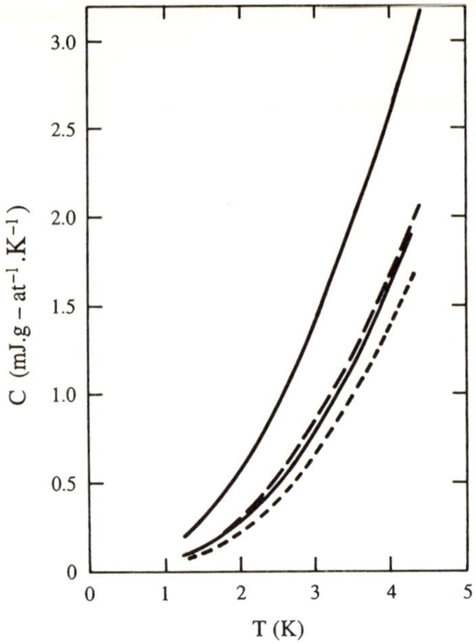

Fig. 5.24. $C_P(T)$ for representative samples of graphite below 5 K [DeS53, DeS58]. —(top curve): graphitized lampblack, crystallites ~ 10nm. — —(middle): pile graphite, crystallites ~ 25 nm. - - - -(bottom): natural crystals ≥ 100μ m.

Note that the low temperature values of Θ^C ($\sim \Theta_0$) are not a good guide to the heat capacity of graphite (or boron nitride) at higher temperatures; the high frequency modes do not become excited and increase C_P until much higher temperatures than predicted by the Debye model. For example, measured values of C_P at about 100 K and 300 K respectively correspond to Debye temperatures of 1050 K and 1500 K.

The thermal expansion of a highly oriented pyrolytic graphite sample from 30 to 270 K [Bai70] is shown in Fig. 5.25 and Table 5.8. A sample of hexagonal BN (of graphite structure) showed similar and even greater anisotropy, see [Bar80, p. 676]. There are very many high temperature expansion measurements on graphites from many sources (not single crystal) because of its importance as a high temperature refractory, but relatively few at low temperatures [Tou77].

Bailey and Yates [Bai70] calculate approximate values for the principal Grüneisen parameters with accuracy limited by lack of low temperature elastic moduli. They find $\gamma_\perp \simeq -5$ at lowest temperatures rising to $\simeq -1$ at 270 K and becoming positive at higher temperatures. In the soft direction, $\gamma_\parallel^0 \simeq 3$ and $\gamma_\parallel^{270} \simeq 1$.

Values of α and $\Delta l/l_{293} = (l_{293} - l_T)/l_{293}$ are given in the American Institute of Physics Handbook [Kir72] for graphite. Values of α (in units of 10^{-6}K^{-1}) may

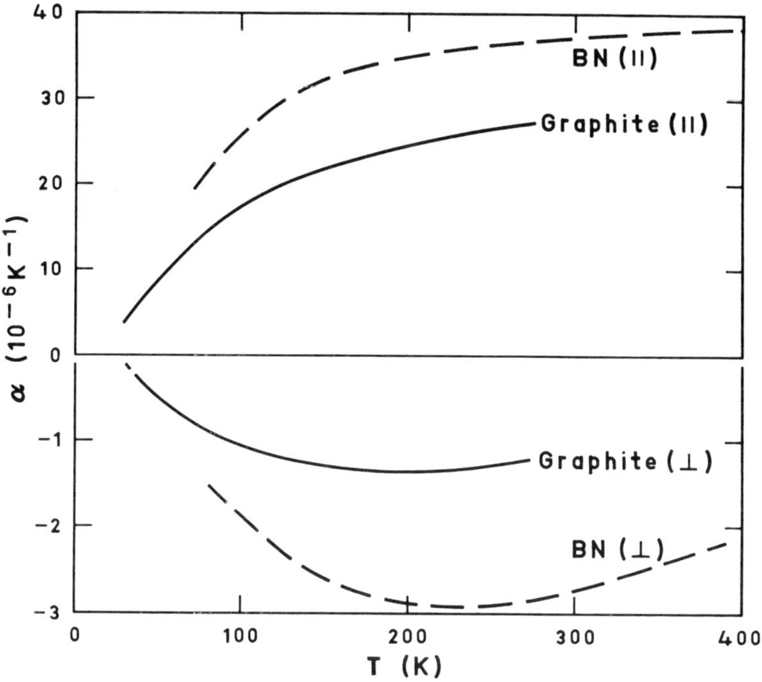

Fig. 5.25. $\alpha(T)$ for pyrolitic graphite and boron nitride. Note scale change above and below the horizontal axis. From [Bai70] and [Bar80, p. 676].

be compared with those for 'glassy' carbon GC-20 (crystallites $\leq$ 10 nm):

T	293 K	200 K	100 K
$\alpha_\perp$	−1.2	−0.8	−0.4
$\alpha_\parallel$	25.9	22.6	15.4
α_{av}	7.8	7.0	4.9
$\alpha_{\text{GC-20}}$	1.9	$\simeq$ 1.6	1.3

Graphite is the classic example of a simple layered crystal, in which the anisotropy of both the elasticity and the Grüneisen functions combine to give large positive expansion perpendicular to the planes and small negative expansion in the planes. But this is not necessarily true of crystals with more complex layers. Thus arsenic has Grüneisen functions which are almost isotropic, and the anisotropy of its thermal expansion stems largely from its elasticity (see Section 6.3.1). And InBi, with compound layers consisting of $Bi_{0.5}InBi_{0.5}$ sandwiches, has reversed anisotropy in thermal expansion, with positive expansion in *ab* directions and contraction perpendicular to the planes (see Section 8.4.6).

Table 5.8. Data for some layer and chain-like crystals. Sources are reviews [Bar80, Tou77]

Material	Θ_0 (K)	B_0 (GPa)	$\alpha_{\perp}^{293}$ (10^{-6}/K)	$\alpha_{\parallel}^{293}$ (10^{-6}/K)	$\gamma_{\perp,293}$	$\gamma_{\parallel,293}$	$\gamma_{\perp,0}$	$\gamma_{\parallel,0}$
Graphite	413	36	−1.3	27	−1	1	−5	3
BN	—	485	−2.8	38	—	—	—	—
Se	160 ±10	12.5 approx	69	−13	1.5	−1.6	0.85	1.0
Te	152	22	29.5	−2.3	1.8	−0.7	1.0	1.1
POM	271	11	75	2.4	1.35	−1.2	0.7	0.7

5.8.3. Chain Structures

Selenium. Se crystallizes in the trigonal system in the form of spiral chains of atoms arranged in a hexagonal array, so that each chain has six neighbors: the space group is D_4^3 ($D3_121$) or D_3^6 ($D3_231$). There are also amorphous and monoclinic forms as for sulphur.

Low temperature values of heat capacity for trigonal Se show some variation depending on the sample preparation. Values for C_P/T^3 below 10 K lie within about 10% of 0.5 mJ·mol^{-1}·K^{-4}, corresponding to $\Theta_0 \simeq 160$ K [Mei78, Las69]. Their values of Θ_0 for monoclinic Se are about 128 K. The thermal expansion data from 10 to 300 K [Gro78] show the expected large anisotropy, with $\alpha_{\perp} = 69.8 \times 10^{-6}$ K^{-1} and $\alpha_{\parallel} = -13.4 \times 10^{-6}$ K^{-1} at 300 K. Other data are given in Table 5.8.

Tellurium. Te crystallizes in the same trigonal structure of spiral chains as Se. There are more consistent thermal data for Te than for Se. The heat capacity, measured from 1.5 to 20 K [Lea73] and above 14 K [Sla39b], gives a $\Theta^C(T)$ curve similar in shape to those of other crystals represented in Figs. 5.4 and 5.7: $\Theta_0 = 152$ K, $\Theta^C_{min} = 131$ K at $T = 10$ K ($\Theta_0/15$) and $\Theta^C_\infty \simeq 180$ K. This contrasts with the shape of the $\Theta^C(T)$ curves for graphite and BN mentioned above.

The thermal expansion of single crystals determined down to 2 K (Fig. 5.26, refs. in [Bar80, p. 691]) shows anisotropy but less marked than for selenium: at 293 K, $\alpha_{\perp} = 29.6 \times 10^{-6}$ K^{-1} and $\alpha_{\parallel} = -2.3 \times 10^{-6}$ K^{-1}. Principal γ values are included in Table 5.8. A theoretical model indicates that the anisotropy may be reduced by radial contraction of the spiral chains with increasing temperature [Gib73].

5.9. POLYMERS

5.9.1. Introduction

Polymers are widely used in cryogenic applications, particularly when fiber reinforced, because of their strength/weight ratio and nonmagnetic nature. They are generally in the amorphous (glassy) or semicrystalline condition. If the crystallites

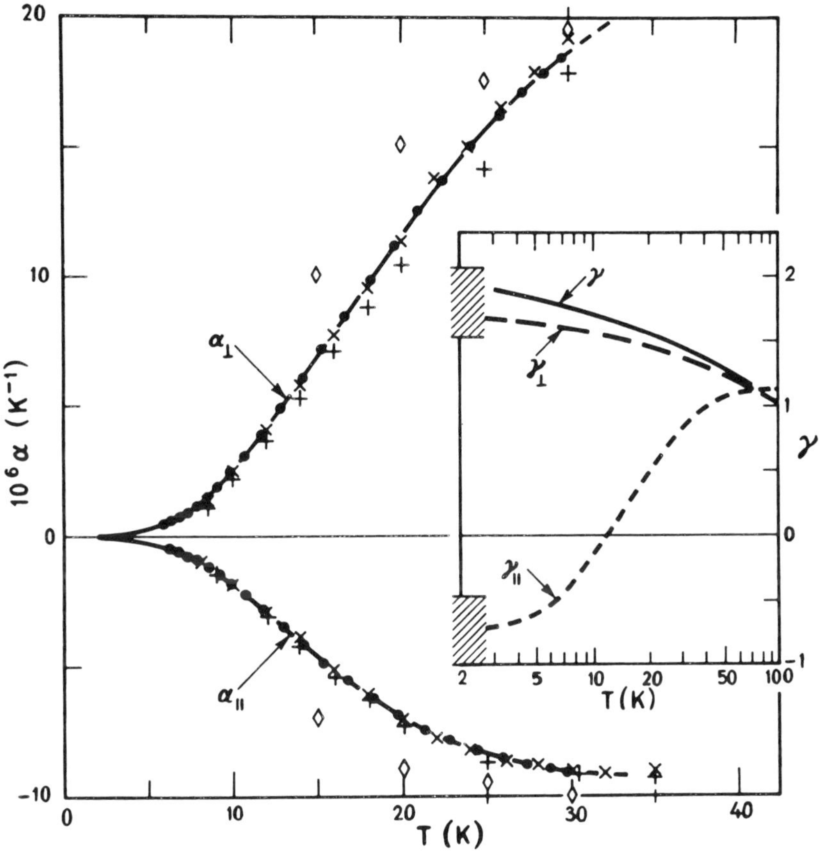

Fig. 5.26. $\alpha(T)$ for Te with $\gamma(T)$ in inset. From [Bar80, p. 692].

are oriented by the growth process or by drawing, the properties are then anisotropic. In the fully crystallized state, which is not usually achieved with bulk samples, the polymers consist of chains (linear, planar zigzag, helical, etc.); the crystals can have axial, orthorhombic or lower symmetry. Whatever the structure, the covalent intrachain linkage (i.e., within chains) is much stronger than the interchain linkage (i.e., between chains), so that we may expect the lattice dynamics of crystalline polymers to resemble that of tellurium or selenium, with the added complication of side groups of atoms. In some polymers such as epoxy resins and vulcanized rubbers the interchain linkage is strengthened by cross bonding. At low temperatures, the thermal properties are largely determined by low frequency modes governed by interchain forces and weak intrachain torsional forces; at higher temperatures the high frequency intrachain modes become important. However, polymers normally melt before the highest frequencies are excited.*

For both isotropic and oriented samples, the heat capacity and thermal expansion depend on the *degree of crystallinity*, i.e., the fraction of the material that is crystalline rather than amorphous, and also on whether the samples are isotropic or have crystallites and/or chains oriented preferentially (*texture*). We shall discuss first the experimental behavior of the few examples available of single or quasi-single crystal specimens, before examining the properties of the partially crystalline and amorphous materials.

Data on heat capacity have been reviewed and tabulated in a series of papers by Gaur *et al.* [Gau81] in the *Journal of Physical and Chemical Reference Data*. A Cryogenic Monograph by Hartwig [Har94] entitled *Polymer Properties at Room and Cryogenic Temperatures* includes a review of heat capacity and thermal expansion, as well as dielectric and elastic behavior and thermal conductivity. Hartwig tabulates values for linear thermal expansion and heat capacity of many common polymers. His values for c_P (per gram) correlate closely with C_P values (per mole) given by Gaur *et al.* [Gau81]. Hartwig does not include data for the powder or fiber-filled polymer composites. Some of the latter are included in the chapters by Clark and others in *Materials at Low Temperatures* [Cla83]. Other good sources of thermal data on polymers and composites are the proceedings of the International Cryogenic Materials Conferences published by Plenum Press as *Advances in Cryogenic Engineering Materials* and specialist ICMC conferences on *Nonmetallic Materials and Composites at Low Temperatures* [Cla79, Har82, Har88b, Oka95a].

Table 5.9 gives selected values for c_P and $\Delta l/l_{293} = (l_{293} - l_4)/l_{293}$ for the following polymers in amorphous (a), crystalline (c) or semicrystalline (sc) state [Gau81, Har94]:

- POM (polyoxymethylene ... $(CH_2O)_n$)
- PE (polyethylene ... $(CH_2)_n$)
- PTFE (polytetrafluorethylene ... $(CF_2)_n$)

*Vulcanized rubbers continue to be solids above the glass transition region and exhibit interesting thermal expansion [Bar98], but this does not usually occur at cryogenic temperatures.

Table 5.9. Values of c_P ($J \cdot g^{-1} \cdot K^{-1}$) and $\Delta l/l = (l_{293} - l_4)/l_{293}$ for some polymers in crystalline (c), semi- crystalline (sc) and amorphous (a) forms [Gau81, Har94]

Material	c_P 10 K	c_P 20 K	c_P 50 K	c_P 100 K	c_P 150 K	c_P 200 K	c_P 250 K	c_P 300 K	$\Delta l/l$ 10^{-4}
POM (c)	0.014	0.082	0.337	0.556	0.73	0.905	1.08	1.28	150
PE (c)	0.008	0.051	0.33	0.68	0.86	1.03	1.22	1.44	160
PE (a)	0.025	0.098	0.35	0.70	0.94	1.19	2.07	2.27	290
PTFE (sc)	0.026	0.079	0.210	0.39	0.55	0.68	0.78	0.87	180
PVC (a)	0.029	0.095	0.260	0.43	0.56	0.69	0.82	0.94	120
PS (a)	0.031	0.102	0.266	0.45	0.62	0.80	0.99	1.21	150
PP (c)	0.017	0.087	0.304	0.61	0.99	1.11	1.31	—	?
PP (a)	0.025	0.120	0.354	0.68	0.94	1.23	1.42	—	130
PET (a)	0.065	—	0.25	0.44	0.62	0.80	0.99	1.17	130
PMMA (a)	0.017	0.080	0.28	0.55	—	0.92	—	—	105

- PVC (polyvinylchloride … $(CH_2 \cdot CHCl)_n$)
- PS (polystyrene … $(CH_2 \cdot CH \cdot C_6H_5)_n$)
- PP (polypropylene … $(CH_2 \cdot CH \cdot CH_3)_n$)
- PET (polyethylenetetraphthalate … $(CO \cdot C_6H_4 \cdot CO \cdot O \cdot (CH_2)_2 \cdot O)_n$).

5.9.2. Crystalline Polymers

There have been many simple theoretical models devised to explain the properties of polymer crystals (see for example [Bau73, Gib74, Bar88]). The basic ideas are simple (Section 2.6.3). The heat capacity depends at low temperatures on the weak forces controlling torsion of the chains and interactions between them, with additional contributions at higher temperatures from modes governed by stronger intrachain forces. The bonds between successive atoms in the chain are very stiff and expand very little, so that their rotational motion (libration) contracts the mean length of the chain, giving negative expansion in that direction (an extreme example of the tension effect), while the anharmonicity of the interchain potentials produces positive expansion in directions normal to the polymer chains. However, the quantitative behavior, especially for the anisotropic expansion normal to the chains, is intricate, requiring detailed models for specific substances. Several such models have been developed for crystalline PE (e.g., [Kob79, Lac94b, Bru98]), and single models for a few other polymer crystals, notably by Lacks, Rutledge and colleagues (e.g., [Lac94a]), but most of the calculations do not extend to very low temperatures.

Polyoxymethylene(POM). A close approach to a single crystal among the non-conducting polymers was made by Anderson *et al.* with their samples of poly-oxymethylene produced by radiation polymerization of tetroxane [And82]. They obtained rods of '... single crystal texture ... fiber bundles are well aligned with respect to one another ... and low angle diffraction measurements indicate that the bundles vary from a few hundred to many thousands of Å in thickness ... and diffraction peaks of width never greater than a couple of degrees in all crystallographic directions ....' They used such 'single crystals' for inelastic neutron scattering determinations of the principal elastic moduli. Measurements on a similar crystal were made of C_P (2–20 K) and the principal linear expansivities (2–100 K) [Whi76b]. Below 10 K, $c_P = 0.013T^3$ mJ·g^{-1}·K^{-1}, corresponding to $\Theta_0 = 271$ K. Above this the temperature dependence changes to an approximate T-law between about 25 and 100 K.

The $\alpha_\lambda(T)$ curves are qualitatively very similar to those of Te (Fig. 5.26). $\alpha_\perp$ is positive and relatively large, reaching a value of 75×10^{-6}K^{-1} at 293 K. $\alpha_\| \approx 2.4 \times 10^{-6}$K^{-1} at 293 K, becoming negative below 100 K and reaching a minimum of -2.0×10^{-6}K^{-1} at 40 K. The $\gamma_\lambda(T)$ curves resemble those of Te and Se, as the values in Table 5.8 indicate, but are less smooth (see [Bar80, Fig. 5.22]).

Polysulphur nitride. $(SN)_x$ is a semimetallic chain polymer which can be produced as an 'imperfect' single crystal, i.e., as an oriented bundle of fibers with diameters between 200 and 1200 Å. Measurements of C_P from 1.5 to 80 K [Har77a] showed that the lattice contribution varied as T^3 below about 5 K ($\Theta_0 = 143$K) with a T-dependence from about 40 to 80 K.

5.9.3. Amorphous and Semi-Crystalline Polymers

When a polymer solidifies from the melt, crystalline regions are formed which may consist of long chains folded into lamellae; these are randomly oriented and joined by amorphous regions and sometimes by chains which enter more than one crystallite. If the polymer is then drawn, the degree of the crystalline orientation increases rapidly and results in well oriented blocks of folded chains separated by amorphous regions and some bridges. This is then a semicrystalline oriented polymer.

Polyethylenetetraphthalate. PET can be produced in both amorphous (a) form and, by heat treatment, in semicrystalline (sc) form. Two series of measurements of C_P at low temperatures are in good agreement and show the effect of changing from 60% crystallinity (by volume) to 90%[Cho75, Col87b]. Figure 5.27 shows one series which was fitted to a linear 'tunnelling' term (dominant below 1 K and varying with crystallinity), plus a Debye term with $\Theta \simeq 138$ K, plus a 1-dimensional Debye term, plus an Einstein term to represent the 'bumps' shown near 4 K [Col87b]. The decrease in the height of the bump with increase in crystallinity (and density) is reminiscent of the change in the magnitude of C_P/T^3 for the silica family in the region near 10 K (e.g., Figs. 5.14 and 5.22): as the density increased in progressing from vitreous silica to quartz due to thermal crystallization or irradiation, the bump decreased in magnitude.

Table 5.10. Values of coefficient, A_1, of linear terms for various materials. Refs: PET [Col87b], PTFE [Boy83], others from Pohl, see [Phi81, Ch. 3]

Material	A_1 (μJ/g·K^2)	Material	A_1 (μJ/g·K^2)
PET(a)	2.28	PTFE	0.36
PET(58%c)	5.77	PE	0.94
PMMA	4.8	silica	1.2
PS	5.3	Zerodur	6.1
Nylon	2.1		

Values of the tunnelling term from the C_P data (down to 0.4 K) were $2.3T$ μJ·g^{-1}·K^{-1} for amorphous PET and $5.8T$ μJ·g^{-1}·K^{-1} for crystalline PET [Col87b], compared with $1.2T$ μJ·g^{-1}·K^{-1} for silica and other polymers in Table 5.10.

Polytetrafluoroethylene (Teflon). PTFE or Teflon is a good prototype of a long-chain polymer for low temperature study of the heat capacity because it is available pure, with well defined structure and with no cross links or side chains. Measurements of c_P from 0.3 to 20 K at pressures from zero to 0.5 GPa [Boy83] gave the results shown in Fig. 5.28. The relatively small tunnelling term at zero pressure, $\simeq 0.36T$ μJ·g^{-1}·K^{-2}, is compared with values for PET, PMMA, PS in Table 5.10.

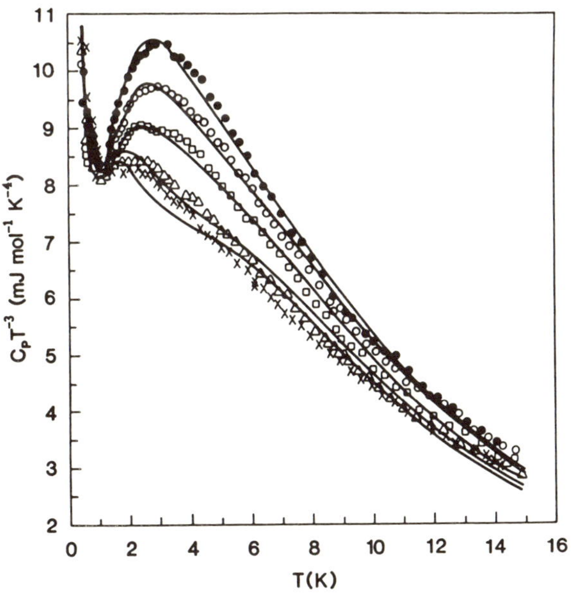

Fig. 5.27. C_P/T^3 against T for five samples of PET varying from zero crystallinity (top, •) to 58% (bottom, ×) [Col87b].

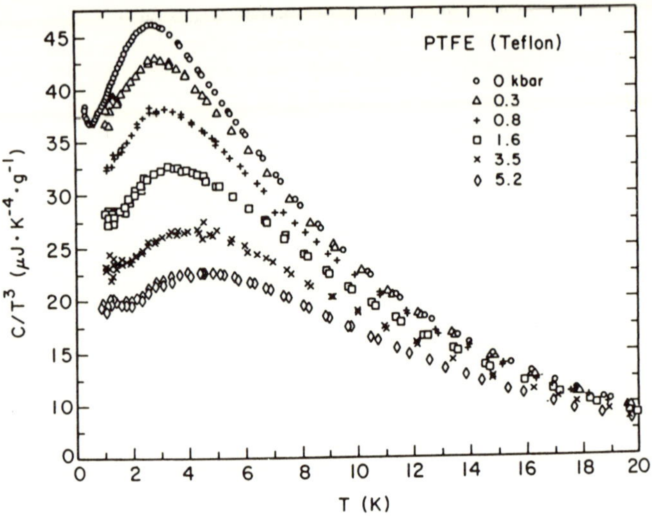

Fig. 5.28. c_P/T^3 (per gram) for PTFE at various pressures [Boy83].

The 'peak' in c/T^3 near 3 K is very obvious at low pressure and decreases by a factor of two at 0.5 GPa. Assuming a bulk modulus of 6.5 GPa, the increase in pressure from 0 to 0.16 GPa (1.6 kbar) corresponds to a density increase of about 2.5%. There is an associated decrease in c/T^3 near 3 K of about 30%, which corresponds to a 10% increase in Θ^C. Therefore the effective Grüneisen parameter is $\gamma = d\ln\Theta_0^C/d\ln\rho \simeq 4$.

Polyethylene. The thermal properties of isotropic PE in various degrees of crystallinity and anisotropic samples have been widely studied, but values for nearly perfect crystals can only be inferred by extrapolation, except at higher temperatures ($\geq$ 100 K) from x-ray diffraction. The values in Table 5.9 for fully crystalline (c) and fully amorphous (a) material were extrapolated linearly from data on a number of samples [Gau81]. At temperatures in the region of 100 to 150 K, the values of c_P (amorphous) exceed those for c_P (crystalline) by only a few percent, but below 20 K the excess increases considerably: by a factor of 2 near 20 K and a factor of 4 at 10 K.

The values of α for isotropic PE (Fig. 5.29) show similar differences, e.g., α^a/T^3 is roughly 5 times greater than α^c/T^3 near 5 K [Whi84a]. Choy and collaborators [Cho81] have measured the effect of orientation by drawing various polymers, including PE, through dies. For a sample with a draw ratio of 11 (Table 5.11) the anisotropy in expansion is like that of a crystal, although it is hard to estimate how much the numerical values of $\alpha_\parallel$ and $\alpha_\perp$ would differ from those of an ideal perfect crystal of PE. Above 100 K, $\alpha_\parallel$ is consistent with α_c as given by x-ray diffraction [Dad81]; $\alpha_\perp$ is in fair agreement with $(\alpha_a+\alpha_b)/2$, but increases more rapidly so

Table 5.11. Values of α (units of 10^{-6}K^{-1}) for isotropic polyethylene extrapolated to fully amorphous (a) and fully crystalline (c); also values measured on rod (draw ratio 11, ca. 80% crystalline) in directions perpendicular and parallel to draw direction [Cho81, Whi84a]

T(K)	PE^a_{iso}	PE^c_{iso}	$PE^{drawn}_{\perp}$	$PE^{drawn}_{\parallel}$
10	3.18	0.68	2.2	−0.175
20	11.6	5.0	11.3	−0.79
55	35	30	53	−4.2
100	57	49	86	−6.0
150	96	65	107	−6.8
200	143	77	123	−8.0
250	203	80	141	−10
300	324	83	158	−11

that at room temperature $\alpha_\perp$ for the drawn sample is about 25% larger. The x-ray measurements reveal marked anisotropy in the ab plane at higher temperatures, with α_a about double α_b, but force-constant models of the perfect crystal suggest that at the lowest temperatures the thermal expansion is approximately isotropic in this plane[Bru98].

γ values for the isotropic PE change from about 3 below 10 K to 1 at room temperature. For the anisotropic drawn rods, $\gamma_\perp \sim 3$ at low temperatures and falls to about 1 at 300 K; in the axial direction, $\gamma_\parallel$ is negative.

Polymethylmethacrylate (PMMA). Measurements of linear expansion between 1.5 and 4 K show a negative T-term and positive T^3-term, fitting the equation $\alpha = (-1.06T + 1.15T^3) \times 10^{-8}\text{K}^{-1}$ [Lyo79]. Using a value of $B_S = 8.2$ GPa and unpublished heat capacity data gives for the linear term, $\gamma_T = -16$. The α measurements extend up to 300 K with $\alpha_{293} = 79.3 \times 10^{-6}\text{K}^{-1}$ and $\Delta l/l_{293} \approx 1.2\%$.

Later measurements of α for PMMA extending down to 0.4 K [Ack84] give values about 25% smaller than [Lyo79], and analysis of the data below 1 K gives a very different value for γ_T, namely -1, while $\gamma_{T^3} \sim 1$.

Epoxy resins. Epoxy resins, with and without fillers, are widely used in cryogenic equipment. Measurements of linear thermal expansion indicate a range of values of α_{293} from 60 to $70\times10^{-6}\text{K}^{-1}$ for unfilled samples and 20 to $40\times10^{-6}\text{K}^{-1}$ for some filled samples [Esc95, Wal94]. Overall contractions, $\Delta l/l$, on cooling from 293 to 4 K range from 1.0 to 1.4% for unfilled samples and from 0.3 to 0.7% for samples with fillers.

A filled epoxy of low expansion which is often used for sealing electrical leads through copper at low temperatures is Stycast 2850FT. Measurements of heat capacity

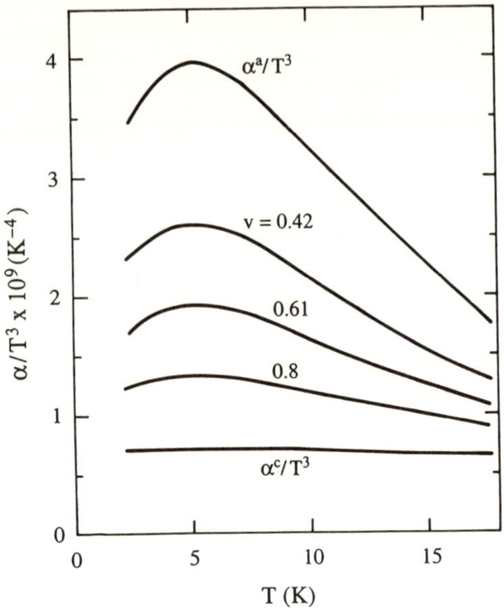

Fig. 5.29. α/T^3 versus T for various isotropic PE samples. α^a and α^c denote coefficients of amorphous and crystalline regions respectively [Whi84a].

(1 to 90 K) and linear expansion (1.5 to 300 K) [Swe97] give $\alpha_{300} = 30 \times 10^{-6}\mathrm{K}^{-1}$ and $(l_{300} - l_{1.5})/l_{300} \approx 0.45\%$. This sample had a maximum in C_P/T^3 and in α/T^3 near 5 K and a linear tunnelling term was apparent below 2 K.

Polymer composites. The thermal expansion and elastic properties of composites can be tailored to meet engineering requirements by addition of powders such as silica, alumina and magnesia or fibers of graphite, silica, kevlar, etc.; the general aim is to produce structural components of light weight, sufficient strength, and thermal expansion compatible with other components. They are also electrically non-conducting and non-magnetic. The electrically non-conducting property is important in ac applications to stop eddy currents and associated magnetic fields. The 'tailoring' consists of choosing suitable proportions of powder or suitable alignment of fibers. These composites are discussed in Ch. 7.

5.10. HIGH T_c SUPERCONDUCTORS

5.10.1. Introduction

High-T_c superconductors (HTS) are included in this chapter because they are ceramics, and because their heat capacity and thermal expansion vary with temperature in much the same way as do those of other ceramics except in the immediate vicinity

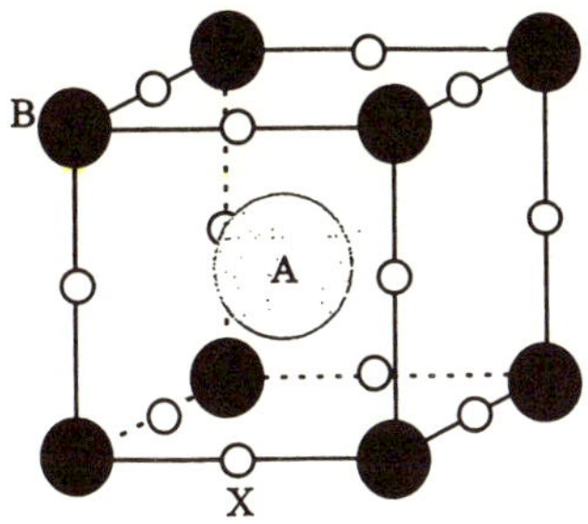

Fig. 5.30. The basic perovskite structure with large metal ion, A, at the center, smaller metal ions, B, at corners and anions (e.g., O) at mid-edge.

of the superconducting transition T_c. At T_c there is sometimes a bump and sometimes a sharp peak in C_P (and also in α) which is relatively small compared with the lattice vibrational background: relatively small because T_c is larger compared with Θ than for metallic superconductors. For example, $T_c/\Theta_0 \sim 0.2$ for $YBa_2Cu_3O_{6+s}$ compared with ~ 0.02 for Sn.

As far as electron transport is concerned, the electrical resistivity of most HTS above T_c is rather like that of graphite: $\rho_{ab} \sim 100\ \mu\Omega{\cdot}\text{cm}\ (= 10^{-6}\Omega{\cdot}\text{m})$ and increases roughly linearly with T, while ρ_c is orders of magnitude larger and is semiconductor-like.

The crystal structures of the HTS are related to the perovskite ($CaTiO_3$) structure, shown in Fig. 5.30. The larger metallic ion A sits at a cube center, the smaller metallic ion B (Cu in most HTS) is at a cube corner, and the non-metallic anions (oxygen in the HTS) are at mid-edge sites. As discussed in more detail below for YBCO, the HTS structure is generally distorted to tetragonal or orthorhombic symmetry, with layers of CuO_2 normal to the principal axis performing a vital function in the superconductivity. When samples are twinned with a common c-axis, α_a and α_b cannot be measured separately but only an average denoted by α_{ab}.

The following sections present data for selected HTS showing how well (or badly) C_P and α fit the Debye-Grüneisen models, and also discuss the anisotropy in expansion; both these properties are important in cryogenic applications. The exact form of the anomalies near to T_c depend critically on the doping or oxygen content and are of fundamental interest. The relative magnitudes of the 'discontinuities' in C and α at T_c are also useful as a means of calculating the pressure and uniaxial strain dependences of T_c via the Ehrenfest relations.

The thermal properties of HTS have been the subject of a number of reviews, including (in chronological order) [Fisc88, Fish88, Jun90a, Phi92, Jun96] for heat capacity and [Gme89, Whi93a] for thermal expansion. Reviews of high pressure studies by Schilling and Klotz [Sch92] and of ultrasonic data by Dominec [Dom93] are sources for compressibility (and bulk modulus) data. However, the ultrasonic data include many measurements on polycrystalline samples of relatively low density which lead to values of χ (and B) which are at variance both with each other and with values from pressure measurements and single crystal data. It appears that the shape of the voids in the porous samples makes it impossible to apply realistic density corrections, and therefore may lead to low values of B. Table 5.12 gives some physical parameters for selected HTS, including: approximate molecular weight (M.W.); 'best' estimates of T_c, Θ_0 and 'linear' term in C_P ($\Gamma(0)$) for optimal stoichiometry; normal-state Γ_n (observed above T_c); linear expansion coefficient, α_{293}^{av}, measured on a polycrystal without preferred orientation; and the volumetric Grüneisen gamma at room temperature, γ_{293}. These data are based on information in the review articles [Fisc88, Fish88, Jun90a, Phi92, Whi93a, Jun96, For97]; also see [Lor94] and expansion data of Meingast and collaborators [Mei91, Mei93, Mei96, Gug94, Pas98]. Abbreviations for HTS in Table 5.12 are as follows:

- Y123 for $YBa_2Cu_3O_{6+s}$ or $YBa_2Cu_3O_{7-\delta}$
- Y124 for $YBa_2Cu_4O_8$
- Bi2212 for $Bi_2Sr_2CaCu_2O_8$
- Bi2223 for $Bi_2Sr_2Ca_2Cu_3O_{10}$ or $(Bi,Pb)_2Sr_2Ca_2Cu_3O_{10}$
- TB2212 for $Tl_2Ba_2CaCu_2O_8$
- LSC for $La_{1.85}Sr_{0.15}CuO_4$
- BKB for $(Ba_{0.6}K_{0.4})BiO_3$.

Perovskites. The perovskite structure plays an important rôle in the HTS. The forerunner of the ceramic HTS was $SrTiO_3$, which was found (in the early 1970s [For97]) to exhibit superconductivity below 1 K if oxygen deficient. Minerals of the perovskite structure are also of great interest to the geophysicist because of their rôle in the earth's mantle. The elastic and thermal properties have been widely studied at high temperatures and pressures (see, for example, recent monograph by Orson Anderson [And95a]). They are of less concern to the low temperature physicist, with the exception of the HTS and some perovskites which display soft ferroelectric modes below room temperature. Some like $SrTiO_3$ remain cubic in structure, while others are distorted to tetragonal or orthorhombic symmetry. Fischer *et al.* [Fisc93] give elastic data on polycrystals of the Ca, Sr and Ba titanates. Values of B_S at room temperature given below are from [Fisc93] and Webb [Web98]. The α values are from [Tou77]. We can compare them with values for the HTS in Table 5.12.

- $CaTiO_3$ ... $B_S = 172 - 176$ GPa (o-rh); $\alpha_{293}^{av} = 11 \times 10^{-6}K^{-1}$; $\gamma_{293} = 1.7$
- $SrTiO_3$ ... $B_S = 172 - 185$ GPa (cub); $\alpha_{293} = 10.3 \times 10^{-6}K^{-1}$; $\gamma_{293} = 1.7$
- $BaTiO_3$ ... $B_S = 136$ GPa (tetr.); $\alpha_{293}^{av} = 6.3 \times 10^{-6}K^{-1}$; $\gamma_{293} = 1.0$.

Table 5.12. Physical parameters for some major members of the HTS family. M.W. is approximate molecular weight. (Sources discussed in text)

Property	Y123	Y124	Bi2212	Bi2223	TB2212	LSC	BKB
M.W.(approx)	667	746	888	1024	979	398	355
T_c (K)	92	81	85	107	110	37	31
Θ_0(K)	430	350	240	285	255	440	330
	± 10	± 10	± 20	± 5	± 5	± 10	± 10
$\Gamma(0)$ (mJ/mol·K^2)	≤ 4	5	≤ 0.1	$\simeq 0$	< 7	<1.5	≤ 0.1
Γ_n (mJ/mol·K^2)	18	11	—	11	—	7–10	1.5
$\Delta C(T_c)/T_c$ (mJ/mol·K^2)	40–60	12–16	20–25	20–30	$\simeq 30$	~ 8	$\simeq 2$
dT_c/dP (K/GPa)	0.7	5.5	1.5	1.1	1.2	3	0.7
$10^6 \alpha_{293}^{av}$/K	11.5	11	12	12–13	—	12	—
$10^6 \alpha_{293}^{c}$/K	$18_{.5}$	18	18	—	—	13	—
$10^6 \alpha_{293}^{a}$/K	10	8(ab)	$8_{.5}$	—	—	11	—
$10^6 \alpha_{293}^{b}$/K	10	8(ab)	11	—	—	11	—
B_S (GPa)	125	120	70	80	120	130	—
γ_{293}	1.6	1.6	1.5	1.1	—	1.6	—

Lawless [Law80] has measured C_P on a number of these ferroelectrics from 2 to 35 K where there are large departures from Debye-like behavior due to the optic or ferroelectric modes. Values of Θ_0^C are rather uncertain because of these soft modes. Their rôle in producing a very large pressure dependence of the dielectric constant and an associated large value of γ ($\sim$ 100) at low temperatures has been discussed by Samara [Sam71].

$KTaO_3$ is one cubic perovskite for which there are low temperature data on both heat capacity and thermal expansion. The C_P measurements extend from 1.5 to about 30 K and depart considerably from the Debye function based on $\Theta_0^{el} = 594$ K [Whi87]. Below 3 K, the difference in C_P from Debye can be represented by an Einstein term based on $\Theta_E \simeq 15$ K; Θ_E increases to about 20 K near 4 K. Measurements of α from 2 to 30 K combined with value of $B_S = 235$ GPa leads to values of Grüneisen γ of about 100 at 3–4 K decreasing to 13 at 30 K and 1.4 at room temperature. If the optic-mode contributions are separated from the total measured values, then γ lies between 100 and 200 at liquid helium temperatures, comparable with the values of $\sim$ +400 calculated from dielectric data [Sam73].

5.10.2. $YBa_2Cu_3O_{6+s}$

Y123 is the most studied of the HTS but is not the simplest in terms of structure and oxygen content. Relating the structure to the perovskite of Fig. 5.30, the copper sites all remain occupied, and every third A site (up the c-axis) is occupied by a Y atom and the intervening two A sites by Ba. But some of the oxygen sites are vacant.

Consider first $YBa_2Cu_3O_6$, which is tetragonal. The sites in the *ab* planes containing Y atoms are all vacant, and so are the sites in *ab* planes midway between adjacent Ba atoms, so that these planes contain only Cu atoms. The other O sites are occupied, and in particular there are CuO_2 planes normal to the *c* axis above and below each Y site. Increasing oxygen content starts to fill vacancies in the planes midway between the Ba ions, until in $YBa_2Cu_3O_7$ half of the mid-edge oxygen sites are filled, thus giving CuO planes. In these planes there are quasi 1-dimensional CuO chains running in the *b* direction, leaving unoccupied sites along the *a* direction. The occupied sites expand the lattice in the *b* direction, producing the orthorhombic distortion. Optimal oxygen content for superconductivity corresponds to $x = 0.92$, with $T_c \approx 93$ K. The minimum value of x for superconductivity to occur at all is around 0.4, giving 6.4 oxygens per molecule, (see, for example, the reviews [For97] and [All97, Fig. 10.5]). There is also oxygen deficiency in the CuO_2 planes, which are those responsible for superconductivity.

Measurements show that C_P generally follows the same form of departure from the Debye model as most other dielectric crystals, except in the regions near T_c and below 4 K. $\Theta^C(T)$ falls from a limiting value of about $\Theta_0 = 430$ K to a minimum of ca. 350 K at $T \simeq \Theta_0/15$ and then rises to ca. 500 K at room temperature. Figure 5.31 from a review by Fisher *et al.* [Fish88] shows the general form of heat capacity up to 100 K, plotted as C/T to illustrate the 'bump' near T_c more clearly; the regions below 10 K and near T_c are amplified in insets.

At temperatures below 4 K, graphs of C/T^3 show a marked upturn, which varies from sample to sample and has been attributed partly to magnetic impurity effects — a Schottky bump due to some Cu^{2+} ions — and partly to a linear T-term (e.g., Fig. 5.31 inset). The question remains whether this 'linear' term (written as $\Gamma(0)T$) arises from tunnelling states (as in glasses), normal regions, or unpaired carriers in CuO chains and/or whether it is an intrinsic part of the superconducting interaction. The lowest value of $\Gamma(0) \simeq 2$ mJ·mol^{-1}·K^{-1} has been observed for an overdoped untwinned single crystal [Fisc88, Fish88, Jun90a, Phi92].

It is difficult to determine the normal electronic component of C_P in HTS because the critical fields are too high to allow destruction of superconductivity at low temperatures. However, careful analysis of C_P near T_c has been applied to separate C_{en} from the much larger phonon component: Loram *et al.* [Lor94] used precise differential calorimetry to compare C_P for the same sample (after different oxygen doping) with a non-superconducting sample in which 7% of the Cu was replaced by Zn, giving a fully oxygenated $YBa_2(Cu_{0.93}Zn_{0.07})_3O_{6.92}$. They had previously established that the phonon and electronic terms for the latter are very similar to those of $YBa_2Cu_3O_{6.92}$ material. Their results showed that $\Gamma_n \approx 18$ mJ·mol^{-1}·K^{-2} = 1.4 mJ·g-at^{-1}·K^{-2} for optimally doped Y123 on samples with $x \geq 0.9$ and $T_c \approx 91$ K. Values of C_P also showed that a degree of oxygen depletion (provided that $x \geq 0.4$) had little affect on Γ_n, but that the height of the step at T_c was substantially reduced for values of $x \leq 0.9$ (see Fig. 5.32). At optimal doping the step height $\Delta C/T_c$ was about 4 mJ·g-at^{-1}·K^{-2} or 50 mJ·mol^{-1}·K^{-2}.

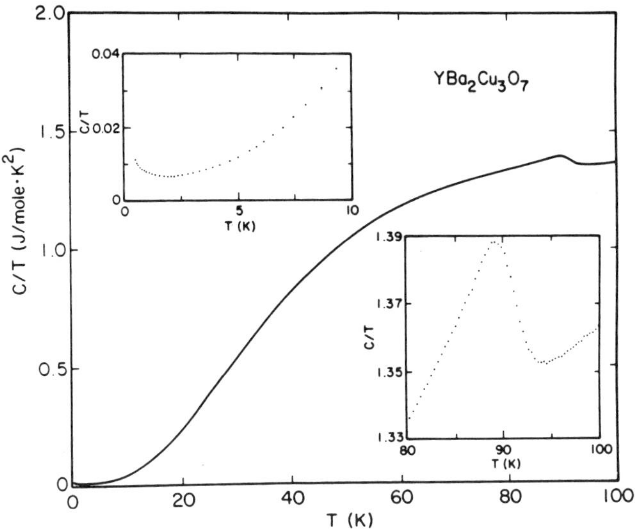

Fig. 5.31. C/T for Y123 with top inset showing upturn below 4 K and lower inset expanded region near T_c. From [Fish88, p. 233].

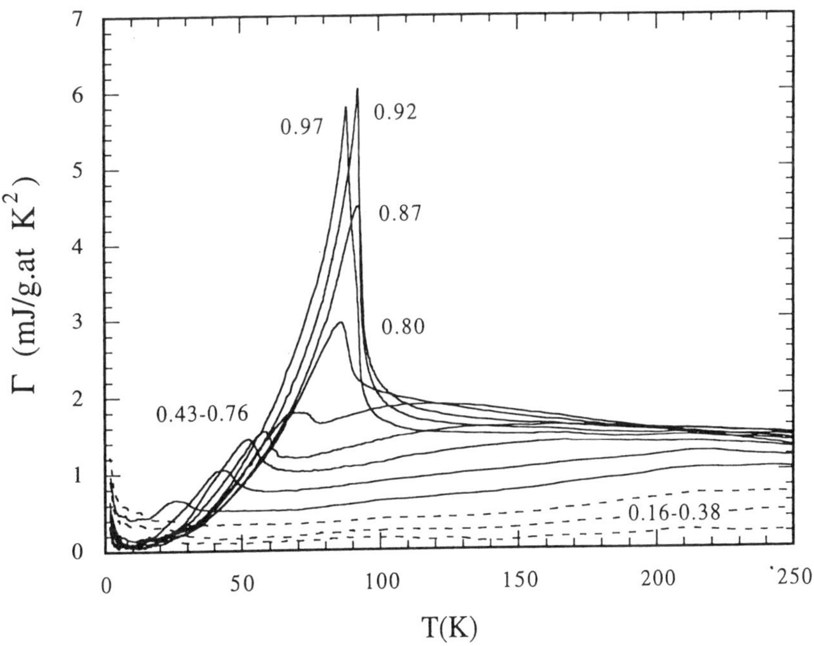

Fig. 5.32. $\Gamma(x,T)$ for $YBa_2Cu_3O_{6+s}$ relative to $YBa_2Cu_3O_6$ [Lor94].

Measurements of the shape and size of the anomaly in C_P near T_c in fields up to 16 T show that T_c is not changed by field but that the height of the peak is reduced; also it is reduced much more drastically by a magnetic field parallel to the c axis than by a field in the ab plane. Junod [Jun96] has reviewed these studies and their implications for various theoretical models.

The expansion coefficient determined on sintered samples, $\alpha^{av}(T)$, follows a similar curve to C_P with variations of a few percent from sample to sample, reflecting differences in oxygen content, microcracks, preferred orientation, etc. There is a peak or bump near 90 K of 1–2%, and indications of departures from T^3 behavior below 4 K as for C_P. Using a value of 125 GPa for B_S leads to $\gamma \approx 1.6$ from 20 to 300 K (e.g., [Whi93a, Swe89]).

Capacitance dilatometry [Mei91] on an untwinned single crystal confirmed earlier X-ray evidence that $\alpha_c > \alpha_a > \alpha_b$ (Fig. 5.33). Near T_c there are obvious differences in the expansion coefficient along the different crystal axes: there is a positive 'jump' (or 'step') in α_a, (so that $\alpha_a^s < \alpha_a^n$), a negative jump in α_b (as in C_P), and no obvious change in α_c. From these jumps the Ehrenfest relations lead to values for $dT_c/dP_{a,b,c}$ of -1.9 K·GPa^{-1} (a-axis), +2.2 K·GPa^{-1} (b-axis), and 0 (c-axis). Adding these gives a net positive sum for the hydrostatic pressure derivative of (0.3 ± 0.4) K·GPa^{-1}. A value of about 0.7 K·GPa^{-1} is obtained from polycrystals and from direct pressure measurements [Sch92, Whi93a] and leads to $d \ln T_c / d \ln V \approx 1.1$.

Further measurements were made on a larger single crystal with near-optimal doping by Pasler *et al.* [Pas98] to investigate evidence for critical fluctuations near T_c. They show very clearly the difference between α_a and α_b near T_c with a large positive 'spike' in $\alpha_b - \alpha_a = \Delta\alpha_{b-a}$ of height 2 to 3 $\times 10^{-6}$ K^{-1}; this 'spike' is centered near 91.5 K and leads to values for critical exponents.

5.10.3. $YBa_2Cu_4O_8$

Y124 has attracted less attention than Y123 because T_c ($\approx$ 81 K) is lower and close to the boiling point of nitrogen. However it has potential interest in that it does not adopt the possible tetragonal structure but retains orthorhombic symmetry, the oxygen content is more stable, and there is less twinning and microcracking. Structurally the difference from Y123 is that Y124 has a double layer of CuO chains, still running in the b direction. Calcium can be added to raise T_c to about 90 K.

C_P per gram atom above 20 K differs by less than 2% from that of Y123, but the step at T_c is much smaller [Jun90b]: $\Delta C/T_c$ is 12 to 16 mJ·mol^{-1}·K^{-2} or about 1 mJ·g-at^{-1}·K^{-2}. Θ^C increases with temperature from the minimum value of about 350 K (near 20 K) to about 550 K at room temperature.

Expansion measurements on a single crystal give values of α_{293} in units of 10^{-6}K^{-1} of about 18 for α_c and 8 for α_a and α_b [Mei93]. Below T_c, $\alpha_b < \alpha_a$. At T_c the jumps in α_b and in α_c are very small, but that in α_a is relatively large: $\Delta\alpha_a = \alpha_a^s - \alpha_a^n = (0.45 \pm 0.05) \times 10^{-6}$ K^{-1} — or about 5%. This leads to a value of the uniaxial pressure dependence of $dT_c/dP_a \geq 4$ K·GPa^{-1}, and a similarly large value for the hydrostatic pressure dependence.

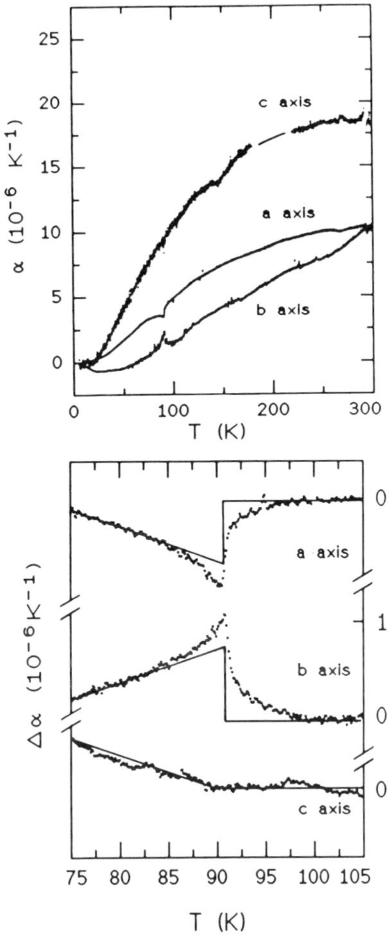

Fig. 5.33. Principal expansivities for Y123, showing $\alpha(T)$ in the upper figure and an expanded view of change in α near T_c in lower figure [Mei91].

5.10.4. $Bi_2Sr_2CaCu_2O_8$

C_P measurements on Bi2212 below 4 K are affected by the presence of magnetic phase material as they are for Y123 and Y124. However, the coefficient of the linear term in T, $\Gamma(0)$, appears to be less than 0.1 mJ·mol^{-1}·K^{-2}, and is possibly zero in the best single crystals. Θ_0^C at the lowest temperatures is in the range 220 to 260 K and increases to about 450 K at 100 K and 550 K at 200 K (see review [Jun96]). At T_c ($\approx$ 85 K) there is no sharp step in C_P as in Y123, but a symmetrical logarithmic 'cusp'; this decreases when a field is applied parallel to the c-axis, but is not affected by fields (up to 10 T) normal to the c-axis.

Meingast *et al.* [Mei96] measured the principal expansion coefficients between 10 and 300 K on a single crystal $9 \times 5 \times 1$ mm^3 with $T_c = 88.5$ K. Values of $\alpha^{293}_{a,b,c}$ are in Table 5.12. Near to T_c, they observed a positive cusp or 'spike' in α_a and α_b of $\sim 0.5 \times 10^{-6}$ above the background and a negative spike in α_c. The Ehrenfest relation leads to the following values of the uniaxial pressure dependence (in K/GPa):

$$\frac{dT_c}{dP_a} = 1.6, \quad \frac{dT_c}{dP_b} = 2.0, \quad \frac{dT_c}{dP_c} = -2.9$$

The resulting hydrostatic pressure dependence is $dT_c/dP \simeq 0.8$ K·GPa^{-1} which compares with values of 1 to 1.5 K·GPa^{-1} from direct pressure measurements.

Kierspel *et al.* [Kie96] measured the cusps in both C_P and α_{ab} near T_c on an untwinned single crystal, finding them to have similar shapes but differing magnitudes: 1.8 J·mol^{-1}·K^{-1} or 1.3% for ΔC_P and 0.16×10^{-6}K^{-1} or 4% for $\Delta\alpha$, leading to a value for dT_c/dP_{ab} of 0.9 K·GPa^{-1}.

5.10.5. $(Bi,Pb)_2Sr_2Ca_2Cu_3O_{10}$

The pure Bi2223 phase is usually achieved by addition of some Pb and has $T_c \approx 107$ K. Data for C_P indicate that $\Gamma(0)$ is close to zero and that Θ_0 is in the range 280 to 290 K, with evidence of dispersion and optic modes affecting Θ^C above 4 K. The bump in C_P near T_c is rather symmetrical like that in 2212, and $\Delta C_P/T_c \simeq 16$ mJ·mol^{-1}·K^{-2} [Phi92, Jun96]. The coefficient α for polycrystals is similar to that of 2212 near room temperature but appears to be smaller at low temperatures, reflecting a higher value of Θ^C.

5.10.6. $Tl_2Ba_2CaCu_2O_8$

Thermal data are scarce for the thallium family, whether 2201 (T_c = 85 K), 2212 (T_c = 110 K) or 2223 (T_c = 125 K). C_P data for TB2212 indicates $\Theta_0 \simeq 250$ K with a linear coefficient $\Gamma(0)$ less than 7 mJ·mol^{-1}·K^{-2}. The bump centered on T_c is symmetrical and rounded, leading to a value of $\Delta C/T_c = 35 \pm 10$ mJ·mol^{-1}·K^{-2} [Jun90a]. A sharper symmetric bump has been observed for a 2201 ceramic, centered at 85 K and with a height of $\Delta C/T_c = 10$ mJ·mol^{-1}·K^{-2} [Jun96]. There are no data for $\Delta\alpha$, but there are direct pressure measurements giving $dT_c/dP = 1.2$ K·GPa^{-1}.

5.10.7. $La_{1.85}Sr_{0.15}CuO_4$

Measurements of C_P and α below 20 K on a polycrystal show a similar pattern (Fig. 5.34 and [Col87c]), with a linear term dominating below 5 K and a T^3 term giving $\Theta_0 \simeq 440$ K. Θ^C falls to a minimum of about 320 K near 25 K ($\sim \Theta_0/15$) and then increases to about 500 K at room temperature. Near 35 K, there are small 'discontinuities' in C_P and α of $\sim 2\%$. Because the transition is in a temperature region where the lattice energy and therefore C_P and α are changing rapidly, the shape of the anomaly is not easily distinguished unless data are plotted as C/T^2 and α/T^2 (or $\Delta l/T^3$) [Jun90a].

Later measurements of $\alpha_{a,b,c}$ were made from 5 to 300 K on an untwinned single crystal, doped at Sr levels of $x = 0.10, 0.15, 0.20$ [Gug94]. For the optimally doped sample near $T_c = 36$ K, there were positive jumps in α_a and α_b respectively of 0.32 ± 0.05 and $0.63 \pm 0.05 \times 10^{-6}$ K^{-1} and in α_c a (negative) jump of $-0.87 \pm 0.1 \times 10^{-6}$ K^{-1}.

This leads to the following values of the uniaxial pressure dependence (in K/GPa):

$$\frac{dT_c}{dP_a} = 2.5, \quad \frac{dT_c}{dP_b} = 4.9, \quad \frac{dT_c}{dP_c} = -6.8$$

The sum of these, the hydrostatic derivative, is zero within rather large limits of experimental error. Direct pressure measurements of dT_c/dP on polycrystalline samples give values in the range 2 to 3 K·GPa^{-1} [Sch92]. Using a value of 120 GPa for B_S gives $\gamma_{293} = 1.6$.

5.10.8. $(Ba_{0.6}K_{0.4})BiO_3$

BKB is of interest because it is a ceramic HTS which is copper-free, i.e., a cubic perovskite without the CuO planes of the cuprate HTS. The low temperature measurements of C_P show that any linear term must be ≤ 0.1 mJ·mol^{-1}·K^{-1}. Θ^C falls from $\Theta_0 \simeq 320$–340 K to a minimum near 25 K and then increases to about 380 K at 90 K. The step $\Delta C/T_c$ at $T_c = 31$ K was observed to be about 2 mJ·mol^{-1}·K^{-2} [Jun90a]. Lacking data for the thermal expansion at the transition, Table 5.12 lists a value $dT_c/dP = 0.7$ K·GPa^{-1} based on pressure measurements [Sch92].

5.10.9. Fullerenes

The fullerenes are members of the carbon family, each looking like a 'soccer ball' made from folding up a graphite sheet. They are closed cage polyhedral molecules of carbon atoms, the most common being C_{60} (see review [Dre94]). When doped with alkali metals to form M_3C_{60} (M=K, Cs, Rb, Na) they show superconductivity at temperatures up to 33 K ($RbCs_2C_{60}$). Most of these compounds have a fcc structure, that is the C_{60} soccer balls form an fcc lattice with the alkali metal ions taking up octahedral or tetrahedral sites in the cube.

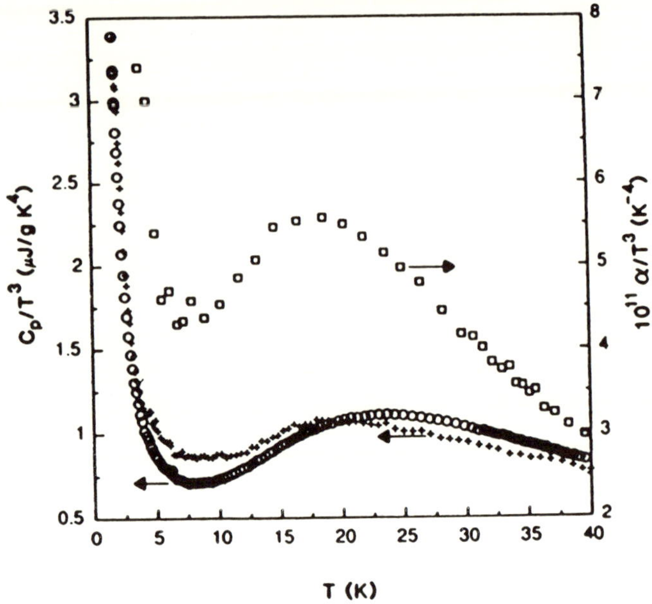

Fig. 5.34. A plot of c_P/T^3 (LSC: ○ and Y123: +) and α/T^3 (LSC: □) [Col87c].

Thermal data are limited. The C_P values for K_3C_{60} [Ram92] at low temperatures fit a Debye term (using $\Theta_0 = 70$ K, estimated from the bulk modulus) plus linear and Einstein terms. The step in $\Delta C/T$ at $T_c \approx 19$ K is $\simeq 30$ mJ·mol^{-1}·K^{-2}. Measurements of C_P on C_{60} from 1 to 20 K give values similar to K_3C_{60} except for the absence of a linear term [Bey92].

Burkhardt and Meingast [Bur96] measured the expansion of K_3C_{60} and Rb_3C_{60}, finding small *positive* steps about 3% high at respective temperatures of 19 and 29 K (Fig. 5.35). The steps in α and C_P at T_c lead via the Ehrenfest relations to *negative* values of dT_c/dP of between -8 and -10 K·GPa^{-1}, consistent with results of direct pressure measurements. Room temperature values of α were about 28 $\times 10^{-6}$K^{-1} for each compound, about twice that for C_{60}.

5.11. NON-METALLIC MAGNETIC CRYSTALS

5.11.1. Introduction

Magnetic moments, electronic or nuclear in origin, contribute additively to the free energy and a component S_m to the entropy. These result in contributions C_m and

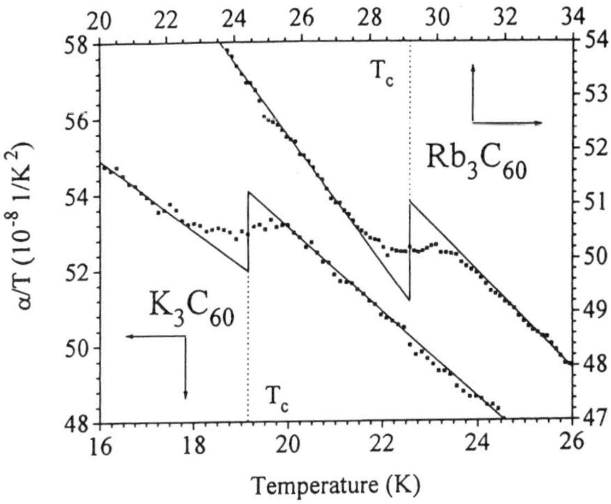

Fig. 5.35. α/T of K_3C_{60} and Rb_3C_{60} near to T_c [Bur96].

β_m to the heat capacity and thermal expansion coefficient:

$$C_m = (\partial S_m/\partial \ln T)_V \tag{5.2}$$

$$\beta_m = \chi_T(\partial S_m/\partial V)_T \tag{5.3}$$

They are linked by a Grüneisen function

$$\gamma_m = (\partial S_m/\partial \ln V)_T/C_m = \beta_m V/\chi_T C_m \tag{5.4}$$

If S_m can be expressed as a function of (E_m/T) where E_m is a magnetic interaction energy depending only on volume, then there is a single Grüneisen parameter [see Eq. (2.42)]:

$$\gamma_m = -(d\ln E_m/d\ln V) \tag{5.5}$$

If we further assume that E_m alone determines the magnetic ordering temperature, T_m, it follows that

$$\gamma_m = -(d\ln T_m/d\ln V) \tag{5.6}$$

To a first approximation these assumptions are usually valid, and so we can get a reliable estimate of the volume or pressure dependence of T_m from the magnetic contribution to thermal expansion. Independently of models, thermodynamically for first order transitions we have the Clapeyron equation Eq. (2.26)

$$dT_m/dP = \Delta V/\Delta S \tag{5.7}$$

and for second order transitions, the Ehrenfest relation Eqs. (2.28)

$$dT_m/dP = VT_m\Delta\beta/\Delta C \tag{5.8}$$

In insulators we are largely concerned with local magnetic moments, generally arising in d- or f-electron shells. Nuclear moments are orders of magnitude smaller, and so their effect on C or β is usually apparent only at very low temperatures (mK to μK region) in high magnetic fields. Exceptions are some rare earth salts where large internal fields from the electron shells can split the nuclear energy levels sufficiently to be comparable with kT at liquid helium temperatures.

We distinguish between cooperative and non-cooperative ordering (Section 2.5) among magnetic spins. Cooperative ordering, whether first order (showing a discontinuity in S at T_m) or second order (showing a discontinuity in the derivative (dS/dT) at T_m) occurs via direct or indirect exchange interactions, and is characterized by a well defined 'cusp' (logarithmic singularity) or λ-shaped peak in C and in β (e.g., Fig. 2.5). In contrast, the non-cooperative ordering of Schottky is characterized by broad 'bumps,' with an exponential shape as $T \to 0$ and a $1/T^2$ tail above the maximum (e.g., Fig. 2.2). Such transitions occur for example in diluted electron paramagnets (e.g., the cooling salt ferric ammonium alum — $FeNH_4(SO_4)_2 \cdot 12H_2O$), nuclear paramagnets, 'dilute' magnetic impurity systems (e.g., ZnS with low levels of Fe impurity), and in linear chain systems (e.g., $CsNiCl_3$, centered around 30 K).

Those systems which order cooperatively do so with the spins either parallel (ferromagnetic ... F and ferrimagnetic) or antiparallel (antiferromagnetic ... AF). At $T = 0$ all spins are aligned, but at finite temperatures spins are excited with spin wave spectra described respectively by $\omega \propto q^2$ (F) and $\omega \propto q$ (AF), which should give rise to $T^{3/2}$ (F) or T^3 (AF) contributions to C and β at low temperatures. In the AF case there can be an energy gap in the spin wave spectrum which causes a departure from T^3 as $T \to 0$. Some examples of these effects are given below.

We do not give details of the phase transitions in this monograph. A recent book by Cyril Domb entitled *The Critical Point* [Dom96] includes a fascinating historical introduction to modern theory of critical phenomena and refers to volumes 1–6 of *Phase Transitions and Critical Phenomena* (eds. Domb and Green, 1972-1976) and later volumes 7–16 of *Phase Transitions and Critical Behavior* (eds. Domb and Leibowitz, 1983-1994) published by Academic Press. We shall refer below to the review of experimental data on 'simple magnetic model systems' by de Jongh and Miedema [deJ74]. General treatments of the heat capacity of magnetic systems are given in the monograph by Gopal ([Gop66, Ch. 4]) and the chapter by Miiller ([Mii88, Section 1.4]).

5.11.2. Ferromagnets

EuO and EuS. Examples of ferromagnetic crystals which have cubic symmetry are EuO and EuS (rocksalt structure), with Curie temperatures of 69.2 and 16.4 K respectively. They have attracted experiments on their thermal properties because of the simple structure and approximation to a cubic Heisenberg ferromagnet (e.g., [deJ74, p. 198]). The linear expansivity α of EuO is shown in Fig. 5.36 [Arg67]; the variation of C_P with T is qualitatively similar. The magnetic components are

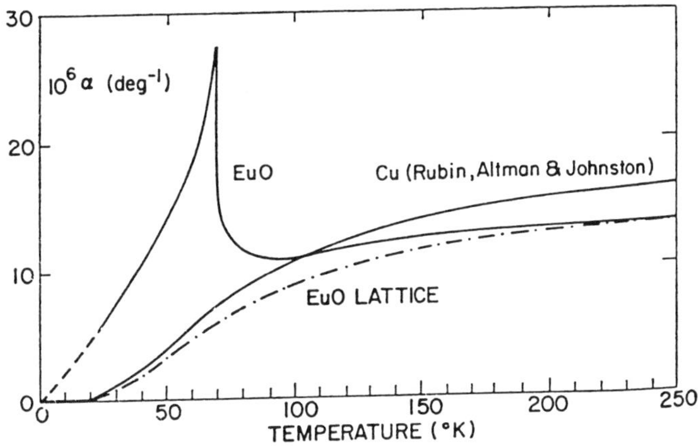

Fig. 5.36. The measured $\alpha(T)$ for EuO, compared with the estimated lattice component (– · –) and α for Cu [Arg67].

extracted by estimating the lattice background using a value of Θ_D determined well above T_c, e.g., above 150 K for EuO. The magnetic component α_m varies linearly with C_m over the temperature ranges 25 to 64 K and 74 to 140 K, that is excluding the region near T_c where C and α change rapidly and are difficult to measure accurately. The ratio $\alpha_m/C_m \simeq$constant, giving $\gamma_m \approx 5.3$, compared with $\gamma_{vib} \approx 1.9$ (measured at temperatures well above T_c).

Magnetite. The low temperature behavior of C_P for a ferromagnet well below T_c is exemplified by magnetite, Fe_3O_4. Figure 5.37 shows that $C_P = AT^{3/2} + BT^3$ in the region 1.5 to 4 K. Another good example of the $T^{3/2}$ dependence of C_m is from data on yttrium iron garnets, see [Gop66, p. 91].

We are not aware of measurements which show a $T^{3/2}$ term in the expansivity at low temperatures. The probable magnitude of a $T^{3/2}$ spin wave term in α for EuO was estimated [Lor67] by assuming that $\gamma_m \sim 5$. The conclusion was that such a term would cause a length change in a 10 mm long rod of about 40 Å (4nm) between 1 and 4 K. This should be observable by capacitance dilatometry.

5.11.3. Antiferromagnets

Uranium dioxide. UO_2 has a first order transition to antiferromagnetism at 30.4 K. It crystallizes in the fluorite structure and appears to remain cubic in the AF ordered state. There is a convenient non-magnetic isomorph, ThO_2, of similar density, molar volume and Debye temperature (see Table 5.5) from which a reliable estimate can be made of the lattice components for UO_2 in the AF region. Data for the elastic moduli of UO_2 give $\Theta_0^{el} \approx 380$ K. Using measurements of C_P [Hun71] and α [Whi74, Bar80], we can calculate the magnetic contributions, leading to $\gamma_m \simeq$

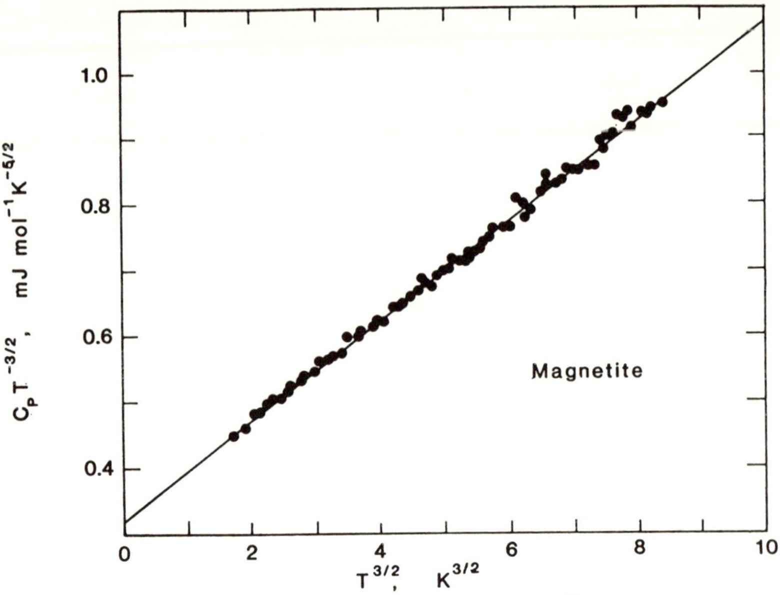

Fig. 5.37. Temperature variation of C_P for magnetite (Fe_3O_4) showing $T^{3/2}$ dependence [Cez88].

4–5 below 35 K, while $\gamma_{293} = 2.2$. Fig. 5.38 shows $\alpha(T)$ for a single crystal of UO_2 and a series of UO_2/ThO_2 mixtures (cold-pressed and sintered ceramics). For the 100% and 90% UO_2 samples the transition is in fact first order, and a distinct jump in length Δl is observed at T_N; but the cusp is broadened for the other samples, due perhaps to non-random distribution of U ions [Whi74].

Below 15 or 20 K, C_P and α both appear to vary more rapidly than T^3 — as about $T^{3.5}$ for the expansivity. The magnitude of C_P in this range is a factor of about two larger than for ThO_2, while the factor is about ten for α, reflecting a more significant magnetic contribution.

MnF_2. Another AF crystal which has been well studied is MnF_2, with $T_N = 67$ K, see [deJ74, p. 154]. It has the tetragonal rutile structure, and α is anisotropic: α_c has a large sharp (positive) cusp at the Néel Point, while α_a is small and negative below 150 K with a less obvious anomaly at T_N [Gib59]. Many hydrated chlorides of Co, Ni, Mn have also been measured with values of T_N between 1 and 6 K. C_P shows logarithmic singularities similar in form for each, but the observed α_i are anisotropic and complex due to the low symmetry of the crystals [Bar80, deJ74].

5.11.4. Schottky Anomalies

Examples of Schottky anomalies include the hydrated paramagnetic salts used for magnetic cooling. In these salts, spins are associated with Ni, Fe, Cr, Mn, etc., and are diluted with nonmagnetic ions and water of crystallization. They have energy level

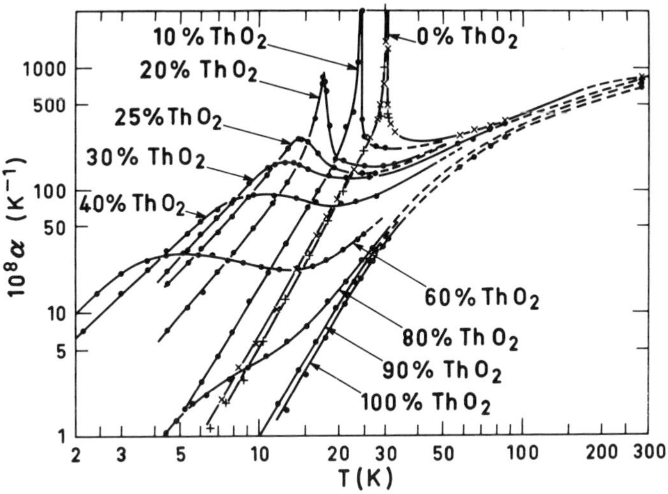

Fig. 5.38. $\alpha(T)$ for UO_2 and UO_2/ThO_2 mixtures [Whi74] (see also [Bar80, p. 705]).

splittings of ~ 1 K, largely from Stark splitting. With application of a magnetic field of a Tesla or less, the splitting is increased sufficiently so that at temperatures of $\sim$1 K enough spins are aligned to reduce the magnetic entropy substantially: examples are potassium chrome alum, ferric ammonium alum, manganous ammonium sulphate, gadolinium sulphate, cerous magnesium nitrate.

The theoretical form of Schottky bumps in C_P (e.g., Fig. 1.5) was discussed in Ch. 2. Characteristic is the $1/T^2$ 'tail' above the maximum. Most of the hydrated cooling salts mentioned above are anisotropic and of low symmetry, and they have not attracted expansion measurements.

In these salts the interaction between the diluted spins is very weak, but it does produce a cooperative ordering peak in C_P at sufficiently low temperature (~ 0.1 K for many of them and $\sim$ mK for hydrated cerous magnesium nitrate).

Figure 5.39 shows magnetic components for both C and α of a cubic (rocksalt) structure compound, TmSb, one of a series of rare-earth antimonides measured by Ott and Luthi [Ott77] giving values of γ_m (Schottky) ranging from -0.5 to -4. The rare earth ions have a 4f ground state which is partially split by the crystal electric field; the observed negative values for γ show that this is not adequately represented by a point charge model, which would predict $\gamma \simeq 5/3$, see [Bar80, Section 9.3].

Magnetic impurities. Small concentrations of magnetic impurity ions such as Fe or Cr in a host lattice can produce localized energy levels resulting in Schottky peaks at low temperatures in the heat capacity and in the expansion coefficient. For two-level defects of splitting energy E_m, the contribution to β is proportional to $\gamma_m C_m$, where C_m is the Schottky heat capacity of the impurities and $\gamma_m = -(d \ln E_m / d \ln V)$ is the Grüneisen parameter. The combination of measurements of C and β can be

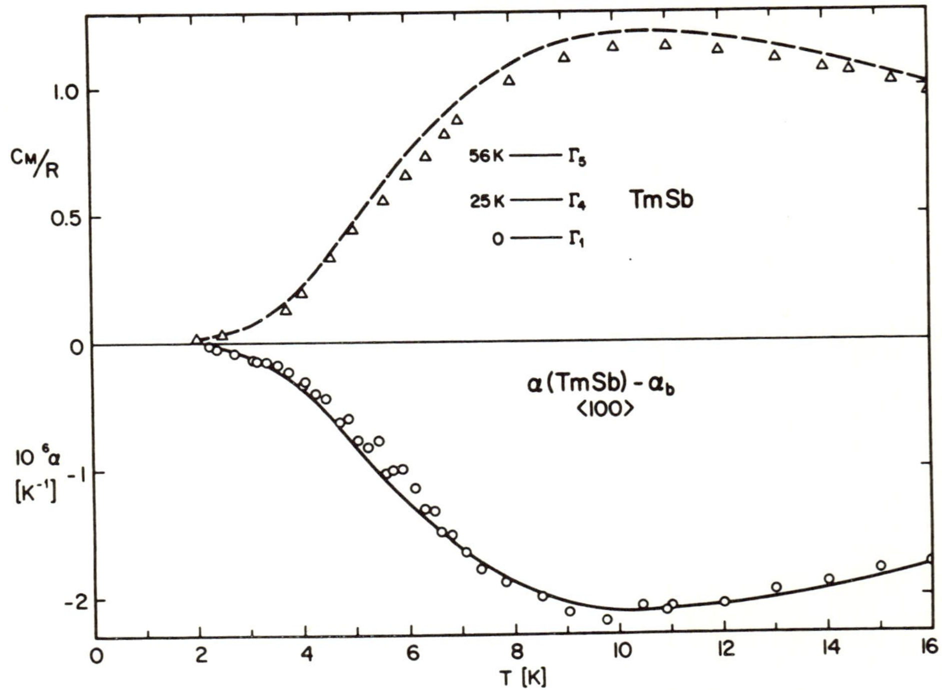

Fig. 5.39. Measured magnetic components of C_P (△) and α (○) for TmSb. Curves are calculated from the scheme of levels shown, with fitted $\gamma_4 = \gamma_5 = -1.2$. From [Ott77].

a useful supplement to spectroscopy in determining the nature of the splitting, for example whether it arises from Jahn–Teller effect or from spin–orbit coupling.

An example is the effect of Fe^{2+} ions in a cubic crystal, ZnS. Measurements of C_P and α on crystals containing 0.1 and 1% Fe give bumps which are centered near 10 K and are roughly similar in shape to those shown in Fig. 5.39, the component α_m being negative. Values of γ_m for the more dilute sample are distributed around the value of $-5/3$ predicted by the point-ion model for second-order spin–orbit coupling, see [She72] and [Bar80, pp. 706-708].

Linear-chain antiferromagnets. Examples of quasi one-dimensional antiferromagnets are the hexagonal ABX_3 crystals such as $CsNiCl_3$. For the latter the Ni atoms form chains parallel to the c-axis, characterized by an intrachain exchange parameter J which is stronger than the interchain parameter J' by a factor of 100. Short range AF order develops within the chains in the vicinity of 30 K, with the result that the acoustic attenuation, magnetic susceptibility and magnetic components of C_P and β show broad maxima centered around 30 K. At much lower temperatures (~ 5 K) the thermal energy becomes comparable with the smaller interchain coupling and three-dimensional order develops, shown by sharp cusps in both C_P and β at 4.85 K.

The individual linear expansion coefficients are highly anisotropic: at $T_N = 4.85$ K, $\alpha_\parallel$ has a sharp positive cusp while $\alpha_\perp$ shows no anomaly. In $CsNiCl_3$ there is also a spin reorientation at $T_R = 4.4$ K, where C_P, β and $\alpha_\perp$ show strong positive cusps and $\alpha_\parallel$ a small minimum [Ray81, Col87d]. The elastic stiffness c_{33}, measured by the longitudinal wave velocity along the c-axis chains, reflects similar anomalies — a fall around 30 K, a smaller drop near T_N, and a sharp minimum at $T_R = 4.4$ K.

The volume coefficient $\beta(T)$ of $CsNiCl_3$ (Fig. 5.40) has a similar form to $C_P(T)$ and leads to values of $\gamma_m \simeq 8$ [Ray81].

C_P has also been measured for $RbNiCl_3$ ($T_N = 11.0$ K) and $CsCuCl_3$ (10.4 K) [Col87d] and the Néel temperatures then related to the relative strengths of the intrachain exchange parameter J, and the interchain exchange J'. Using C_P data for the non-magnetic 'isomorph' $CsMgCl_3$, the lattice contributions can be estimated and subtracted from the total to give the T^3 antiferromagnetic contributions to heat capacity at low temperatures.

5.12. MIXED SYSTEMS, DIPOLES ETC.

5.12.1. Introduction

How do we expect the heat capacity of a polycrystalline mix of say KCl and NaCl to relate to that of their constituents? Does C_P (or β) vary linearly with concentration? The lattice dynamics will certainly be affected by the changes in atomic masses, the binding energy and the molar volume, with consequent changes

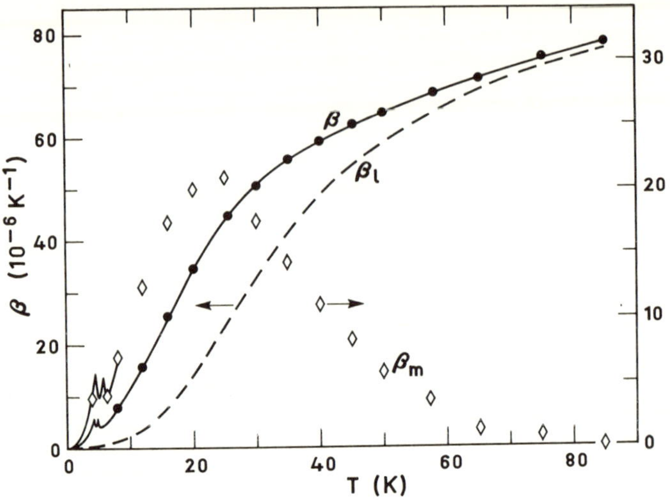

Fig. 5.40. Volume expansion coefficients for $CsNiCl_3$ [Ray81]: —•—smoothed values of $\beta = 2\alpha_\perp + \alpha_\parallel$; β_l is the vibrational (lattice) component; ◇—magnetic component, $\beta_m = \beta - \beta_l$.

in Θ_0 and Θ_∞^C. Changes in the expansion coefficient will also reflect changes in the anharmonicity of the interatomic potential.

Data on homogeneous mixed systems are limited compared with data on composites or aggregates. Low temperature interest has been largely confined to the effect of impurities at liquid helium temperatures, where magnetic ions or electric dipoles may have considerable effect on C_P and β because of the small lattice background at these temperatures. These impurities include well separated Fe and Cr ions in alumina (sapphire) and Fe in ZnS (discussed under Schottky anomalies in Section 5.11.4 above); also electric dipoles resulting from hydroxyl ions (OH^-) and Li^+ ions in alkali halides.

5.12.2. Polycrystalline Mixtures

There do not appear to be systematic studies of the heat capacity of mixed polycrystals to see whether averages based on molar fraction give a good guide to the values of, say, a 30/70 or 50/50 mix of two compounds. Nor are there many data for their expansion coefficients. Figure 5.41 shows the variation of α near room temperature for a series of sintered compacts of UO_2 and ThO_2 (see also Section 5.11.3), the components being each cubic, of similar molecular weight and miscible. Not surprisingly α varies roughly linearly with concentration. It is less clear why the end values for the sinters are significantly smaller than the single crystal values: the samples were made by coprecipitation, powdering, cold pressing and sintering with final densities varying from about 75 to 95% of theoretical [Whi74].

The thermal expansion of an equimolar mixture of Ca and Ba fluorides, measured

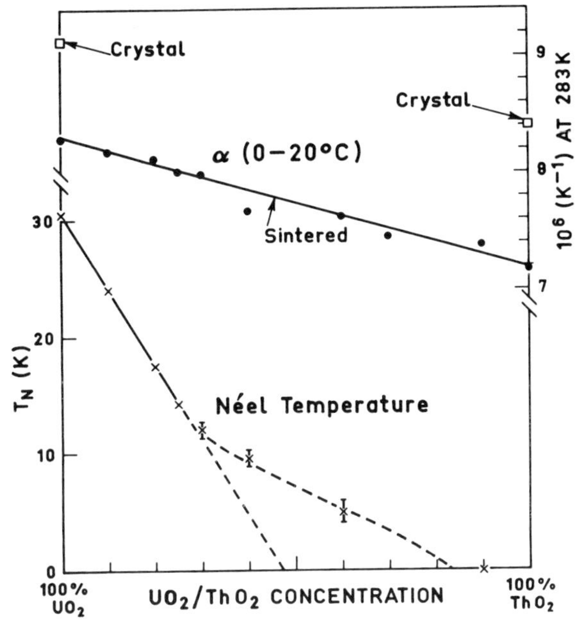

Fig. 5.41. Upper curve shows average α measured between 0 and 20° C for mixtures of UO_2 and ThO_2 [Whi74].

from 80 to 300 K, gave values which are a few percent higher than either of the components [Bar80, p. 662].

A rather different system referred to in Section 5.7.2 are the sodium silicate glasses for which there are some data on both C_P and α. These data cover concentrations up to 40 mol% Na_2O in SiO_2, chiefly at low temperatures (below 30 K) but including some at liquid nitrogen temperatures and room temperatures (Fig. 5.42 [Bar80, Whi77]). The heat capacity of a number of alkali silicate glasses has been measured, including $SiO_2 \cdot xNa_2O$ ($x = 0.14, 0.2, 0.3, 0.4$) from 2–95 K [Kru72] and mixed Cs–K, Cs–Na, K–Na silicates from 77–300 K [Hir70]. The differences in C_P at 300 K are a few percent, generally reflecting an increase with addition of alkali and with increase in atomic mass of the alkali. Differences are larger at 100 K and are qualitatively consistent with change in Debye Θ. In the mixed alkalis the changes are not linear in concentration. The structure and lattice dynamics of these glassy silicates do not lead to any useful conclusions about simpler mixed polycrystals.

5.12.3. Substitutional Impurities — Heavy Ions

When a low concentration of foreign atoms is introduced into a crystal lattice, there are three factors affecting the lattice dynamics:

1. The mass difference, $\Delta M/M$;

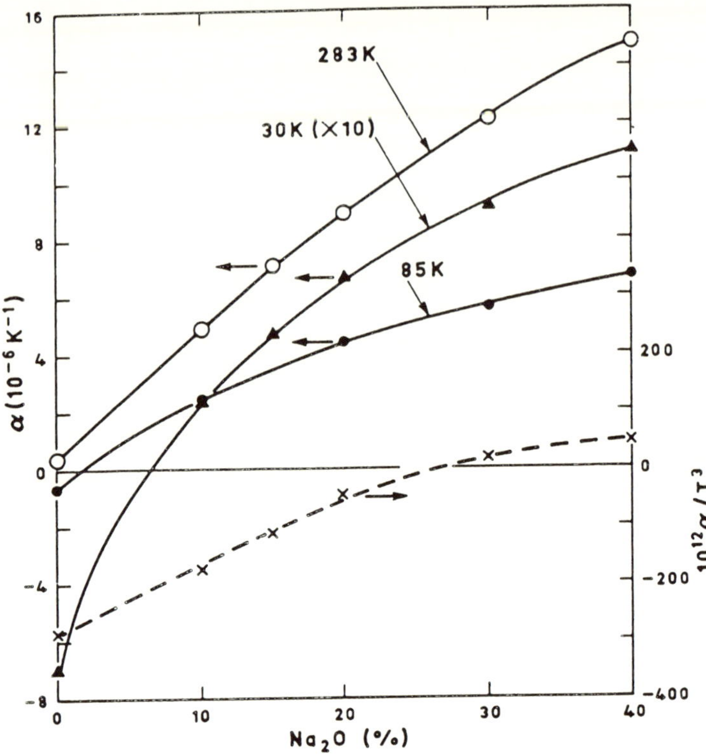

Fig. 5.42. Sodium silicate glass. Upper three curves show variation of α with composition x mol% soda at 283 K, 85 K and 30 K; dashed curve shows values of α/T^3 as $T \to 0$ [Whi77].

2. The coupling constant K determining whether binding is weak or strong;

3. Change in the anharmonicity of the potential, $\Delta A/A$, which affects α rather than C_P.

Timmesfeld and Elliott [Tim70] analyzed the likely effect of adding a heavy ion to an alkali halide. They concluded that if loosely bound, the heavy ion would have a resonant frequency in the lower end of the spectrum and therefore produce observable changes in C_P and β at low temperatures. The change in β could be negative or positive depending on $\Delta A/A$. If the impurity was tightly coupled the change in the phonon spectrum would be less localized and less obvious.

The difference is born out by observations below 30 K on two samples of NaCl doped respectively with 1% Ag and 1% Rb. Each of these has a large mass difference from Na but only the Ag produced a significant (localized) 'bump' in C_P (Figure 5.43) and in α (some values in Table 5.13). Both C_P and α showed a maximum increase of about 10% for the Ag additive over the range 10 to 15 K.

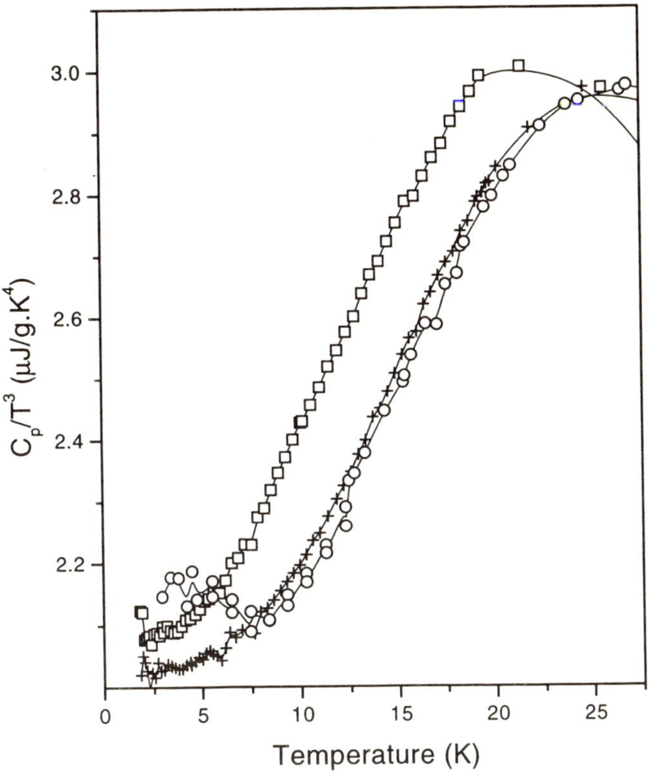

Fig. 5.43. Plot of C_P/T^3 versus T for pure NaCl (○) and NaCl containing 1% Ag (□) or 1% Rb (+) [Col84b].

Table 5.13. Values of linear coefficient α in units of 10^{-8} K^{-1} for pure NaCl, NaCl + 1% Rb and NaCl+1% Ag [Col84b]

T(K)	NaCl	+1%Rb	+1% Ag	T(K)	NaCl	+1%Rb	+1% Ag
4	0.37	0.35	0.39	20	61.2	61.3	65.9
	±0.01			24	115.4	116.1	121.9
5	0.72	0.71	0.82	28	184.9	195.3	202.7
6	1.24	1.24	1.32	32	299	299.5	307.5
8	3.05	3.07	3.27	35	392	392	401
10	6.02	6.14	6.64	40	571	571	579
12	10.8	10.9	11.95	65	1555	1570	1560
14	18.0	18.1	20.0	75	1890	1880	1895
16	28.1	28.35	30.95	85	2180	2200	2210
18	42.4	42.5	46.1	285	3910	3910	3910
					±5		

5.12.4. Impurity Dipoles — Tunnelling Effects

Another class of substitutional impurities which have been studied with an alkali halide as host are those which form an electric dipole which has a number of equivalent orientations along symmetry directions in the crystal. The orientations are separated by potential barriers through which quantum tunnelling may occur, with energy splitting corresponding to temperatures of the order of 1 K. One such dipole is OH^-, often present as an unwanted impurity in alkali halides. Others are the cyanide ion, CN^-, with the added complication of rotational and librational states; and Li^+, a small ion which sits off center from its substitutional site and forms a dipole which can tunnel between 8 equivalent directions.

For an isolated impurity there are a small number of energy levels giving rise to a Schottky contribution to the thermodynamic properties. The spacing between these levels decreases as the height of the potential barriers increase, so that the Grüneisen parameter and consequent contribution to the thermal expansion will be negative or positive depending upon whether the barriers are enhanced or reduced by pressure. However, even for low impurity concentrations electric dipole–dipole interactions can be expected to broaden the distribution of Schottky levels. Increasing concentrations of impurities in random sites gives rise to the formation of an 'orientational glass,' with an apparent linear T region in C_V and α.

Much early work on these tunnelling states was done at Cornell University, involving spectroscopic, thermal conductivity and heat capacity measurements. This arose from their interest in thermal properties of alkali halides, and was reviewed by Narayanamurti and Pohl [Nar70]. Where tunnelling states are involved, the barrier height is expected to be very sensitive to changes in strain or volume and therefore to make a much larger contribution to the thermal expansion than to C_P.

NaCl + OH⁻. When the thermal expansion of pure NaCl and NaCl crystals doped with 6 and 80 ppm of hydroxyl ion were compared at temperatures from about 2 to 30 K, the increase in α below 15 K was very obvious: there was $1/T^2$ tail, characteristic of a Schottky anomaly around 10 K. The Grüneisen parameter associated with the anomaly was large and positive — $\gamma \simeq 40$, see [Cas72] and [Bar80, p. 706] — indicating that pressure reduces the barriers between orientations.

KCl + Li⁺. Similar measurements on α for KCl containing about 100 ppm of Li^+ between 1.3 and 30 K showed the 'tail' of a Schottky bump, expressible as $\alpha \simeq 5 \times 10^{-8} T^{-2}\ \mathrm{K}^{-1}$ below 5 K [Cas72]. Thus again pressure reduces the barriers, presumably because it reduces the distances between the off-center Li sites. Later measurements of C_P and α extended down to 0.1 K, showing the fuller extent of the Schottky bump (see Fig. 5.44), which was broadened and had a value of $\gamma_{sch} \approx 160$ [Dob86a, Col87a]. Using a simplified model but taking account of dipole–dipole interaction, M. W. Klein has derived $\alpha(T)$ and $C_V(T)$ for a very low concentration of 8-oriented $\langle 111 \rangle$ tunnelling dipoles dissolved in alkali halides, obtaining qualitative agreement with the experiments on KCl+Li [Kle87].

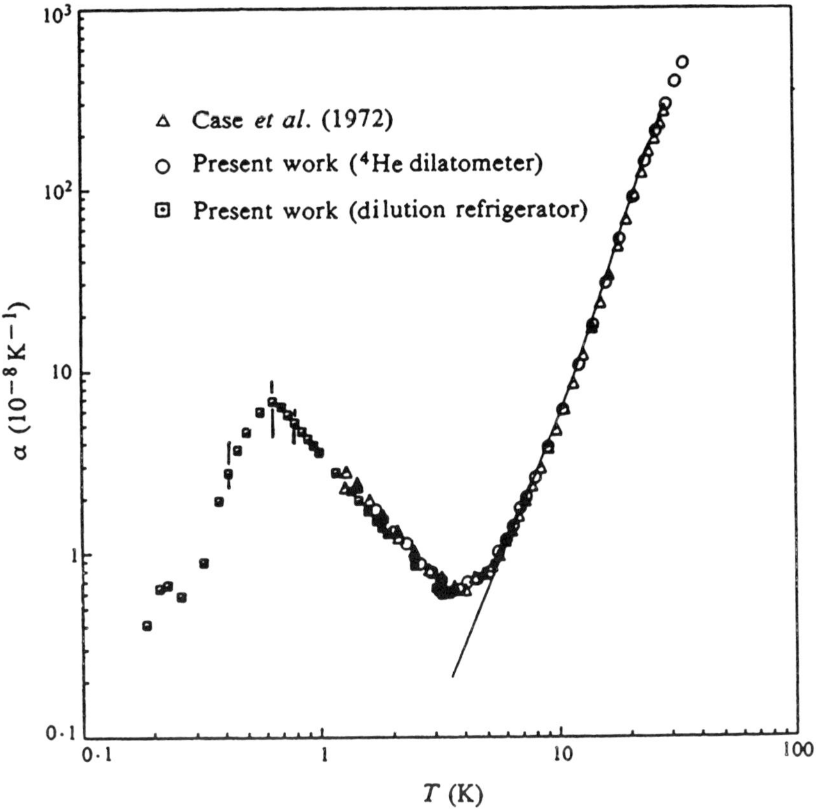

Fig. 5.44. Linear coefficient $\alpha(T)$ of KCl containing 10^{24} Li^+ ions per m^3. Solid line shows α for pure KCl [Col87a].

CN^- ions. The rod-shaped CN molecule has been studied in a number of alkali halide host lattices. In some hosts such as NaCl, the low temperature properties show complex ageing effects which are not easily interpreted. A clearer picture emerges from the measurements on KBr:CN. At low enough concentrations the isolated CN ions occupy $\langle 111 \rangle$ localized orientational tunnelling sites. Dobbs *et al.* [Dob86b] measured a sample with a 0.03% concentration of CN from 0.08 to 10 K, finding a rather broadened Schottky bump in C_P centered around 0.5 K, with α becoming negative below 0.5 K: the Grüneisen parameter for the tunnelling states, γ_{CN}, reached nearly -300 at 0.1 K, indicating strong enhancement of the barriers under pressure.

With higher concentrations, the CN ions 'freeze into an orientational glass' and display the T-term (in C_P) and T^2-term in heat conductivity characteristic of amorphous solids. CN levels up to 60 mol% have been measured [Wat89, Dob86b].

Chapter 6

Metals

6.1. INTRODUCTION

6.1.1. The Separation of Electronic and Vibrational Effects

Chapter 5 was concerned with non-metallic solids for which the bonding forces are predominantly ionic, covalent or Van der Waals. For them the thermal energy responsible for the specific heat and expansion derives from the collective modes of vibration (and rotation and libration) of the atoms. There are also magnetic exchange interactions in many crystals containing Ni, Cr, Fe, Mn or rare earth ions, and tunnelling levels in some imperfect crystals and glasses.

The electronic structure of metals is characterized by the presence of mobile 'conduction' electrons, which in general are distributed throughout the volume of the metal rather than being concentrated near ion cores or in directional bonds between atoms; they then act as a jelly or glue to bind the ions together. This kind of bonding usually results in a more closely packed type of structure than that in most insulating crystals. The usual structures found in pure metals are close-packed (cubic or hexagonal, each having 12 nearest neighbors) or body-centered cubic (8 nearest neighbors). The structures of alloys can also often be understood in terms of packing, but may then depend upon the relative ion sizes.

The conduction electrons play a major rôle in determining the binding energy, and hence also the compressibility and other elastic moduli, as well as the vibrational frequencies; but their contributions to these properties are not separable from those of the ions, any more than are those of the valence electrons in insulators; they are inextricably mixed. However, there is a crucial difference between metals and insulators: in metals there is a continuum of electron energy levels extending above the ground state. The excitation of these levels as the temperature increases gives rise to specifically electronic contributions to the entropy, heat capacity and thermal expansion coefficients. We have seen in Ch. 1 that these are easily distinguished from the vibrational contributions, at least at low enough temperatures, by their linear dependence on T, and that a most convenient separation is achieved by simply plotting data points obtained below $T \sim \Theta_D/25$ as C/T versus T^2 or α/T versus T^2 (e.g., Figs. 1.6 and 1.7). The intercept at $T = 0$ then gives the coefficient of the electronic term and the initial slope gives the coefficient of the T^3 vibrational term. Computer fitting may achieve the same result, but can be misleading unless it is done with awareness and careful assessment and weighting of the data. As with insulators, a value for Θ_0 can also be obtained from the averaging of ultrasonic wave velocities derived from elastic moduli (see Section 2.9). Denoted by Θ_0^{el}, it should agree with Θ_0^C obtained from heat capacity data within experimental error (for metallic elements see [Phi71, Table 5]). In addition a value of γ_0^{el} can be calculated from the pressure derivatives of the elastic moduli or from third order stiffnesses, for comparison with the value of γ_0^{th} obtained from the ratio of the T^3 terms in α and C.

As discussed in Section 1.3.4, the electronic term in C_V at low temperatures is a measure of the density of states at the Fermi surface, $n(\epsilon_{F0})$, while the corresponding term in the expansion coefficient gives the volume derivative of this density of states, $dn(\epsilon_{F0})/dV$. At higher temperatures the separation into lattice and electron terms is more difficult and its interpretation less certain, for various reasons. We have already seen that C_{vib} varies with T in a complicated way, mainly because of vagaries of the frequency spectrum and also because of increasing anharmonic effects as the temperature rises. Another reason is that the electronic component C_e does not generally remain linear in T as the temperature increases; even in simple metals 'phonon enhancement' of C_e due to electron–phonon coupling is reduced from its low-temperature value (see Section 6.1.2).

6.1.2. Description of Electronic Structure of Metals

In the following sections we present the salient features of the heat capacity and thermal expansion, first for cubic metals and alloys, including technical alloys used in cryogenic construction, and then for anisotropic metals. We next deal with the effect of magnetic spins, which are particularly important in determining the thermal expansion of materials such as chromium, manganese and many alloys; and we conclude with superconducting metals and heavy electron (fermion) metals. To understand qualitatively what is going on in these increasingly complex materials, and in particular the nomenclature that is applied to them, we must say a little more about the electronic structure of crystals. To go further, readers are referred to *Electron — a centenary volume* [Spr97a], which includes general reviews of many aspects of solid state theory, including the 'electron glue,' see [Gyo97].

Electronic Band Structure. In Section 1.3.4 we introduced the independent particle model. Each electron moves in an averaged potential field produced by the totality of all the nuclei and electrons. Because electrons are fermions, no two electrons can exist in the same single particle quantum state; and at $T = 0$ the N electrons of the system fill up the N single particle states of lowest energy. When the highest occupied state falls within a continuum of allowed energies, its energy is the *Fermi energy* ϵ_{F0}.

The assumption that each electron moves in the averaged potential field is called the *Hartree–Fock* approximation. When it is applied to individual atoms, the field has spherical symmetry; the single particle wave functions are called *orbitals*, which have characteristic symmetries labelled s,p,d,f — for example, s orbitals have spherical symmetry. The energies of atomic orbitals are discrete, and successive shells of orbitals (called $1s, 2s, 2p, 3s$, etc.) are occupied by the available electrons.

In a crystal the atoms are so close to each other that the outermost atomic orbitals would overlap considerably. Under these circumstances they merge to form orbitals extending throughout the solid, and the discrete energy levels of the atomic orbitals widen into 'bands.' The governing symmetry is now the periodic symmetry of the crystal structure, with the consequence that the orbitals take the form of waves. These can be labelled by their wave-vectors, denoted by $\mathbf{k}$ (in distinction from the phonon wave-vectors $\mathbf{q}$). This description of the total electronic wave-function of the crystal in terms of bands of single electron states is called the *band structure*.

For atoms of large atomic number the overlap of inner shells is negligible, and it is only for the outer shells that band-width becomes appreciable. In general, s- and p-shells overlap more than d-shells, which themselves overlap more than f-shells. Consequently, outer s- and p-bands are broad, d-bands are narrow and f-bands narrower still. The bands also may themselves overlap in energy, so that for example a narrow d-band with a rapidly varying density of electron states comes in the middle of a broad, low density s/p-band. If the Fermi-level falls in this region, the density of states $n(\epsilon_{F0})$ and other crystal properties can depend critically on the total number of available electrons (Fig. 1.10).

The Hartree–Fock approximation is drastic, in that it neglects any correlation between the motion of different electrons, despite the strong Coulomb interaction between them; and at first sight it might seem incredible that it provides a useful description of the electronic structure. However, further investigation reveals that the effective interaction between two electrons falls rapidly with distance, because of screening by the other electrons; and although there still remains appreciable correlation, theory has been developed so that band structure calculations can take this into account: an effective independent particle picture is retained, but with a 'renormalized' band structure.

Phonon Enhancement of C_e. Correlation also occurs between the motion of the electrons and that of the ions (electron–phonon interaction). This normally has a negligible effect on the vibrational contributions to thermodynamic properties (although it may cause subtle changes to phonon dispersion curves), but gives rise to an indirect attraction between pairs of electrons which affects the electronic contributions at low temperatures — spectacularly so when the attraction is strong enough to cause electron pairing and consequent superconductivity (Section 6.5). Even in the normal (non-superconducting) state, analysis shows that at low temperatures the attraction has a marked effect on the energies of effective single particle states in the immediate neighborhood of the Fermi energy, in such a way as to enhance the effective density of states that should be used in Eq. (1.19):

$$n_{eff}(\epsilon_{F0}) = n_{bs}(\epsilon_{F0})(1+\lambda) \tag{6.1}$$

where the subscript *bs* refers to the renormalized electronic band structure. Estimates of λ for different non-transition metals are tabulated in [Phi71, Table 6] and [Gri81], and vary from about 0.10 for Be and 0.15 for Na and K to unity or larger for Pb. The electronic heat capacity is proportional to $(1+\lambda)$, and the thermal expansion depends on its volume dependence as well as that of $n_{bs}(\epsilon_{F0})$. At higher temperatures ($T \approx \Theta$) this enhancement has fallen off, both because only a narrow range of states is affected and because the effect on them is itself temperature dependent.

Magnetic and Other Complex Materials. Band theory has proved very fruitful in the interpretation of metallic properties and their calculation, especially for the simpler metals. However, electronic structure becomes increasingly complex for heavier metals, especially when *d*- and *f*-electrons participate. Some of these materials are discussed later in this chapter.

6.1.3. References for Data

Much of the data on metals and alloys which we use to illustrate the sections of this chapter are from the following compilations:

- Heat capacity from volume 4 of *Thermophysical Properties of Matter* [Tou70a], Hultgren *et al.* [Hul73], Corruccini and Gniewek [Cor60], *American Institute of Physics Handbook* [Fur72] and review by Phillips [Phi71].
- Thermal expansion data from volume 12 of *Thermophysical Properties of Matter* [Tou75], Corruccini and Gniewek [Cor61], *American Institute of Physics Handbook* [Kir72], Clark [Cla68], and Barron *et al.* [Bar80].
- Elastic Moduli from volumes of *Landolt–Börnstein* [Hea66, Hea69, Hea79, Hea84] and Simmons and Wang [Sim71].
- Other more recent sources which are referenced explicitly.

6.2. CUBIC METALS

6.2.1. General

We may get an idea of the variation of behavior of non-magnetic materials from Table 6.1, which takes a slice across the Periodic Table listing the cubic metals beginning with Rb, Sr and ending with Rh, Pd, Ag. This row includes the 4d elements (see Periodic Table) and has been preferred because many of the elements in the 3d row are magnetic. Table 6.1 lists their structure (bcc or fcc), atomic volume

V, adiabatic bulk modulus B_S (293 K), Θ_0^C, C_P (293 K), α (293 K), Γ (electronic heat capacity coefficient), and the Grüneisen parameters γ_e, γ_0 and γ_{293}. The first obvious difference is the larger molar volume of Rb and Sr compared with the transition metals. This is due largely to weaker binding by the electron gas, which is also reflected in the low values of bulk modulus and Debye temperature.

Heat capacity values at room temperature are not too dissimilar because the majority of the lattice vibrations are excited at room temperature which for all these elements is comparable with or greater than the Debye temperature. At lower temperatures C_V (and C_P) decrease with T roughly in accord with the Debye function, departures being illustrated in Fig. 6.1 by the variations of Θ^C with temperature. The shallow minimum in $\Theta^C(T)$ indicates that in this vicinity C_V is larger than the Debye model would predict. Generally the variations in $\Theta^C(T)$ are smaller for the cubic metals than for highly anisotropic metals or for the non-metals of more open structure. At very low temperature, the effect of the $(T/\Theta)^3$ dependence amplifies the effect of differences in Θ: for example at 10 K, $C_P \approx 7$ J·mol^{-1}·K^{-1} for Rb and 0.2 J·mol^{-1}·K^{-1} for Nb. Above the Debye temperature C_V usually exceeds the Dulong and Petit value, because of electronic contributions and possible anharmonic effects.

The differences in expansion coefficient at 293 K (Table 6.1) arise from the interplay between thermal pressure and bulk modulus (or compressibility):

$$\beta = \gamma \chi_T C_V / V = \chi_T (\partial P/\partial T)_V$$

For example, the molar volume of Rb is larger than that of Nb by a factor of 5, so that the heat capacity per unit volume is about 5 times smaller for Rb than for Nb. But this difference is far outweighed by the difference in compressibility—Rb being about 70 times more compliant than Nb. Figure 6.2 shows the temperature variation of α for these cubic metals, except that K replaces Rb because data are lacking for Rb below room temperature. At lower temperatures the variation of α (or β) with T is roughly similar to that of $C_P(T)$, departing from the Debye function partly due to the nature of the frequency spectrum with its transverse and longitudinal branches, but also due to variations in γ for the different parts of the spectrum. Table 6.1 shows that the low temperature average of mode gammas, γ_0, is usually smaller than the average at ambient temperature, γ_{293}. Also at temperatures of 10 K or less, the electronic contributions to α (and to C_V) are noticeable (e.g., Figs. 1.6 and 1.7), particularly for the transition elements.

6.2.2. Group 1A—Alkali Metals

For these body-centered cubic metals, Phillips's review [Phi71] gives the following values of Θ_0, noting that Li and Na can transform martensitically from bcc to hcp: Li ... 355 K (hcp) and 335 K (bcc); Na ... 159 K (hcp) and 153 K (bcc); K ... 91 K; Rb ... 56 K; Cs ... 40 K. The progressive decrease in Θ_0 from Li through to Cs reflects a decrease in vibrational frequencies due mainly but not entirely to the

Table 6.1. Data for a selection of cubic metals

Property	Rb	Sr	Nb	Mo	Rh	Pd	Ag
structure	bcc	fcc	bcc	bcc	fcc	fcc	fcc
V^{293}(cm^3/mol)	55.9	33.7	10.8	9.40	8.30	8.87	10.28
B_S^{293}(GPa)	2.3	11.8	168	260	270	189	104
Θ_0(K)	56	147	277	470	500	274	226
C_P^{293}(J/mol·K)	30.5	26.8	24.6	23.9	24.6	25.8	28.5
C_V^{293}(J/mol·K)	28.0	26.4	24.3	23.7	24.2	25.2	27.4
$\alpha^{293}(10^{-6}$/K)	90	22.6	7.1	5.1	8.4	11.7	19.0
Γ(mJ/mol·K^2)	2.5	3.6	7.8	1.84	4.7	9.4	0.65
γ_e	—	−4.4	1.5	1.1	2.8	2.2	1.1
γ_0	—	0.44	0.9	1.3	1.9	2.2	2.2
γ_{293}	1.1	1.1	1.6	1.6	2.3	2.4	2.38

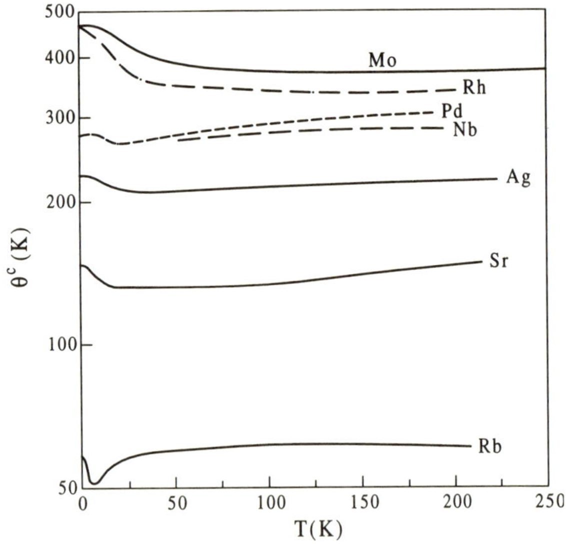

Fig. 6.1. Variation of Θ^C with temperature for Rb, Sr, Nb, Mo, Rh, Pd and Ag.

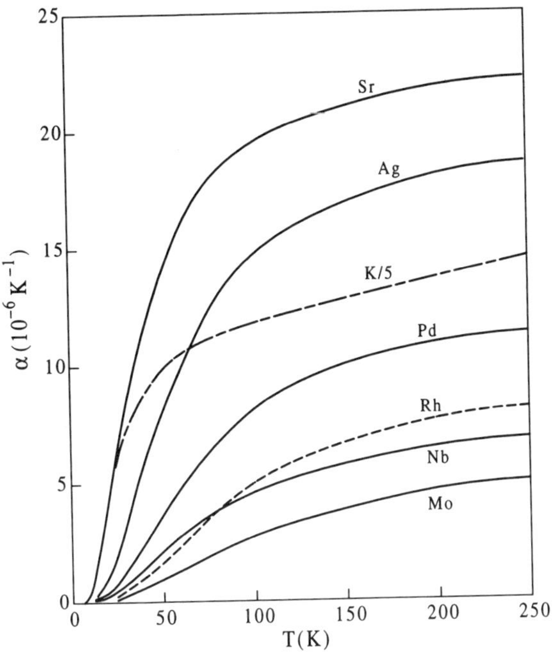

Fig. 6.2. Variation of α with temperature for K (note different scale), Sr, Nb, Mo, Rh, Pd and Ag.

increase in atomic mass. Experimental values of C_P (and enthalpy data) at high and low temperatures are tabulated in an extensive review [Alc94].

Experimental data for thermal expansion at low temperatures are confined to potassium. There are equation of state PVT measurements for other members of the alkali family which give values of the isothermal bulk modulus B_T [And83b, And85]. The γ values at room temperature lie between 0.9 and 1.2 [Bar80].

6.2.3. Group 1B—Noble Metals; also Al, Pb

The fcc metals Cu, Ag, Au and Al have been more thoroughly studied than most, particularly Cu as it is both a common cryogenic material and a 'standard' reference material. The critical analysis of the C_P data by Furukawa [Fur68] and a later review [Phi71] both give similar limiting values of Θ_0, *viz.* 344.5 K for Cu, 226.0 K for Ag and 162.3 K for Au. The Furukawa analysis tabulates 'best' values for C_P and enthalpy from 0 to 300 K, and a IUPAC publication [Mar87a] includes polynomial expressions for C_P for Cu from 1 to 300 K. For Al and Pb Phillips gives Θ_0 as 430 K and 105.0 K respectively. These latter two elements are superconducting below 1.1795 K (Al) and 7.1999 K (Pb); this has no measurable effect on the lattice component, but gives rise to a λ-shaped anomaly in C_e at T_c, below which C_{es} falls exponentially (see Section 6.5).

The elastic moduli lead to values of B_S at 293 K and at 4 K respectively of 137 and 142 GPa (Cu), 103.5 and 109 GPa (Ag), 173 and 180 GPa (Au), 76.1 and 79.4 GPa (Al) and 44.9 and 48.8 GPa (Pb).

For all these metals the temperature dependence of α is qualitatively similar to that of C_P, as shown for Ag in Fig. 6.2. The Grüneisen function falls a little for Ag and Cu at temperatures below 50 K (Fig. 6.3) [Whi72b, Bar80]; but for practical engineering purposes of calculating dimensional changes this could be ignored, and heat capacity data or the Debye function used as a good guide to $\alpha(T)$. Aluminium is a little unusual in that γ_0 (lattice) =2.6 is larger than the room temperature value of 2.16, because the mode gammas are larger for many low-lying transverse acoustic modes than for the longitudinal modes [Col73, Bar80]. For Al and Pb in the superconducting state the electronic component α_e falls more rapidly with temperature than in the normal state, the difference, $\alpha_{en} - \alpha_{es}$, being of basic interest as it is related thermodynamically to the strain dependence of T_c and H_c (Section 6.5).

An important factor in cryogenic design is whether mechanical strain or traces of chemical impurity have any significant effect on the thermal or other properties of a material. Copper is one element for which there is a considerable body of thermal data for samples of varying degrees of physical and chemical purity. Measurements of heat capacity and expansion both indicate that these properties which depend principally on the lattice vibrational energy are not significantly affected by trace impurities or strain effects. For example, measurements of C_P from 20 to 300 K for commercial and 99.999% pure samples, cold-worked and annealed, showed that differences are much less than 1% [Mar60]. Similarly thermal expansion measurements from 273 to 293 K and from 2 to 100 K on various coppers (6N, OFHC, electrolytic tough-pitch, phosphorus-deoxidized, free-machining tellurium copper and deformed rod) give differences of less than 1% [Whi72a].

There have also been thorough intercomparisons of thermal expansion between high purity Cu and various oxygen-free coppers which have been used in precision gas thermometry. No measurable differences were found. Only dissolved magnetic impurities (particularly Mn and Fe) have *relatively* large effects on both C_P and β at low temperatures (see [Bar80, p. 700] and [Whi72a]). Impurity levels of $\sim 0.1\%$ of Fe or Mn may produce changes of $\sim 10^{-8}$ K^{-1} in α of Cu at 10 K. This is equivalent to more than a 20% change at this low temperature (see also Section 6.4.1).

6.2.4. Group II—Alkaline Earths

Best estimates of Θ_0 for the alkaline earths are [Phi71]:

Ca (fcc) ... 230 K; Sr (fcc) ... 147 K and Ba (bcc) ... 111 K.

These agree well with the compilation by Alcock *et al.* [Alc93] except for Ca, for which $\Theta_0 = 250 \pm 5$ K based on more recent data than that available to Phillips.

The few measurements of thermal expansion which extend to liquid helium temperatures were made on rather impure specimens ([Bar80, p. 685]). The most

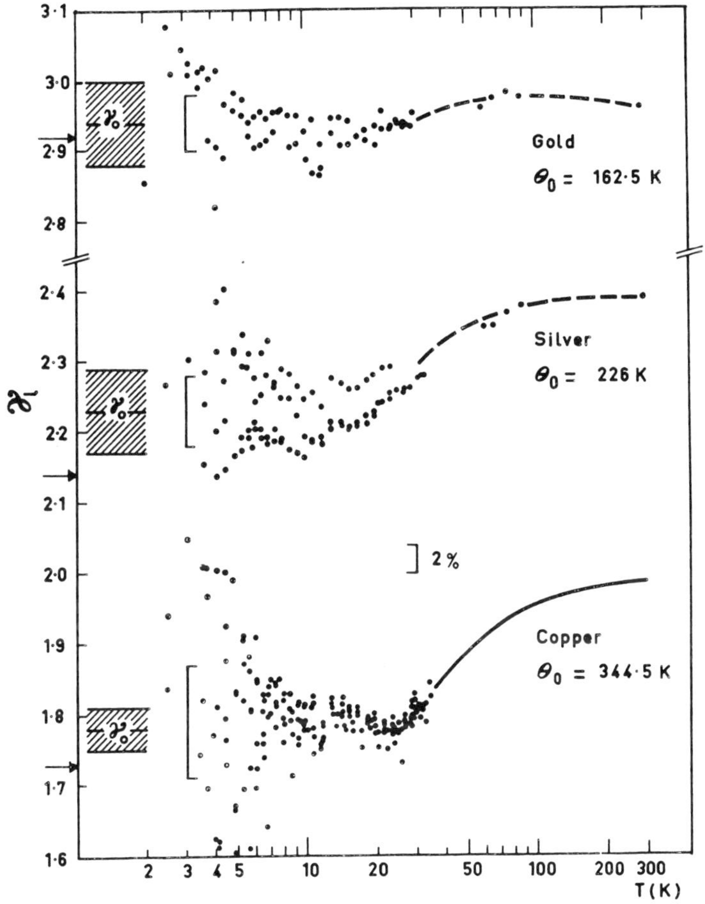

Fig. 6.3. Variation of the lattice Grüneisen function γ_{vib} with temperature. Arrows show the best estimates of γ_0^{el}; shaded areas show γ_0^{th} and error bars at 3 K represent $\Delta\alpha = \pm 10^{-10}\mathrm{K}^{-1}$ [Whi72b, Bar80].

interesting aspect of the results are the values for the electronic contribution, which is near zero for Ca and negative for Sr (see $\gamma_e = -4.4$ in table 6.1). These appear to arise from the overlapping electron energy bands causing the density of states to *increase* under compression.

Bulk modulus values are available from the experimental equations of state measured at pressures up to 2 GPa (20 kbar) and temperatures from 4 to 295 K [And90a]. Room temperature γ values are about 1.1 for Ca and Sr and 0.9 for Ba.

6.2.5. Transition Metals

These are an interesting family, with structures delicately balanced between close-packing (fcc or hcp) and body-centered cubic. Molar volumes generally lie in the narrow range of 8 to 11 cm^3 for the 4d and 5d rows and 6 to 10 cm^3 for the 3d elements. The additional bonding due to the d-electrons shows itself in the variation of bulk modulus, as for example in the 4d elements, Nb to Ag, listed in Table 6.1. Consider also a column in the Periodic Table, such as VIB: B_S =157 GPa for V, 168 GPa for Nb and 190 GPa for Ta. For these three elements Θ_0 decreases from 400 K (V) to 277 K (Nb) and 258 K (Ta). The decrease in characteristic temperature (and vibrational frequencies) is less than that expected from the increase in the atomic mass, owing to increase in elastic stiffness.

The electronic contributions to the low temperature heat capacity vary as we move along a row of the Periodic Table. This variation is revealed clearly for the 4d elements by a plot of the electronic coefficient Γ as function of electron–atom ratio (Fig. 6.4). For the 3d and 5d elements the pattern is similar, but magnetic ordering in the 3d elements Cr, Mn, Fe, Co and Ni adds a little more irregularity to Γ as a function of electron–atom ratio.

As a practical guide, α (and β) varies with temperature for the cubic elements roughly like the Debye function, because Θ_D and γ are fairly constant down to $T \sim \Theta/5$. Below this γ_{vib} decreases for most elements (see γ_0 in Table 6.1). An interesting feature of the electronic component of expansion is that $\gamma_e \sim 2$ for most of the transition elements, which is much larger than the free electron value of 2/3 but comparable with the value of 5/3 predicted by the Heine–Ehrenreich model of an electron density of states controlled by a narrow d-band [Shi74, Fle79].

6.2.6. Cubic Alloys (Non-Magnetic)

Dilute Alloys and Local Modes. At low temperatures, heavy solutes of much greater mass than the solvent atoms can significantly affect the low-frequency end of the lattice spectrum, producing local mode effects akin to those discussed in Section 5.12.3 for Ag in NaCl. For metals the effect on C and α was clearly shown by measurements at Kharkov (see [Pop75, Pop76] and [Bar80, p. 698]) using Al or Mg as solvents and 1% or less of Pb or Cd as additives. These produced broad peaks in C and α centered around 40–50 K. The *relative* increases, $\Delta C/C$ and $\Delta\alpha/\alpha$, reached maximum values at 10–15 K; e.g.,

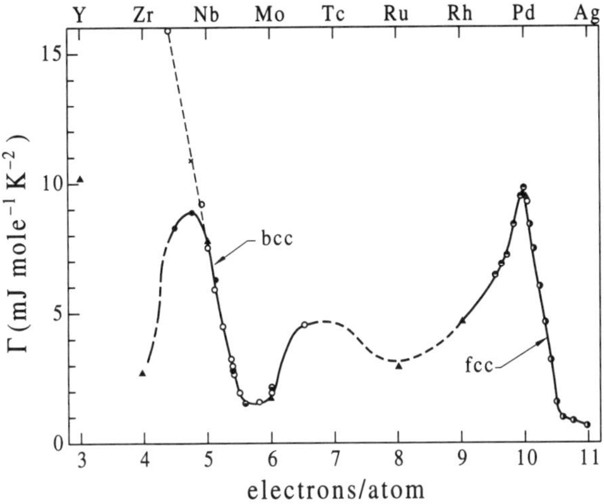

Fig. 6.4. Variation of electronic heat capacity coefficient Γ with number of valence electrons per atom in 4d elements and alloys [Hei66].

AlPb, mass ratio of 8.0, $\Delta\alpha/\alpha \simeq 90\%$ per at%

MgCd, mass ratio of 4.5, $\Delta\alpha/\alpha \simeq 30\%$ per at%.

The relative increases were usually similar in magnitude for C and α, indicating that the Grüneisen γ for the local modes was similar to that for the solvent.

Light Impurities. Solutes or impurities of relatively small mass have little effect at low temperatures, but may contribute a band of optic modes at higher temperatures. PdH is an example of a system (not necessarily dilute) where significant amounts of H can be absorbed into octahedral interstitial sites in the fcc Pd lattice, increasing α at ambient temperatures through optic modes which have been shown by neutron scattering to be centered near 650 K, see [Bar80, p. 699].

Alloys. For alloys in general, mass differences will have more direct effect on the phonon spectrum $f(\omega)$ than on the electron density of states, whereas differences in the electron–atom ratio will affect $n(\epsilon_F)$ (see Fig. 6.4) more than the lattice vibrations. In this section we restrict discussion to 'simple' non-magnetic disordered binary alloys of the same structure as the constituents.

One example for which we have elastic and heat capacity data are the α–brasses [Ray59, Vea63]. Ultrasonic measurements from 4.2 to 300 K on crystals containing Cu + 0, 4, 9, 17, and 23% Zn showed that c_{11} and c' fell by about 10% with addition of 23% Zn, while c_{44} fell by rather less. The calculated values of Θ_0^{el} fall from 345 to about 325 K over this range of concentration. Measurements of C_P below 4.2 K were made on samples containing up to 33% Zn, giving values of Θ_0^C and Γ shown in Figs. 6.5 and 6.6.

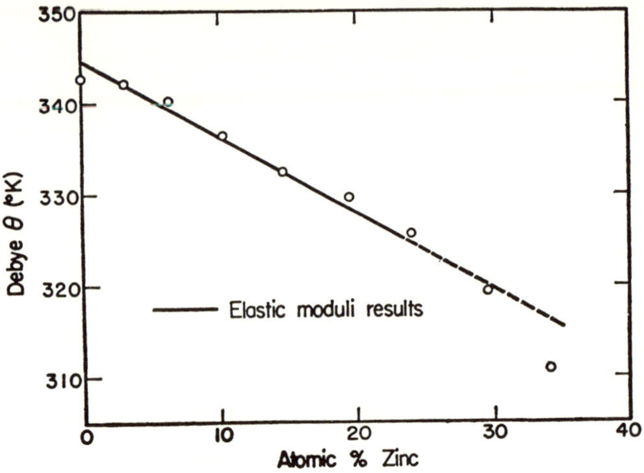

Fig. 6.5. Measured variation of Θ_0^C for CuZn alloys compared with Θ_0^{el} [Vea63].

The addition of Zn to Cu increases the lattice spacing by about 0.07% per atomic percent, so that 10% Zn would increase molar volume by about 2%. We might expect this expansion to produce a decrease in Θ_0 of about 3.5% or 12 K per 10% Zn, based on the Grüneisen parameter for Cu,

$$\gamma_0 = -d\ln\Theta_0/d\ln V = 1.7$$

The calculation ignores the effect of mass difference (small between Cu and Zn) and details of changes in local geometry and bonding. The rate of change agrees roughly with the observed changes of Θ_0^C and Θ^{el} (Fig. 6.5).

We may attempt the same sort of correlation between Γ and volume change, assuming that changes in Γ are controlled by (Section 1.3.4)

$$\gamma_e = d\ln n(\epsilon_F)/d\ln V = d\ln\Gamma/d\ln V \tag{6.2}$$

where $\gamma_e \approx 0.9$ for Cu. The increase in Γ with addition of Zn correlates roughly with that expected from volume change (Fig. 6.6), but this ignores the change in the number of electrons, which on the rigid band model would lead to a *decrease* in Γ. To account for this discrepancy, modifications to the rigid band model are discussed in [Vea63].

A similar picture comes from the heat capacity data on Ag-rich alloys with added Cd, In, Sn and Sb [Mon67, Phi71]: a decrease in Θ_0^C with added solute correlates roughly with increase in electron–atom ratio and volume, while Γ increases slightly with these parameters. This increase, again at variance with the rigid band model, is discussed by Phillips in [Phi71, p. 498].

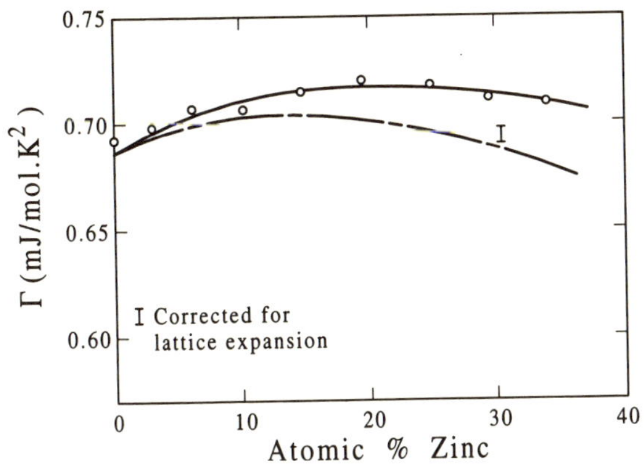

Fig. 6.6. Measured variation of Γ for CuZn alloys [Vea63].

Montgomery *et al.* [Mon67] also measured C_P for the Ag/Pd system, finding that small additions of Ag to Pd reduced Γ markedly as the d-band was filled but had little effect on Θ_0.

We expect that the thermal expansion of these binary alloys would reflect the same sort of changes in lattice spectrum and electronic density of states as the heat capacity, with the added effect of change in the compressibility modulus. There are relatively few expansion data over a wide range of temperature and composition. Measurements of a range of Pd–Ag alloys from 30 K to 270 K [Bai69] show that the expansion coefficient varies monotonically with alloy content (Fig. 6.7) roughly as predicted by the 'mixture' model of Turner,

$$\alpha = \sum_i \alpha_i B_i V_i \Big/ \sum_i B_i V_i \tag{6.3}$$

where V_i is the volume occupied by the ith component and α_i and B_i are the corresponding expansion coefficient and bulk modulus. The measurements did not extend to low enough temperature to see how γ_0 and γ_e varied with composition.

6.2.7. Technical Materials

When a cryogenic engineer is faced with designing a storage vessel or a motor or a magnet, two important considerations are (i) the total heat content of material that is to be cooled and (ii) the stresses that may be caused by thermal contraction. For this purpose the input data may only need to be reliable to within a few percent, and can be obtained from tables such as C.1 and C.2 in Appendix C. If such data are not available, then estimates may often be made based on the room temperature values and the Debye approximation, particularly if the Debye Θ near room temperature is known from the heat capacity. The heat capacity at room temperature for most

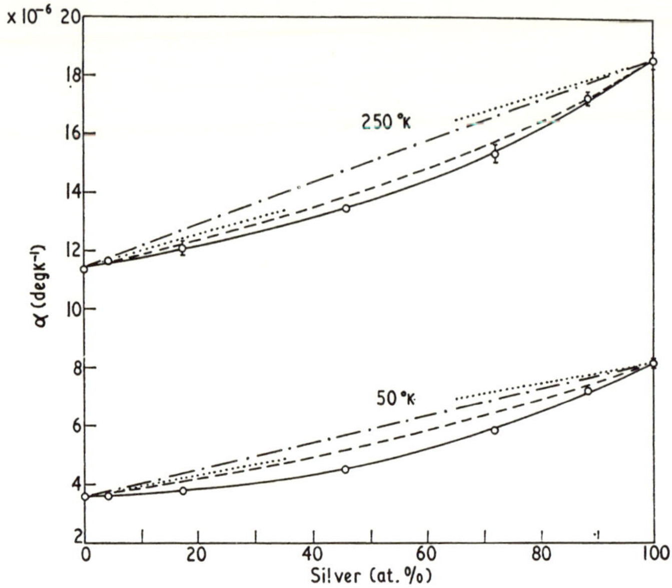

Fig. 6.7. Experimental values (o) for α of Pd-Ag alloys compared with theoretical predictions of 'mixture' models, including that of Turner (—) [Bai69].

metals will approach the Dulong and Petit value of ~ 25 J.(g-at·K)$^{-1}$ and fall with temperature roughly as the Debye function, so that for an unlisted alloy the energy change between 293 K and 80 or 4.2 K can be estimated within say 10% with the help of Debye Tables (Table C.5 in Appendix C).

As an example, we assume the validity of the Debye approximation for the change in internal energy U per gram atom between 300 and 4 K and calculate the ratio $\Delta U/C_V = (U_{300} - U_4)/C_V^{300}$ for solids of different Debye temperatures (600 K, 500 K, 450 K, etc.) From the Debye tables we obtain the following values:

Θ_D (K)	$\Theta_D/300$	ΔU_{300-4} (J/g-at)	C_V^{300} (J/g-at · K)	$\Delta U/C_V^{300}$ (K)
600	2	3300	20.6	160
500	1.67	3810	21.8	175
450	1.5	4103	22.4	183
400	1.33	4400	22.8	192
300	1	5043	23.7	212
200	0.667	5766	24.4	236

Most structural metals have Θ_D values in the range from 300 to 450 K, so that if they behave in a Debye-like fashion, the change in U from 300 to 4 K is from 180 to 210 times the value of C_V at room temperature. The change in total heat or enthalpy,

H, cannot be estimated from the Debye function, but it does not differ from U by more than a few percent below room temperature.

This approximation for $\Delta U/C_V$ is also useful for estimating the length change in metals, provided that Debye and Grüneisen formulae are a good guide for the temperature ranges involved. Estimates of the overall length change in cooling would appear to be difficult because the room temperature values of α, unlike C_V, vary considerably from one element or alloy to another. However, for most solids the temperature dependence $\alpha(T)$ is roughly like $C_P(T)$ or C_V down to 20 or 30 K; and so if a measurement of α can be made or is available at room temperature, a fair estimate of the length change can be made based on the same functional relation as the change in heat capacity. In particular, if the parameters Θ and γ are fairly constant then the overall change $\Delta l/l$ will be roughly proportional to the value of α at room temperature. Clark [Cla68, Cla83] examined this correlation for a number of technical metals and showed that the ratio $[(l_{293} - l_4)/l_{293}]/\alpha_{293}$ lies between 170 and 200. The metals include commercial Al alloys, Hastelloys, Inconels, phosphor bronze and Fe–Ni alloys (excluding Invar composition). Collings (see [Col86a, p. 366]) tabulated a similar ratio for 17 metallic elements over the intervals 293 to 20 K and 293 to 200 K, finding respective mean values of 193±16 and 89±3. He excluded data for Cd (anisotropic), Pb (low Θ), and Cr (magnetic), which gave ratios outside these limits.

Two materials for which there are very reliable expansion data are Cu, for which measurements give

$$(\Delta l_{293-4}/l)/\alpha_{293} = 3257 \times 10^{-6}/16.65 \times 10^{-6} = 196$$

and W, for which the measured ratio is 198.

Figure 6.8 illustrates the fractional length change from 293 K to 0 K for several metallic elements (including polycrystalline Ti and Be) and a common alloy, a $Cu_{65}Zn_{35}$ yellow brass. Although the values are very different, the ratios of length change to the room temperature α lie generally between 180 and 200. Obvious exceptions are Be ($\Theta_D \approx 900$ K, ratio about 117) and Pb ($\Theta_D \approx 90$ K, ratio 244). Such exceptions occur for unusually high and low values of Θ_D because of the nature of the Debye function, even if Θ^C and γ are constant. There may be further discrepancies for some magnetic materials, and for non-metallic solids for which γ changes considerably with temperature, e.g., Si.

6.3. NON-CUBIC METALS

6.3.1. Introduction

In Ch. 2 we introduced the anisotropic nature of the lattice spectrum $\omega(\mathbf{q},s)$, as revealed by inelastic neutron scattering and by the elastic moduli and the ultrasonic wave velocities which vary with direction even in cubic crystals (unless $c' = c_{44}$).

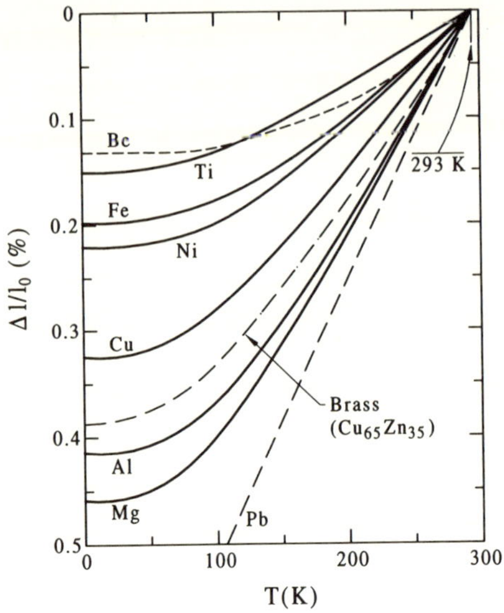

Fig. 6.8. Linear thermal dilation vs. temperature for some common metals [Cla83].

Fortunately the thermal and transport properties of cubic crystals depend on averages over the vibrational spectrum and so do not show this underlying elastic anisotropy. Non-cubic crystals show anisotropy in directional properties such as thermal expansion (and conductivity). The effect is small in some hexagonal close-packed metals which have nearly ideal 'billiard ball' packing, e.g., Mg (hcp) where the *c/a* ratio is 1.623, very close to the ideal ratio of 1.633 (Fig. 6.10). By contrast, in layer structure crystals (e.g., As) or chain-like crystals (e.g., Te), the anisotropy in thermal expansion is very large and usually accentuated at low temperatures.

The heat capacity is a scalar quantity and so cannot show the effects of anisotropy directly, although the departures of C_V from the Debye model at low temperatures may be relatively greater for the more anisotropic metals. For example, compare $\Theta^C(T)$ for Zn and Cd with the more isotropic Ti (see Fig. 6.9). The relatively larger decrease in Θ^C from the limiting value Θ_0 to the minimum (near $T \simeq \Theta_0/15$) is clear, and comes partly from the existence of a lower frequency transverse acoustic branch of the vibrational spectrum. However, from an engineering viewpoint it is enough to know that Θ^C is fairly constant for these metals above about 30 or 40 K, allowing the Debye formula to be used to calculate changes in energy in cooling from room temperature as well for Zn as for Ti or Cu.

Most of the non-cubic metallic elements have crystal structures which are axially symmetric, so that the linear expansion can be expressed by two components, $\alpha_{\|}$ and $\alpha_{\perp}$, respectively parallel and normal to the axis (tetragonal, hexagonal or trigonal). An exception is gallium, which is orthorhombic with three principal axes and nine independent elastic moduli. The two principal coefficients for axial crystals are

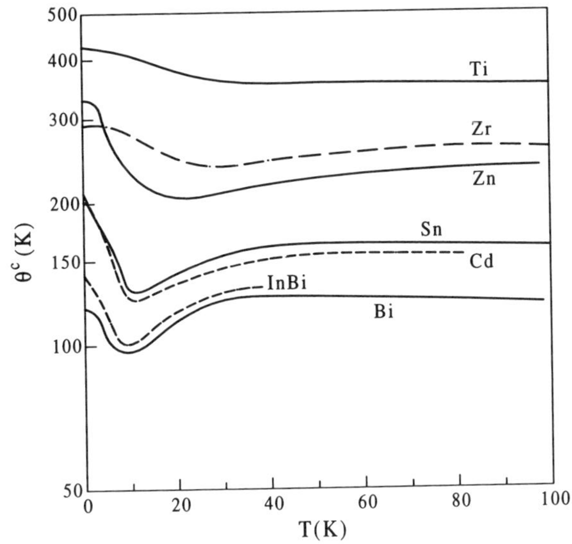

Fig. 6.9. Variation of Debye $\Theta^C(T)$ for some anisotropic metals.

related to the adiabatic values of the elastic compliances, $s^S_{\lambda\mu}$, and the heat capacity at constant stress, C_σ, by the relations (see Section 2.8.4)

$$\alpha_\perp = (C_\sigma/V)[(s^S_{11} + s^S_{12})\gamma_\perp + s^S_{13}\gamma_\parallel] \tag{6.4}$$

$$\alpha_\parallel = (C_\sigma/V)[2s^S_{13}\gamma_\perp + s^S_{33}\gamma_\parallel] \tag{6.5}$$

The cross compliance s^S_{13} is negative for these metals, and the magnitude and sign of the principal coefficients is controlled by the relative magnitude of the compliances as well as by the Grüneisen parameters.

Munn [Mun72] showed that Eqs. (6.4) and (6.5) can be expressed also in terms of the cross-compliance s_{13} and the linear compressibilities χ_λ defined in Section 2.8.3:

$$\alpha_\perp = (C_\sigma/V)[\chi^S_\perp\gamma_\perp + s^S_{13}(\gamma_\parallel - \gamma_\perp)] \tag{6.6}$$

$$\alpha_\parallel = (C_\sigma/V)[\chi^S_\parallel\gamma_\parallel - 2s^S_{13}(\gamma_\parallel - \gamma_\perp)] \tag{6.7}$$

Similar equations may be written for orthorhombic crystals:

$$\alpha_1 = (C_\sigma/V)[\chi^S_1\gamma_1 + s^S_{12}(\gamma_2 - \gamma_1) + s^S_{13}(\gamma_3 - \gamma_1)], \text{ etc.} \tag{6.8}$$

Such equations show more clearly the differing dominant factors which control the α_λ. For example, compare two axial crystals, As (double layered structure — see Section 6.3.4) and graphite (simple layers — see Section 5.8.2). Both are highly anisotropic, with $\alpha_\perp$ negative at low temperatures, but with different anisotropy in their elasticity and Grüneisen functions. Along the c axis both are highly compressible ($\chi_\parallel$ large), while perpendicular to the axis $\chi_\perp$ is small and negative for As and very small and positive for graphite; the cross-compliance s_{13} is very large and negative for As, very small and negative for graphite. Consequently, although both crystals have $\alpha_\perp$ negative at low temperatures, it is for different reasons.

For As, $\gamma_\parallel$ and $\gamma_\perp$ are positive and almost equal at very low temperatures, with a value of about 4; despite the large s_{13}, $s_{13}(\gamma_\parallel - \gamma_\perp)$ is negligible, and the negative expansion is due to the negative value of $\chi_\perp$. At room temperature, however, a small difference $(\gamma_\parallel - \gamma_\perp) \simeq -0.08$ is sufficient to drive $\alpha_\perp$ positive.

In contrast, for graphite the gammas differ in sign: at room temperature $\gamma_\perp \simeq -1$, decreasing to -5 at 20 K, whereas $\gamma_\parallel \simeq 0.4$ near room temperature, increasing to +4 at 20 K. The negative $\alpha_\perp$ at low temperatures is thus due primarily to the negative $\gamma_\perp$, with the two terms in Eq. (6.4) roughly equal.

The reader who wishes to get a better feeling for the lattice dynamics of axially anisotropic crystals could study the rhombohedral lattice model [Bar74a]. This quasiharmonic model employs Mie–Lennard-Jones 6–12 potentials between near neighbors, and explores the effect on the Grüneisen functions and the expansion coefficients of (i) changing the relative bonding within a layer and between layers, (ii) changing the axial ratio. The changes in bonding are seen to change the principal Grüneisen functions qualitatively in the fashion observed experimentally, while changes in axial ratio have more effect on the expansion coefficients. Details are not appropriate in this monograph, but the conclusions are important in understanding anisotropic behavior: quoting from [Bar74a] " ... The forces within planes normal to the symmetry axis are affected strongly by strain $\eta_\perp$ but only weakly by strain $\eta_\parallel$. Forces between the planes are affected more strongly by $\eta_\parallel$ ... ." See also the discussions on mechanisms in Section 2.6.3 and on highly anisotropic non-metals in Section 5.8.

Table 6.2 lists the more common anisotropic metals, giving values for the *c/a* ratio at room temperature, Θ_0^C, electronic heat capacity coefficient Γ [Phi71], and principal linear expansion coefficients at 293 K [Kir72]. It also includes approximate values for Θ_{293}^C based on heat capacity data [Fur72, Tou70a, Whi79] and phonon spectra [Sch81]. The latter should be appropriate when calculating *approximate* values for the heat capacity or internal energy at temperatures from about 50 K to 293 K. They will also be useful in calculating *average* length changes in polycrystalline samples of anisotropic solids, provided that there is not any significant texture (preferred orientation). Clearly single crystal values are more difficult to predict.

6.3.2. Group II—Be, Mg, Zn and Cd

The anisotropy of these hexagonal close-packed metals increases as the *c/a* ratio departs from its ideal 'sphere-packing' value of $(8/3)^{1/2} = 1.633$. For Be, *c/a* =1.58, corresponding to a 'compression' along the *c* (hexagonal symmetry) axis. Mg is almost ideal with a ratio of 1.623, while Zn and Cd are 'stretched' along the symmetry axis by about 15% with *c/a* values of 1.86 (Zn) and 1.89 (Cd) at room temperature.

Table 6.2. Data for some non-cubic metals. Ga is orthorhombic with three principal axes—*a,b,c*. For data sources see [Phi71, Bar80, Kir72] and text (Section 6.1.3)

Element	Structure	c/a	Θ_0^C (K)	Γ $\left(\frac{\mathrm{mJ}}{\mathrm{mol\cdot K^2}}\right)$	Θ_{293}^C (K)	$\alpha_{\parallel,293}$ $\left(\frac{10^{-6}}{\mathrm{K}}\right)$	$\alpha_{\perp,293}$ $\left(\frac{10^{-6}}{\mathrm{K}}\right)$	γ_{293}
Be	hcp	1.57	1470	0.17	920	8.9	12.3	1.0
Mg	hcp	1.623	405	1.23	330	27	25	1.55
Zn	hcp	1.86	327	0.653	240	64	13	2.0
Cd	hcp	1.89	209	0.687	175	54	20	2.2
Ti	hcp	1.59	429	3.34	360	~ 10.5	~ 9	1.2
Zr	hcp	1.59	291	2.80	250	6.5	5.5	0.9
Co	hcp	1.62	460	4.5	380	14.6	11.0	2.1
Ga	orh	—	322	0.598	240	31^c	16.6^a 11.5^b	1.45
Tl	hcp	1.599	78.5	1.47	94	72	9(?)	1.2
In	tet	1.076	110	1.63	110	−10	53	2.35
Sn	tet	0.546	198	1.77	160	32.5	16.5	2.2
As	rhl	—	282	0.192	290	41	1.2	1.3
Sb	rhl	—	210	0.110	210	16.2	8.4	0.95
Bi	rhl	—	120.4	0.008	120	16.2	11.7	1.05

The stretching along the c–axis for Zn and Cd is associated with an increase in the linear compressibility $\chi_{\parallel}$ which is 6 or 7 times larger than $\chi_{\perp}$, with a corresponding increase in $\alpha_{\parallel}$ (see Table 6.2). At low temperatures the differences in α increase: $\alpha_{\perp}$ is negative below 50 K (Fig. 6.10) while $\alpha_{\parallel}$ is large and positive. The volume coefficient and average linear coefficient (for a polycrystalline sample) remain positive and vary with temperature roughly as does the heat capacity. The bulk Grüneisen parameter, γ, has a broad maximum (2.9 for Zn) at 20 to 30 K and values of 2.0 at room temperature and at liquid helium temperature.

The interplay between the electronic and lattice contributions to α at liquid helium temperatures is complex, and discussed in [Bar80, Mun72].

6.3.3. Anisotropic Transition Metals

The more studied of these metals are included in Table 6.2. For others, single crystal data are limited, particularly for the expansion at low temperatures. Phillips [Phi71] gives the following values from heat capacity measurements:

- Hf ($\Theta_0 \approx 252$ K, $\Gamma = 2.15$ mJ$\cdot$mol^{-1}K^{-2})
- Re ($\Theta_0 \approx 415$ K, $\Gamma = 2.26$ mJ$\cdot$mol^{-1}K^{-2})
- Ru ($\Theta_0 \approx 555$ K, $\Gamma = 3.00$ mJ$\cdot$mol^{-1}K^{-2})
- Os ($\Theta_0 \approx 500$ K, $\Gamma = 2.3$ mJ$\cdot$mol^{-1}K^{-2})

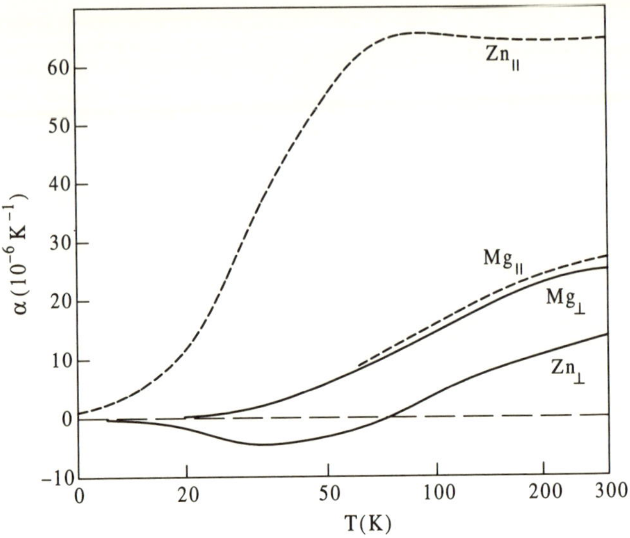

Fig. 6.10. Anisotropy of linear expansion coefficient of Zn compared with Mg; for data sources see [Bar80, p. 686].

Thermal expansion data at low temperatures are restricted to polycrystalline samples of Ti, Zr, Co and Re and may not be representative of the true average due to preferred orientation, see [Bar80, p. 693–5]. Values of α for the principal axes near room temperature can generally be obtained from X-ray lattice spacings measured at relatively large temperature intervals (e.g., *Handbook of Lattice Spacings of Metals and Alloys* [Pea67]). They can be used to calculate volume coefficients, and together with the Debye function to estimate the approximate length changes during cooling of rods or tubes made by powder metallurgy techniques. Single crystal values have not been reported for low temperatures.

6.3.4. Others: Ga, In, Tl, Sn, As, Sb and Bi

The first four of these anisotropic elements become superconducting at liquid helium temperatures, and this feature of their behavior will be considered further under Section 6.5.

The heat capacity of Ga (orthorhombic) is shown in Fig. 1.8. An interesting feature of its thermal expansion is the anisotropy of the electronic component as determined from single crystal data from 1 to 10 K [Gri77]: linear electronic components are negative in the *a* and *c* directions and positive in the *b* direction. Vibrational components are all positive and lead to $\gamma_0 \approx \gamma_{293} \approx 1.5$.

Indium is face-centered tetragonal with no clear-cut soft direction: the linear compliance $s_{33} > s_{11}$, but the linear compressibility $\chi_{\parallel} < \chi_{\perp}$. The linear expansion

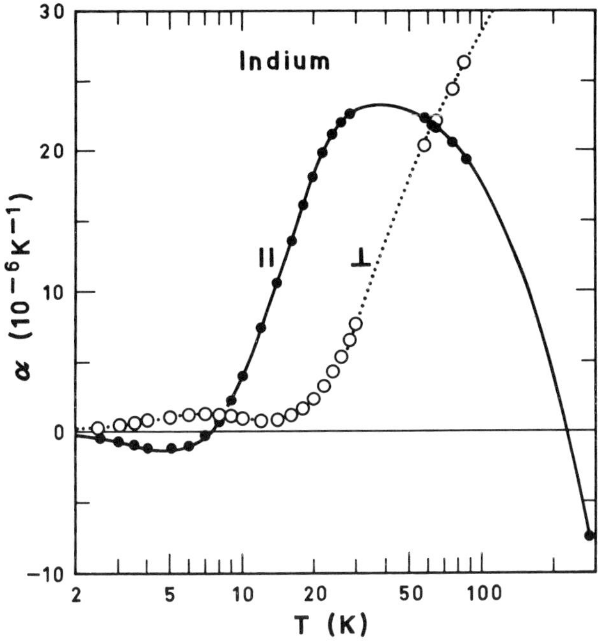

Fig. 6.11. Principal expansion coefficients of In; data sources in [Bar80, p. 689].

coefficients show an interesting oscillatory pattern arising from interplay of thermal and elastic forces (Fig. 6.11). The volume coefficient increases monotonically with temperature, as does the heat capacity, so that the volume change below room temperature can be predicted fairly well from Debye and Grüneisen models, unlike the length changes in a single crystal! The principal Grüneisen coefficients are fairly constant over the temperature range from 5 to 300 K with values of about 2.3.

What about As, Sb and Bi? These group VA semi-metals show an interesting progression with atomic mass. They become less metallic (judging from their conductivity) and more isotropic. The primitive lattice is like an fcc which is stretched along a [111] direction which becomes the trigonal symmetry axis. The rhombohedral primitive cell contains two atoms, and has an axial angle of about 54° for As and 57° for Sb and Bi (cf. 60° in a fcc lattice).

Arsenic crystals cleave easily like mica or graphite. They are layered perpendicular to the trigonal axis with planes of atoms alternately separated by 1.25 and 2.25 Å, thus forming double layers which are only loosely bonded to the adjacent double layers. The thermal expansion of arsenic is highly anisotropic, not unlike Zn (Fig. 6.10), with $\alpha_\perp$ negative below 150 K while $\alpha_\parallel$ is relatively large and positive. The volume coefficient is positive.

Bismuth has positive coefficients in both directions with $\alpha_\parallel$ being about 50% larger than $\alpha_\perp$ from about 100 to 300 K. Antimony is intermediate in behavior, the linear coefficient being negative normal to the symmetry axis below 20 K and otherwise positive.

6.4. MAGNETIC METALS

6.4.1. Introduction

The behavior of non-metallic magnetic solids including diluted magnetic ions in paramagnetic salts has been described in Section 5.11. Metals and alloys exhibit a greater variety of magnetic behavior, because the *d*- or *f*-electrons giving rise to magnetism can interact and mix with non-magnetic conduction electrons in many different ways.

Dilute Magnetic Impurities. To make the transition from non-metallic to magnetic metals, we start with the comparatively simple systems in which magnetic ions are present only as well-separated impurities in non-magnetic metals. They are reviewed in detail by Phillips in [Phi71, Sections 12,13,14,23].

We take as examples dilute solutions of Cr and Fe in Cu. Figure 6.12 shows C_P below 10 K for two Cu–Cr samples with 0.56 and 0.073 at% Cr. Below 1 K the magnetic contribution C_m is seen to be linear in T and to be at least roughly independent of the concentration c. But at much lower concentrations a very different behavior is seen (Fig. 6.13): C_m is now proportional to c, and has the form of a broad bump peaking at about 1 K. It is clear that at low enough c we are seeing the additive effects of isolated impurities. These are believed to arise from the *Kondo* mechanism, in which the magnetic ions form localized zero spin complexes with the sea of the conduction electrons, below a *Kondo temperature* T_K. Theoretical approximations differ in detailed predictions of behavior, but that shown in Fig. 6.13 is in good agreement with experiment with T_K fitted at a value of 2.1 K. At the higher concentrations of Cr in Fig. 6.12 a different mechanism is brought into play, owing to effective interactions between the magnetic ions; the results again are in qualitative agreement with theory. In CuFe the Kondo effect is stronger, with $T_K \approx 28$ K, and is operative even at a concentration of 0.24 at%, except at very low temperatures when ion-ion interactions become important. Several other systems are described by Phillips.

For thermal expansion, rather similar Kondo-like effects have been observed at low temperatures in Cu with small additions (0.2 to 2%) of Mn and Fe and in AgMn, $\Delta\alpha$ reaching a maximum around 6 K with a Grüneisen $\gamma_m \simeq 3$. Figure 6.14 shows the increase for CuMn expressed as $\Delta\alpha/T$ [Bar80, p. 701].

Ordering in Magnetic Metals. We now turn to magnetic metals proper, in which the magnetic ions are close together. Theoretical models show that long range magnetic ordering can occur among localized or among itinerant electrons, in either case due to *exchange interactions* resulting from the anti-symmetry of the electronic

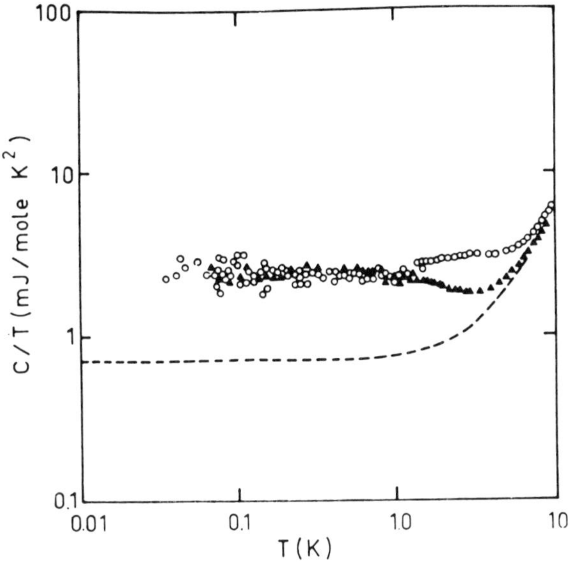

Fig. 6.12. Heat capacities of pure Cu (- - -) and of Cu containing 0.56 (○) and 0.73 (filled triangle) at% Cr [duCh66, Phi71].

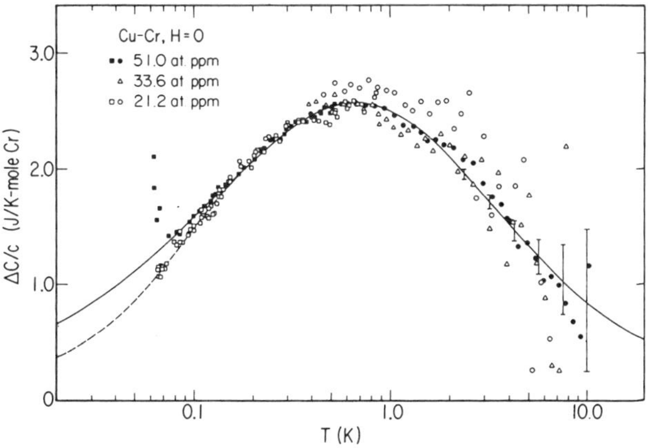

Fig. 6.13. Magnetic heat capacities of very dilute Cu–Cr. The solid curve is obtained by fitting the Bloomfield–Hamann theory to the 51 at ppm data. Error bars denote ±0.1% of total C_P. From [Phi71, p. 510].

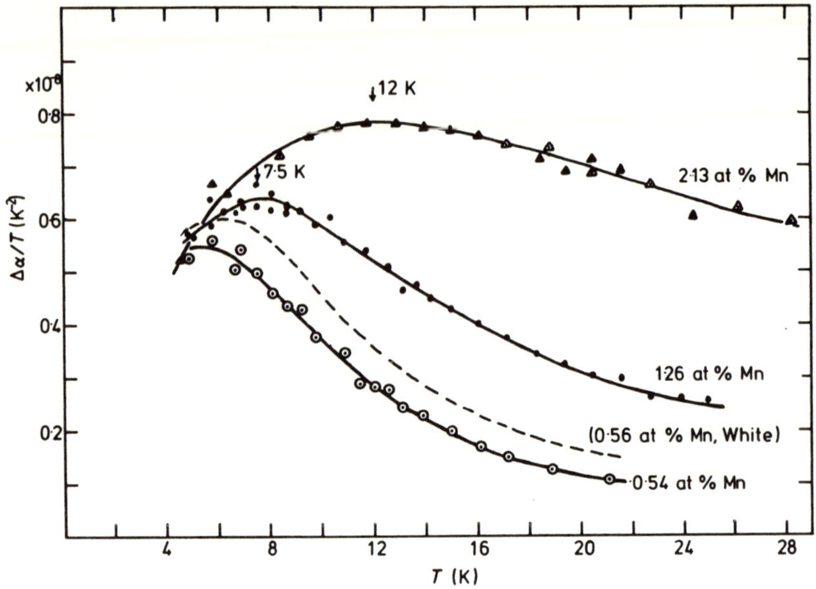

Fig. 6.14. Change in expansion coefficient, $\Delta\alpha/T$, for CuMn [Bar80, p. 701].

wave function with respect to exchange of electrons. The ordering may be ferromagnetic, in which neighboring spins are aligned; antiferromagnetic, in which the spins are ordered in different directions so that there is no net magnetic polarization; or ferrimagnetic, in which the spins are ordered in different directions but with a net magnetic polarization. At sufficiently high temperatures all long range magnetic ordering disappears at a critical temperature, called the Curie temperature T_c for ferro- and ferrimagnetism, and the Néel temperature T_N for anti-ferromagnetism. In some solids the energies of these and other phases are finely balanced, so that transitions occur between them as the temperature or pressure is changed. In particular, many of the rare earths display complex patterns of magnetic order.

The simplest theoretical models comprise interacting spins localized on the magnetic ions. In the Heisenberg model the exchange interaction energy between two neighboring ions A and B is taken to be of the form $-J\mathbf{S}_A \cdot \mathbf{S}_B$, where a positive J favors parallel spins and hence ferromagnetism, and a negative J favors anti-parallel neighbors and hence anti-ferromagnetism. In the ground state of a Heisenberg ferromagnet the spins are all aligned, and above this there is a continuum of energy levels in which the excitations are *spin-waves* in the component of spin normal to the direction of magnetization, the energy of each wave being quantized in *magnons*. The magnons are thus analogous to phonons, but with the important difference that the energy of magnons with long wavelengths is proportional to the square of the wave-number. This has the consequence that magnetic contributions to the limiting heat capacity and thermal expansion, C_m and β_m, are proportional to $T^{\frac{3}{2}}$ at low enough temperatures. A similar result has also been found for models of itinerant ferromagnetism. Low temperature measurements on non-metallic ferromagnets such as magnetite (Fe_3O_4) can therefore be analyzed by plotting $C_P/T^{\frac{3}{2}}$ against $T^{\frac{3}{2}}$.

For metallic ferromagnets, the presence of C_e usually prevents direct identification of the small C_m term, although analysis may be helped by subtracting the vibrational contribution derived from Θ_0^{el}. In this way evidence has been found for the presence of a $T^{\frac{3}{2}}$ term for both Fe [Ray61] and Ni [Ray56]. For Fe the data for C_P below 4 K fitted the relation

$$C_P = 4.90T + (0.08 \pm 0.04)T^{3/2} + 1.944(T/\Theta_0^{el})^3$$

where C_P is in mJ·mol^{-1}·K^{-1}, T is in K, and $\Theta_0^{el} = 477$ K [Ray61]. Any such term in the thermal expansion has so far proved unmeasurable (see below).

Antiferromagnets behave differently at low temperatures. For simple models their magnon energies at long wave-lengths are linear in the wave-vector, so that C_m, like C_{vib}, is proportional to T^3. Other models predict more complex behavior.

At higher temperatures it is usually difficult to identify magnetic contributions until near the critical temperature. In ferromagnetic elements such as Fe, Ni and Co the magnetic contribution to the entropy and energy manifests itself near to T_c (e.g., 1043 K for Fe) by a sharp peak in the heat capacity, but below this region the contribution is not obvious or easily separated from other components. Take nickel as an example: a peak in the expansion coefficient near $T_c = 629$ K has been carefully measured by Kollie [Kol77], but at lower temperatures the expansion coefficient of Ni is much like that of its non-metallic neighbors Pd and Pt.

For the antiferromagnetic elements Cr and Mn, the transitions at T_N are marked by narrow 'delta functions' in the heat capacity (small latent heats); but the anomalies in the thermal expansion are large, indicating strong magnetoelastic coupling (see Figs. 6.15 and 6.16).

Table 6.3 gives some comparative data for ferromagnetic (F) iron, nickel and cobalt, antiferromagnetic (AF) chromium and manganese and paramagnetic (P) palladium and $Cr_{95}V_5$ (an alloy which is similar in density and lattice dynamics to Cr but is non-magnetic). We shall discuss below elements and alloy systems involving Fe, Ni, Cr, Mn and rare earths, etc. Their heat capacities and enthalpies both near to magnetic transitions and at very low temperatures are of fundamental interest; and although these magnetic interactions are usually less important to the engineer, magnetostrictive effects can be large and technically important.

6.4.2. Transition Elements (Magnetic) and Their Alloys

Among the 3d elements from Cr to Ni, it is energetically favorable for the spins of the d-electrons to order magnetically. The spin-wave contribution to the heat capacity at low temperatures is small in the elements, but in some alloys there are significant increases in C_P below about 10 K, variously attributed to ferromagnetic clusters, exchange enhancement and spin fluctuations [Mor81b, Mor85]. Their origin

Table 6.3. Data for a selection of magnetic metals. V is atomic volume [Bar80, Phi71]

Property	Pd	Fe	Co	Ni	Cr	Cr5%V	Mn
structure	fcc	bcc	hcp	fcc	bcc	bcc	cub
mag. struct.	P	F	F	F	AF	P	AF
$T_{c,N}$(K)	—	1043	1380	629	311	—	96
V_{293}(cm^3)	8.87	7.10	6.62	6.60	7.23	7.2	7.39
B_S^{293}(GPa)	189	170	212	184	$\sim$ 170	200	125
Θ_0^C(K)	274	479	460	470	600	600	390
Θ_{293}^C(K)	290	400	380	390	480	480	410
C_P^{293}(J/mol·K)	25.8	25.0	25.1	26.0	23.3	23.3	26.3
$\alpha_{293}(10^{-6}$/K)	11.7	11.8	12.8	12.8	5.0	7.9	$\simeq$ 23
Γ(mJ/mol·K^2)	9.4	4.78	4.5	7.05	1.42	2.3	9.2
γ	2.2	2.4	2.4	2.1	-10	2.0	~ -10
γ_0	2.2	1.3	1.7	1.6	1	1.1	—
γ_{293}	2.37	1.67	2.1	1.87	$\sim$ 1	1.45	2.5

is generally beyond the scope of this monograph and does not play a significant rôle in the overall change in total heat (enthalpy) on cooling from room temperature.

Figures 6.15 and 6.16 illustrate how magnetic interactions may effect the expansion much more than the heat capacity. Differences in C_P between say Cr, Pd and Ni at low temperatures reflect chiefly differences in their Debye temperatures. The much larger differences in α between Pd, Cr and FeNi show the effect of magnetoelastic coupling. Chromium has attracted particular attention from experimentalists and theorists as the "... archetypal itinerant antiferromagnet whose incommensurate spin-density wave (SDW) is characterized by a wave vector determined by the nesting properties of its Fermi surface" (from review [Faw88]). The Nèel temperature is very sensitive to strain and to alloying. If we calculate a magnetic Grüneisen parameter γ_m from dT_N/dV, values of ~ -50 are obtained [Faw88, Faw94]. The magnetovolume effects in some of these systems are also discussed in [Hol84, Kai85].

Corsan and Mitchem [Cor76] measured C_P from 4 to 300 K for a range of stainless steels with nominal Cr/Ni percentages ranging from 12/12 to 24/20. Their tabulated values down to about 40 K lie within an envelope of $\pm 2\%$, but below this there are differences of 20% or more. The alloys of higher Ni content show the largest departure from the $T + T^3$ behavior at liquid helium temperatures. Values of C_P of the cryogenic stainless steels 304 (Cr/Ni contents of 20/9), 316 (19/11) and 310 (28/19) show a similar pattern [Col86b]. The coefficients of expansion of these same three steels agreed within a few per cent above 70 K but diverged below this. Below 40 K, values for the 304 and 316 became negative, while the 310 remained positive down to 6 or 7 K [Col86b]. Other technical metals (see Section 6.2.7) of the NiCrFe family for which there are expansion data include the Hastelloys and Inconel, etc. [Cla68, Cla83].

The most famous (from the technical view) or infamous (as a long- standing puzzle) are the alloys of the Invar family, pioneered by Chevenard and Guillaume around the turn of the century in the search for metals of very low expansion at ambient tem-

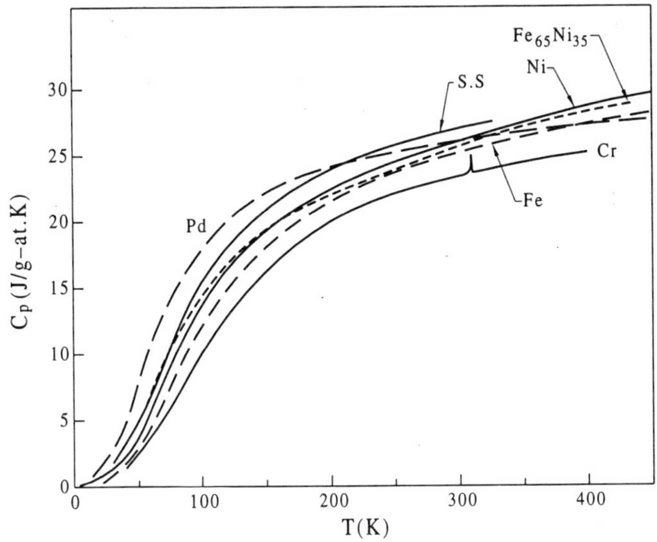

Fig. 6.15. Heat capacity of Pd ($\Theta^C_{293} \approx 290$K), Fe (400 K), Ni (390 K), Cr (480 K), $Fe_{65}Ni_{35}$ (400 K) and an 18:8 stainless steel (S. S.).

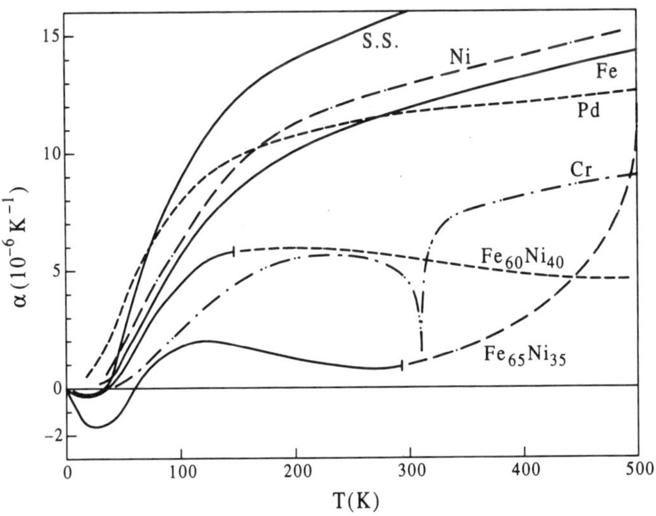

Fig. 6.16. Linear expansion coefficients of Pd, Fe, Ni, Cr, $Fe_{65}Ni_{35}$ and an 18:8 stainless steel (S. S.).

perature (e.g., [Gui897]). These are fcc FeNi alloys with Ni content of about 35 at% (see Fig. 1.10). Alloys with lower Ni content undergo a martensitic transformation on cooling. The shape of the $\alpha(T)$ curve (Fig. 6.16) arises from a balance between the positive lattice contribution to α and a negative magnetostrictive contribution, resulting in a negative overall coefficient below 60 K and a relatively small positive coefficient at room temperature. The exact value near room temperature depends sensitively on the Ni content and heat treatment, but approximates $1\times10^{-6}K^{-1}$.

The Invar effect, the negative expansion at low temperatures and small expansion at ambient temperature, is equally marked in FePd and in FePt alloys (including the *ordered* Fe_3Pt) [Woh75]. It occurs to varying degrees in many alloys of Fe–Cr–Ni–Mn–Co, as discussed in the reviews by Wasserman [Was89, Was90] (see also Fig. 1.10). There have been many theoretical models to explain the effect ranging from earlier attempts based on metallurgy and/or inhomogeneity to latent antiferromagnetism, local order, Weiss two-state models, weak itinerant ferromagetism, etc. These have been reviewed (e.g., [Was89] and [Chi97, Ch. 14]) and discussed at many conferences; see for example the *Proceedings of the International Symposium on Magnetoelasticity in Transition Metals* published in *Physica*, volumes **119 B+C**, **161 B** of 1983 and 1989. Experiments do show that the Fe–Fe interactions on an fcc lattice are antiferromagnetic in nature and are likely to contribute to negative expansion.

The concept of a negative magnetic contribution to the expansion coefficient can seem puzzling. It is equivalent to a positive magnetovolume at low temperatures, that is, the volume near 0 K is larger in the magnetically ordered state than if it were in the disordered state. On warming the magnetic disordering produces a contraction in order to minimize the free energy. For some magnetic systems, the rôle of spin fluctuations is important. Moriya [Mor85] discusses the self-consistent renormalization theory of spin fluctuations in relation to the weak itinerant ferromagnetic compounds MnSi, Ni_3Al, $ZrZn_2$, Sc_3In, which have Curie temperatures at 40 K or lower and exhibit this positive magnetovolume below T_c. Brommer and Franse [Bro90] have reviewed their considerable experimental data on magneto-volume effects in 'enhanced itinerant intermetallics and alloys.'

6.4.3. Rare Earths, Actinides and Their Alloys

Thermal properties of the rare earth elements are interesting, although not usually of concern to cryogenic engineers. For many rare earth elements the intrinsic values have not been accurately measured for a number of reasons. Firstly, most of these elements have one or more magnetic transitions at low temperatures involving planar and spiral rearrangement of the electron spins. These ordering processes complicate the task of fitting electronic and lattice terms in the heat capacity (or expansion).

Further problems are caused at very low temperatures by a nuclear heat capacity in the form of a Schottky 'bump' arising from a very large internal magnetic field acting on the nuclear moment. This bump at liquid helium temperatures makes it difficult to fit T and T^3 terms to the heat capacity data. The presence of impurities in the rare earths, particularly hydrogen, nitrogen, oxygen, and carbon, add to the difficulty. Only recently have electrotransport methods of purification produced samples of rare earths with impurity levels below 0.1 at%.

Analysis of thermal expansion data is complicated by anisotropy. The structure of most rare earths is hexagonal close-packed (ABAB stacking) or double hexagonal close-packed (ABACAB for Pr and Nd), with *c/a* ratios in the range of 1.57 to 1.62. Exceptions are Ce and Yb (fcc), Eu (bcc) and Sm (complex hexagonal).

Because of these difficulties, the values of some parameters in Table 6.4 may have significant errors, e.g., in Γ and Θ. These values are based largely on the reviews [Sun78, Sco78], with some Θ_0 values calculated from ultrasonic elastic data. Expansion coefficients are from the AIP Handbook [Kir72] and CINDAS Tables [Tou75]. Values of Θ^C_{293} are approximations derived from neutron scattering data [Sch81] or specific heats. For Sc and Lu there are recent comprehensive measurements of C_P and α [Swe96]).

The relatively complex temperature dependence of the thermal properties for the magnetic rare-earths is shown by $\alpha_\perp(T)$ for a holmium single crystal (Fig. 6.17). On cooling, a second order magnetic transition to an antiferromagnetic state occurs at the Néel point, $T_N \approx$132 K, a first order transition to ferromagnetism at $T_c = 19.5$ K and a further reordering at $T_R = 17.3$ K. The linear expansion $\Delta l(T)$ shows a discontinuity in length at 19.5 K confirming the first order nature of the transition. A curve of $\alpha_\parallel(T)$ has a qualitatively similar shape to that in Fig. 6.17 but with *negative* sign.

In addition to showing these three transitions, the heat capacity of Ho has an upturn below 4 K revealing a hyperfine nuclear spin term [Ste89] which is not evident in the thermal expansion data.

In contrast to the complexity of holmium (Fig. 6.17) are the smooth curves for the heat capacity and expansion of the non-magnetic elements Sc and Lu [Swe96]. For example, C_V for Lu after subtracting the bare–band electron component is well represented between 20 and 300 K by a Debye function with $\Theta_D = 160$ K.

Alloys. As an example of the separation of magnetic components there are measurements of both C_P and α from about 1 K up to 50 K on a series of polycrystalline samples of $Gd_xY_{1-x}Cu_2$ [Luo85]. The RCu_2 family (where R is a heavy rare earth metal) have an orthorhombic structure and order antiferromagnetically. $GdCu_2$ orders at $T_N = 40$ K whereas YCu_2 is nonmagnetic. Values for C_P of alloys with Gd contents ranging from 0 to 100% are shown in Fig. 6.18. The curves for $\alpha(T)$ are of similar shape to C_P but the magnetic 'peaks' are relatively much higher. C_P for the YCu_2 could be fitted to an electronic term $6.7T$ mJ$\cdot$K$^{-1}\cdot$mol^{-1} plus a Debye phonon term with $\Theta_D = 236$ K. For the other alloys, Θ^C ranged down to 198 K (x=1). Magnetic contributions were obtained by subtracting the electronic and phonon terms, and showed cusps at the respective Néel temperatures. Similar analyses were done on the expansion data. Together they lead to values of the respective Grüneisen parameters

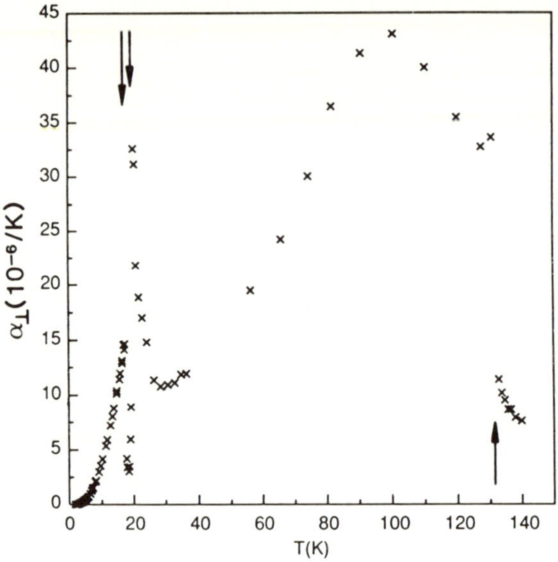

Fig. 6.17. Linear expansivity $\alpha_\perp$ for holmium from 1 to 140 K [Whi89a].

(assuming a value $B_S = 170$ GPa): $\gamma_{vib} = 2$, $\gamma_e = 1.7$ and $\gamma_m \simeq 10$. The relatively larger 'peaks' in α compared with those in C_P lead to the large value of γ_m.

Another important group of intermetallic compounds is the RMn_2 (R= rare earth) family, some of which show very large volume expansions below the magnetic ordering temperature. A notable case is YMn_2 with an observed volume change of about 5% at $T_N \simeq 100$ K. Nakamura [Nak94] has discussed them in terms of the magnetic moment of the Mn atoms above and below the transition temperature.

6.5. TYPE I AND TYPE II SUPERCONDUCTORS

6.5.1. Introduction

Superconductivity is a cooperative ordering process in which the charge carriers (electrons in these metals) 'condense' below a critical temperature into a ground state in which they have zero entropy. Electron pairs in this ground state can circulate without being scattered by lattice waves or other irregularities.

The effect of this ordering on the mechanical properties such as elasticity or on the vibrational spectrum is small: the differences in $c_{\lambda\mu}$ or bulk modulus between superconducting (s) and normal (n) states amount to a few parts in a million and require sophisticated ultrasonic interferometry to measure. Alers and Waldorf [Ale62] measured the changes in ultrasonic velocities in single crystals of Pb, V and Nb with a 'sing-around' technique capable of resolving to a part in 10^7. They did this in

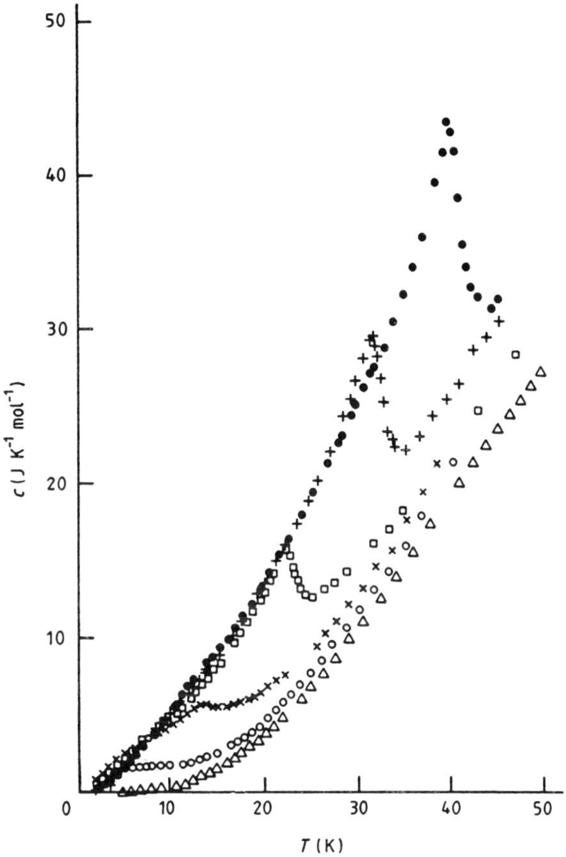

Fig. 6.18. The heat capacity of $Gd_xY_{1-x}Cu_2$ as function of temperature; x=1 (•), x=0.8 (+), x=0.6 (□), x=0.4 (×), x=0.2 (○), x=0 (△) [Luo85].

Table 6.4. Data for Rare earths and actinides from [Sun78, Sco78, Kir72, Tou75, Swe96]. Note that U is orthorhombic with *a*,*b*,*c* axes

Element	T_c (K)	T_N (K)	Θ^C_{293} (K)	Θ_0 (K)	Γ (mJ/mol·K^2)	$\alpha^{293}_{\parallel}$ (10^{-6}/K)	$\alpha^{293}_{\perp}$ (10^{-6}/K)	B^{293}_S (GPa)
Sc	—	—	310	352	10.4	14	8	57
Y	—	—	220	256	8.2	19	5	41
La	—	—	~ 130	154	9.5	15	1	28
Ce	—	13	120	136	12	5.2	5.2	21
Pr	—	—		152	~ 22	14	4	29
Nd	—	12	145	163	—	11	5	32
(Pm)								
Sm	13.7	105	145	169	12	19	9.5	38
Eu	—	90		118	6	25	25	13
Gd	293.4	—	150	182	10	−5 (310 K)	6 (310 K)	38
Tb	221	230	145	177	4.3	13	7.5	39
Dy	85	179	155	183	4.5	15	6	41
Ho	19.5	132	150	190	~ 2	17	3.5	40
Er	19	85	160	189	4.3	16.5	6.0	44
Tm	25	58	—	200	—	~ 18	~ 9	45
Yb	—	—	—	118	2.8	24.6	24.6	30
Lu	—	—	159	190	8.3	20	7	48
Th	—	—	140	165	4.4	11	11	57
U	—	—	—	222	9.8	20[c]	23[a],0.5[b]	—
Pu	—	—	—	161	15.9	~ 50(pc)	~ 50(pc)	—

magnetic fields up to 0.7 T and found that differences between s- and n-states in the elastic moduli ($\Delta c_{44}/c_{44}$, $\Delta c'/c'$, $\Delta B/B$) were at most 100 parts per million. They calculated that this would result in a change in Θ_0^{el} of ≤ 0.02 K.

It follows that differences in the vibrational contributions to heat capacity or thermal expansion should be too small to detect by thermal techniques. Notwithstanding this there was some calorimetric evidence in the early 1960s for In, which indicated an apparent small difference between C_{ls} and C_{ln}, probably arising from calorimetric problems as Θ is small, see [Phi71, p. 504]. For other metals including Pb, later work did not show any measurable difference.

The electronic contributions to heat capacity (and expansion) in the normal and superconducting states, C_{en} and C_{es}, are very different, and this is especially noticeable in those metals for which $T_c \ll \Theta_D$. For example, in Sn and Al the lattice contribution to these thermal properties is small near T_c and falling as T^3, so that electronic components are *relatively* large.

Values of T_c for the more common superconducting elements range down from Nb (9.25 K), Pb (7.200 K), V (5.38 K), Ta (4.48 K), Sn (3.722 K), In (3.415 K), Al (1.180 K), Mo (0.92 K), Zn (0.87 K), Cd (0.56 K), Zr (0.546 K), Ti (0.39 K) to W (0.015 K). The values of T_c are much more sharply defined for the elements which are readily available in high purity strain-free crystals than for 'dirtier' materials.

In the next section we discuss the relation between the heat capacity, length, and critical magnetic field for Type I superconductors. Type I are those metals for which a magnetic field exceeding a critical field, H_c, penetrates completely and destroys the superconductivity with accompanying latent heat and discontinuity in dimensions.

Type II are distinguished by a shorter range of ordering interaction which makes it energetically favorable for flux penetration to be gradual and produce an 'intermediate' state in which normal regions are interspersed with superconducting regions until the magnetic field is large enough to render the whole sample normal.

6.5.2. Type I Superconductors

According to the theory of Bardeen–Cooper–Schrieffer (BCS) for weak-coupling superconductors,

$$T_c = 1.14\overline{\Theta}\exp[-1/\{n(\epsilon_F)V\}] \tag{6.9}$$

where V is an interaction potential between the Cooper pairs and $\overline{\Theta}$ is an average over the lattice spectrum but not in general the same average as Θ_0^C or Θ_{RT}^C. Other forms of the BCS equation for strong coupling due to McMillan, Dynes and others involve the electron–phonon coupling parameter, λ, and screened Coulomb interaction, μ^*, in the exponential. It is clear that the ratio T_c/Θ increases with the strength of the electron–phonon coupling [Col86a, Ch. 8].

In this monograph we are less concerned with the limitations and extensions of the BCS theory than with the thermodynamic relations between the heat capacity and mechanical properties in the normal and superconducting states. Consideration of the Gibbs free energy shows that in zero magnetic field the difference in heat capacity at T_c is ([Sho60, Ch. 3] and [Bar98, Section 1.8.4.1])

$$C_s - C_n = \mu_0 V T_c \left(\frac{\partial H_c}{\partial T}\right)_P^2 \tag{6.10}$$

which is a positive quantity. Note that in the electromagnetic units used by Shoenberg $\mu_0 = \frac{1}{4\pi}$.

According to BCS theory, at T_c an energy gap appears at the Fermi surface which increases from zero at T_c to 3.52 kT_c at $T = 0$. The dependence of C_{es} on temperature is complicated but approximates to T^3 just below T_c and becomes exponential over a limited temperature range below this. From BCS for weak coupling, the discontinuity in C_e at T_c is given by

$$(C_{es} - \Gamma T_c)/\Gamma T_c = 1.43 \tag{6.11}$$

The measured values of this 'jump' ΔC_e at T_c depart from the BCS model to varying degrees for elements and alloys, e.g., [Phi71, Section 20] and [Col86a, Ch. 8]. Figure 6.19 shows the total measured heat capacity for vanadium, for which $\Delta C/\Gamma T_c \simeq 1.43$.

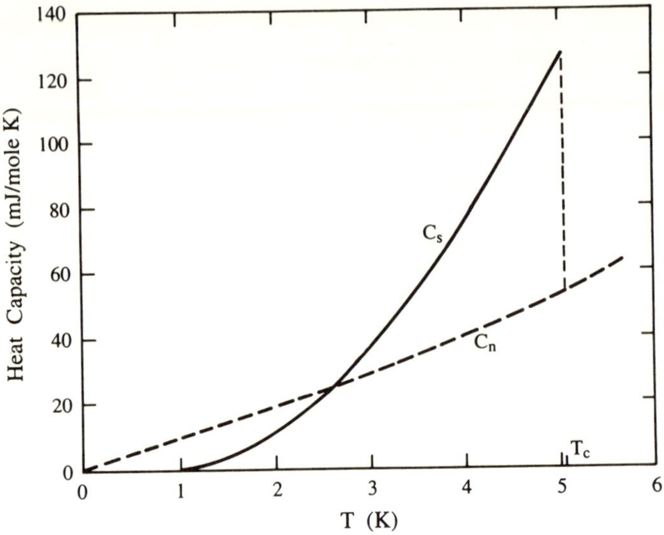

Fig. 6.19. Heat capacity of vanadium in the normal (C_n) and superconducting (C_s) states [Gop66].

The changes in volume and length can also be deduced from the thermodynamic relations [Sho60, p. 74]

$$V_n - V_s(0) = \mu_0 V_s H_c \left(\frac{\partial H_c}{\partial P}\right)_T + \frac{1}{2}\mu_0 H_c^2 \left(\frac{\partial V_s}{\partial P}\right)_T \tag{6.12}$$

$$l_{ni} - l_{si}(0) = \mu_0 l_{si} H_c \left(\frac{\partial H_c}{\partial P_i}\right)_T + \frac{1}{2}\mu_0 H_c^2 \left(\frac{\partial l_{si}}{\partial P_i}\right)_T \tag{6.13}$$

where P_i denotes a uniaxial compression in direction i. The second term on the right of the equations is the smaller magnetostrictive contribution.

Differentiating the above and taking $H_c = 0$, we find

$$\beta_n - \beta_s = \mu_0 \left(\frac{\partial H_c}{\partial T}\right)_P \left(\frac{\partial H_c}{\partial P}\right)_T \tag{6.14}$$

$$\alpha_{ni} - \alpha_{si} = \mu_0 \left(\frac{\partial H_c}{\partial T}\right)_P \left(\frac{\partial H_c}{\partial P_i}\right)_T \tag{6.15}$$

At $T = T_c$, an Ehrenfest relation [see Eqs. (2.28)] gives also

$$\beta_n - \beta_s = \frac{C_n - C_s}{V} \cdot \frac{d \ln T_c}{dP} \tag{6.16}$$

This last equation shows that a combination of measurements of differences in C_P and in α can provide data on the pressure or stress dependence of T_c. More direct pressure or stress measurement can be difficult and sometimes misleading because of

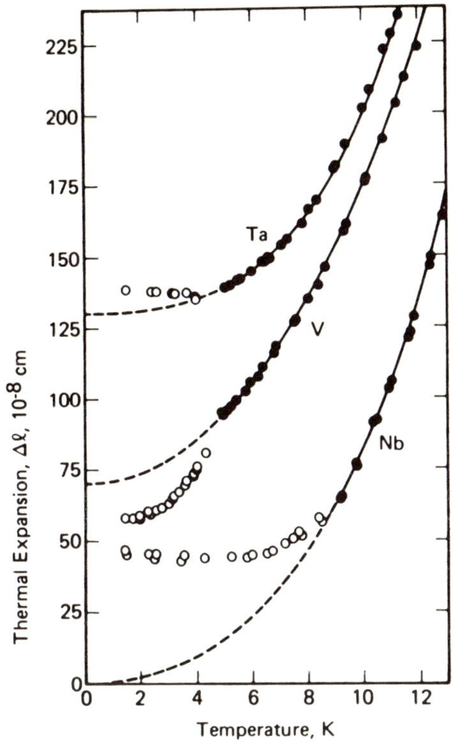

Fig. 6.20. Linear thermal expansion, Δl, of 5 cm long rods of Ta (type I), V (type II), and Nb (type II) in normal (•) and superconducting (○). Dashed curves are extrapolations of the normal state. From [Bar80, p. 711] and [Whi62].

non–hydrostatic conditions, anisotropy and frozen–in magnetic flux, etc., particularly with single crystals.

Figure 6.20 shows the length changes observed for tantalum, vanadium and niobium in normal and superconducting states below 10 K, see [Bar80, p. 711]. Note that $\Delta l = l_n - l_s$ is negative for Ta and Nb (and most other elemental superconductors) but is positive for V. The positive sign indicates that T_c *increases* under pressure for V in contrast to the more usual decrease with pressure. The order of magnitude of dT_c/dP for many type I superconductors is ~ -0.1 K·GPa^{-1}, corresponding to $d\ln T_c/d\ln V \sim 1$.

6.5.3. Type II Superconductors Including Technical Alloys

In type II superconductors below T_c, partial penetration by a magnetic field begins at the lower critical field, H_{c_1}, which may be relatively small, e.g., ~ 0.1 T or less. With increasing field a relatively larger volume of the metal becomes normal

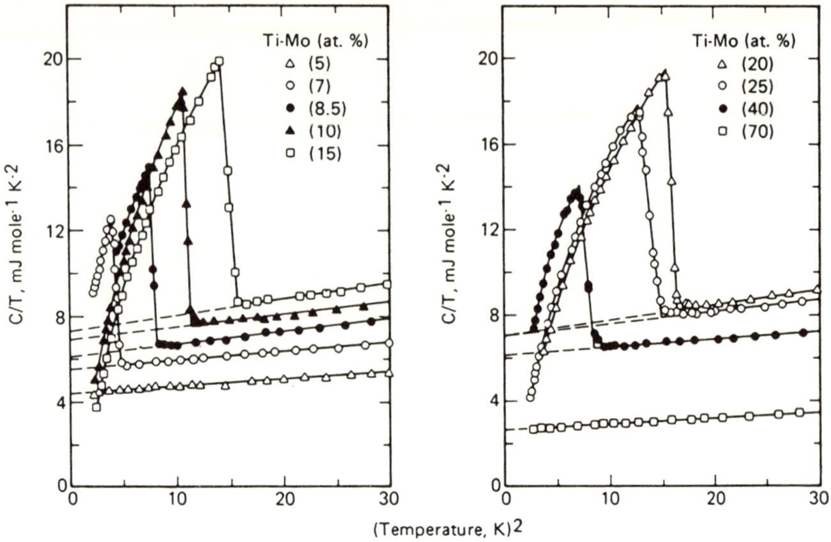

Fig. 6.21. C/T for a number of Ti–Mo alloys [Col86a, p. 317].

until at the upper critical field, H_{c_2}, the whole volume is in the normal state and finite electrical resistance is restored. The thermodynamics of the type II are more complicated than type I [Hak68, Hak69].

Most superconducting alloys and compounds are type II. Some with sufficiently high values of H_{c_2} are useful for construction of high field magnets (Table 6.5 and review [Eki83]) although many are unsuitable for mechanical or metallurgical reasons. The most tractable alloys used for fields up to 12 T have been Ti–Nb of about 50/50 wt% (35 at% Nb).

In zero magnetic field the behavior of $C(T)$ near to T_c for type II is similar to that for type I: see Fig. 6.19, and also the C/T versus T^2 graphs of Fig. 6.21, see [Col86a, p. 317]. At higher temperatures the Debye function is generally a good rough guide to the lattice contribution. The smaller electronic term is more difficult to assess because the electron–phonon coupling decreases with increase in temperature. Analysis of the Ti–Mo alloy data (Fig. 6.21) shows that $\Gamma = 5 \pm 2$ mJ·mol^{-1}·K^{-2} and that the coupling factor $\lambda \sim 0.4$ at low temperatures.

For all superconductors C_P decreases on warming through T_c, but the change in α at T_c may be positive or negative. Figure 6.22 shows the change in α_e for (a) Nb/Zr alloy for which $\alpha_{es} > \alpha_{en}$, and (b) pure Nb for which $\alpha_{es} < \alpha_{en}$ at T_c. The Ehrenfest relation then leads to a positive value of dT_c/dP for the Nb/Zr alloy and a negative value for Nb.

An alloy which is commonly used in constructing superconducting magnets is Ti–Nb, for which the fractional contraction on cooling from room temperature to 4 K is about 0.19%. This compares with a contraction of 0.326% for Cu, which is

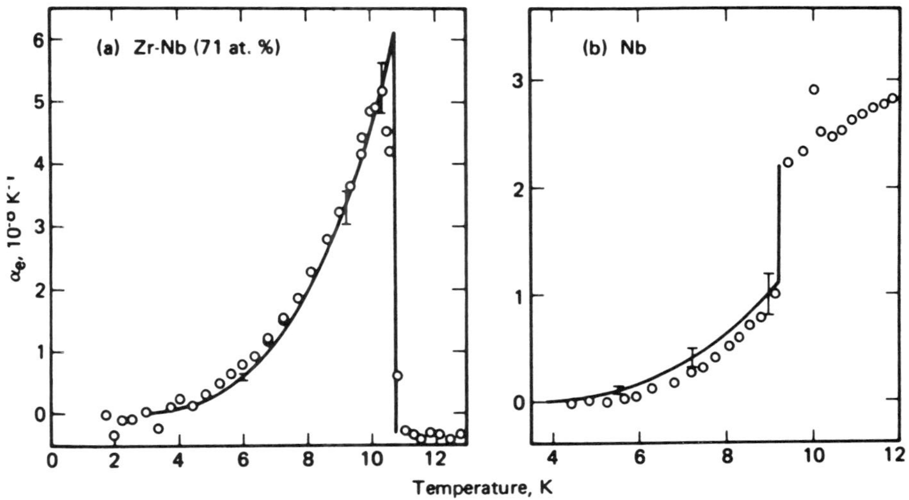

Fig. 6.22. The electronic contribution α_e for Nb + 29 at%Zr and for pure Nb [Sim78].

Table 6.5. Data for practical high field superconducting alloys and compounds, including contraction $-\Delta l/l_{293} = (l_{293} - l_4)/l_{293}$. (Cu/Nb–Ti is a multifilament material with Nb–Ti in a Cu matrix). Data sources [Eki83, Cla81, Cla83] (see also [Wil83, Rob72, Col86a])

Material	Structure	T_c (K)	$H_{c2}^{4.2}$ (T)	Θ_{293}^{c} (K)	α_{293} (10^{-6}/K)	$-\Delta l/l_{293}$ (10^{-4})
Nb	bcc	9.25	0.27	250	7.1	14.3
Ti–Nb	A2	10.2	12	310	9.0	18.9
Nb–Zr	A2	10.8	21	310	—	—
Cu/Ti–Nb	comp	—	—	—	~ 14	~ 25
V_3Ga	A15	16.5	22	400	—	—
V_3Si	A15	17.0	—	500	7.5	—
Nb_3Sn	A15	18.3	22	—	7.2	16.5
Nb_3Al	A15	18.9	32	—	7.8	—
Nb_3Ge	A15	23	30	—	6.7	—

commonly used as a matrix for the multifilament Ti–Nb wires. Such magnets can be used up to about 14 T.

The compound which has been used for magnets producing higher fields (up to 20 T) is Nb_3Sn, usually in the form of tape. It is one of the A15 compounds; these include V_3Si, Nb_3Sn, and V_3Ga, which have relatively high transition temperatures and a martensitic transformation at low temperatures. This transformation is associated with a 'soft' transverse acoustic mode, $c' = (c_{11} - c_{12})/2$. In V_3Si, the cubic-to-tetragonal transformation occurs at $T_M \approx 21$ K, and may give rise to comparatively large anisotropic length changes (positive and negative) of parts in 10^4, with magnitude dependent on the stoichiometry and applied stress (e.g., [Ott85, Liu95]). The rôles of lattice vibrations, electronic density of states and electron–phonon interaction in determining T_c, T_M, and dT_c/dP for A15 compounds are still debated [Liu95].

C_P has been measured for a series of V_3X compounds (where X= Si, Ga, Sn) and corrected for expansion to obtain C_V [Kna75]. The C_V values were analyzed into a Debye–like lattice term (e.g., $\Theta \approx 500$ K for V_3Si) and an electronic term, $C_e = \Gamma T$. At normal temperatures Γ for V_3Si was reduced to about 30% of its low temperature phonon–enhanced value, falling from 16.7 (at low temperatures) to 5 or 6 mJ·g-at^{-1}·K^{-2} (cf. band-structure value of 9.18 mJ·g-at^{-1}·K^{-2}).

6.6. HEAVY ELECTRON METALS

6.6.1. Introduction

Another collective electron system of considerable interest is that of the 'heavy electron' or 'heavy fermion' metals, which exhibit large heat capacity values and even larger expansion coefficients at low temperatures. They are called heavy because the

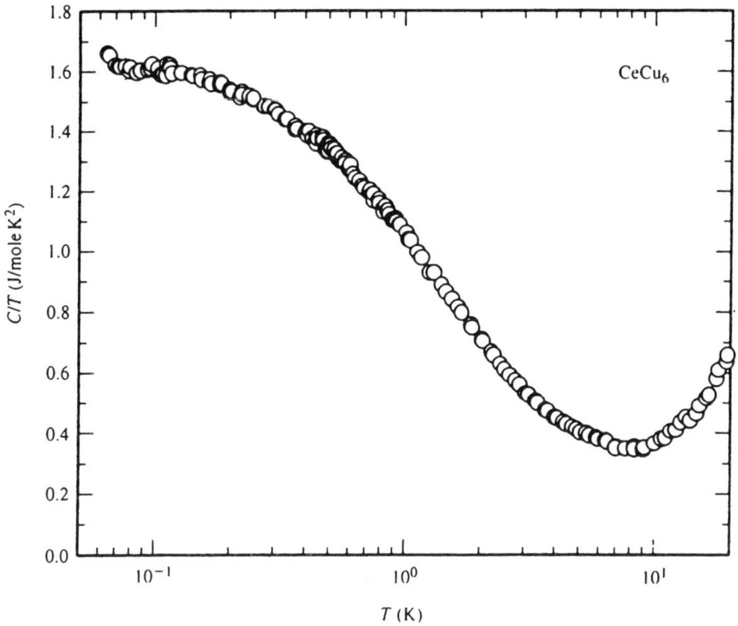

Fig. 6.23. C_P/T versus T of $CeCu_6$. Limiting value as $T \to 0$ gives $\Gamma = 1.67$ J/mol·K^2 (= 238 mJ/g-at·K^2) [Spr97b].

effective masses of the conduction electrons are very much larger than the bare electron mass. To quote Fisk *et al.* [Fisk88] '... at room temperatures and above, heavy electron systems behave as a weakly interacting collection of f-electron moments and conduction electrons with ordinary masses; at low temperatures the f-electron moments become strongly coupled to the conduction electrons and one another, and the conduction electron effective mass is typically 10–100 times the bare mass ....'

The materials are all intermetallic compounds with one constituent atom having a 4f or 5f shell partially filled. A Curie–Weiss type susceptibility at higher temperatures shows that these give rise to localized magnetic moments, which have energies that are split in the crystal field. The coupling at low temperatures of these moments to the conduction electrons is thought to resemble in some way that in the dilute Kondo systems (Section 6.4.1); but the distance of magnetic atoms from each other in heavy fermion materials is much smaller than the radius of an isolated Kondo complex, and also they form an ordered lattice. We have therefore a *strongly correlated system*, which is found experimentally to give an effective very narrow band of conduction 'electrons' with very large effective mass. This has sometimes been called a *Kondo-lattice system*. Moreover, the type of correlation is highly sensitive to strain, as shown by very large Grüneisen parameters that can be either positive or negative, giving rise to marked effects at low temperature in the thermal expansion and even in the elasticity.

Salient features of their physical properties at low temperatures are:

- They show a large linear (T-) term in the heat capacity, with Γ up to 150 times larger per mole than for Pd or Ni
- They show a paramagnetic susceptibility which depends only weakly on temperature but of magnitude 10^2 to 10^3 times that of normal metals
- Values of the expansion coefficient may be 1000 times greater than those for Pd or Ni, with values of $\gamma \sim 100$
- The electrical resistivity is relatively large ($\sim 100 \times 10^{-8}$ μm) at room temperature, and falls at low temperatures as T^2. The coefficient of T^2 is many orders of magnitude larger than for transition metals.
- Some order antiferromagnetically and some show superconductivity.

The present understanding of these materials has been surveyed recently by Springford [Spr97b]. Despite extensive and continuing study, there is as yet no agreed theory, and 'such diverse properties are not easy to accommodate within a general framework.' Nevertheless, the systems are of such interest and potential importance that they merit brief discussion here.

6.6.2. Heat Capacity

The review by Fisk *et al.* [Fisk88] lists 11 heavy electron compounds for which there were low temperature C_P data with Γ values ranging from about 140 to 1600 mJ$\cdot$mol$^{-1}\cdot$K^{-2}. Some of these (for which there are also expansivity data) are listed in Table 6.6, which includes approximate values of the adiabatic bulk modulus, Γ values per mole and per gram atom, the Grüneisen function γ from thermal data at the lowest temperatures, and the temperature at which any ordering occurs (superconducting or magnetic).

The compound $CeCu_6$ for which $C_P(T)$ is shown (Fig. 6.23) was chosen as an example by Springford in his review [Spr97b] for detailed discussion of heat capacity and electron transport because it has been widely studied and remains paramagnetic without onset of ordering down to a few mK. This compound and $CeAl_3$ (which also does not order, at least above 20 mK) are discussed further in the review by Ott [Ott87]. Values of Γ in Table 6.6 are estimates based on different sources [Fisk88, Ott87, deV89, deV90].

6.6.3. Thermal Expansion

Except for UBe_{13}, the compounds in Table 6.6 are highly anisotropic. Values of principal linear coefficients of some of them have been measured by de Visser and colleagues, and reviewed in [deV89]. The observed behavior is very varied and often

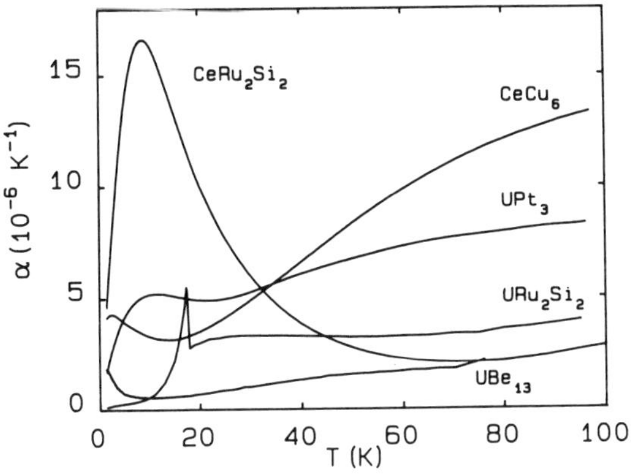

Fig. 6.24. Coefficients of linear expansion, $\beta/3$, for heavy electron compounds listed in Table 6.6 [deV89].

Table 6.6. Selected heavy electron compounds for which there are low temperature C_P and α values. S or AF denote superconducting or antiferromagnetic transition at temperature T_S or T_N [deV89, Fisk88, Ott87]

	Structure	Γ $\left(\frac{\text{mJ}}{\text{mol}\cdot\text{K}^2}\right)$	Γ $\left(\frac{\text{mJ}}{\text{g-at}\cdot\text{K}^2}\right)$	B_S (GPa)	γ	$T_{S,N}$ (K)
$CeAl_3$	hex	1600	400	—	$-200(< 0.3$ K) $60(\sim 1$ K)	0
$CeCu_6$	o-rh	1600–1700	240	83	$200(< 0.3$ K) $60–100(\sim 1$ K)	0
$CeRu_2Si_2$	fct	330	16	~ 100	150	17(AF)
URu_2Si_2	fct	60–180	12–30	137	25	1.5(S)
UBe_{13}	cub	800–1100	60–80	93	30–40	0.9(S)
UPt_3	hex	430	107	208	70	0.5(S)
Pd	fcc	10	10	189	2.2	0

striking. In the absence of a common pattern, we quote here $CeCu_6$ as one example. No superconductivity or magnetic ordering was observed down to 18 mK. The linear coefficients of expansion were measured from 1.4–200 K, and it was found that '... the unit cell expands along the *b* and *c* axes, while it shrinks along the *a* axis up to 17 K ... ' This anisotropy is strongly temperature-dependent. At 2 K α_c is more than double α_b, while α_a is negative; by 25 K α_c is small and negative, α_a small and positive, and α_b is flattening off to a high value of about $1.2 \times 10^{-6}\ K^{-1}$; and by 100 K α_a is rather greater than α_b and α_c, which are roughly equal. Other materials are reviewed in [Bar98, Section 1.8.6]. After subtraction of the phonon contribution, obtained from data on $LaCu_6$, the *f*-electron contributions were identified as (i) a low temperature anomaly centered at 2.5 K, attributed to the 'Kondo-lattice' effect, and a broad contribution centered near 60 K attributed to crystal field effects. Other materials are reviewed in [Bar98, Section 1.8.6].

The averaged linear expansion coefficients ($= \beta/3$) for these compounds are shown in Fig. 6.24. There is further discussion of the Grüneisen parameters by de Visser *et al.* [deV90] and in the review by Ott [Ott87].

Chapter 7

Polycrystals, Composites and Aggregates

7.1. INTRODUCTION

This chapter covers mixtures which are not homogeneous. In a polycrystalline specimen of a single component only the orientation of the crystallites varies, while in composites and aggregates there are also two or more distinct components forming a mosaic throughout the solid.

In such a heterogeneous solid there are usually local stress and strain fields, and these local fields change non-uniformly when the solid is subjected to stress or strain or change of temperature. For example, quoting from Rosen and Hashin [Ros70]: When the bulk C_V (or C_P) is measured 'there is no change in the *average* strains (or stresses), but the local values of strains (or stresses) may change. Hence the composite specific heats are not simple weighted averages of the constituent specific heats.' This can make it difficult to relate the bulk properties to those of the constituents, especially when the crystallites are strongly anisotropic, and (in composites) when the thermoelastic properties of the components differ strongly. Moreover, if the inhomogeneous stresses are large enough slip may occur, and this may also give rise to hysteresis.

We must note, however, an important exception. In a polycrystal of crystallites which have cubic symmetry, change of temperature and pressure do not give rise to inhomogeneous stress and strain fields, so that the bulk modulus, heat capacity and thermal expansion are the same as for a single crystal. Inhomogeneous fields then arise only under shear stress.

In composites and aggregates the two or more components are spacially separated, and thus are not chemically linked except possibly at the interface between them. They include many everyday materials ranging from rocks (mineral aggregates) through fiber-reinforced metals to fiber-reinforced polymers and particle-filled polymers. The term composite therefore does not apply to homogeneous mixtures such as ^{3}He/^{4}He or stabilized zirconia. Those of most cryogenic interest are the polymer composites, mentioned in Section 5.9, in which high strength fibers or low

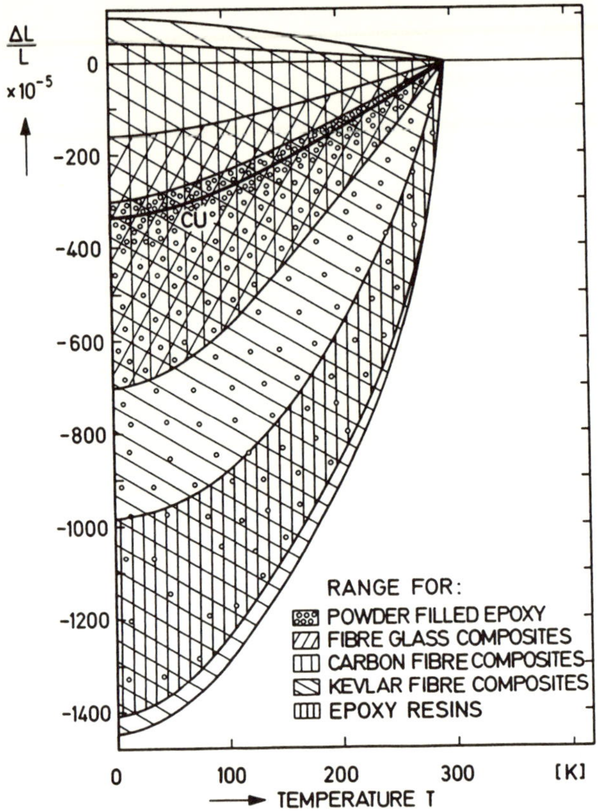

Fig. 7.1. Linear thermal expansion, $\Delta l/l$, relative to 293 K for several composites and epoxy resins, compared with Cu [Har88a].

expansion powders are added to the polymer to give constructional materials of light weight, high strength and compatible thermal expansion (or contraction). Thermal properties of composites can be tailored for many purposes by addition of powders such as silica, alumina, magnesia or fibers of graphite, silica, kevlar, etc. They are also electrically non-conducting and non-magnetic. The electrically non-conducting property is important in ac applications to avoid eddy currents and associated magnetic fields. The 'tailoring' consists of choosing suitable proportions of powder or suitable alignment of fibers. Figure 7.1 [Har88a] illustrates the wide range of linear thermal contractions that can be achieved with different 'mixes,' $\Delta l/l$ (relative to 293 K) varying from zero to about 1.4%. It is not within the scope of this monograph to cover in detail the multitude of 'mixes' and their engineering applications.

Although significant departures from additivity generally arise from inhomogeneous stress and strain due to mismatch of the thermal expansion at interfaces between components, to a first approximation the heat capacity of a composite can be taken as simply the sum of the heat capacities of the individual constituents. But for the expansion coefficient departures from additivity are usually larger and more difficult to estimate, particularly if the one or more of the components is anisotropic. Useful sources of information on composites are

- *Handbook of Composites* edited by Lubin [Lub82], which covers applications, ingredients, and processing rather than physical properties.
- *Mechanics of Composite Materials* by Christensen [Chr91], which is primarily a mathematical treatment of the elastic properties with a short chapter on effective thermal properties of composites.
- Papers on the thermal expansion of composites include those of Rosen and Hashin [Ros70] and Hale [Hal76]. These are theoretical treatments of the thermoelastic properties of composites and their constituents, in which bounds are derived for the effective expansion coefficients and specific heats of multiphase anisotropic composites having anisotropic constituents.
- Papers by Klemens [Kle86, Kle88] on the theory of thermal expansion of composites with spherical or cylindrical inclusions in a matrix.
- On the experimental side, there are many reports in the physics journals, particularly in the Proceedings of ICMC (International Cryogenic Materials Conference) published as volumes of *Advances in Cryogenic Engineering — Materials* (e.g., volume **40B**, 1994) by Plenum Press (New York), and in specialist ICMC conferences on *Nonmetallic Materials and Composites at Low Temperatures* [Cla79, Har82, Har88b, Oka95a, Har98].

It appears that for composite materials there has been insufficient marriage between theory and experiment, particularly for the anisotropic fiber-reinforced polymers. On the one hand there are theoretical upper and lower bounds for values of thermoelastic properties, and various models and approximations for estimating where the true values lie between these bounds. On the other hand there are extensive measurements on practical composites, but usually without any comparison with theoretical models. We shall discuss them separately below.

7.2. THEORY

The complex pattern of local stress and strain fields in a composite depends on the size, shape and orientation of the different grains and their distribution relative to each other. For simple models of the distribution, it may be possible to compute bulk properties using finite element techniques. Much work has also been done on developing approximate general theories. But, to quote [Tay98a]: 'Many simplified relationships have been derived [Cha87]. Unfortunately, different equations predict different values of expansion coefficient for a given composite.' To chart reliable paths through this specialist minefield is beyond the scope of this monograph, but we can indicate the principles on which some of the theories are based.

7.2.1. Reuss, Voigt and Related Approximations

The historic approximations of Voigt [Voi889] and Reuss [Reu29] are still often quoted and used. Together they give limits for bulk properties within which the true values must lie. Reuss assumes that any change in *stress* is uniform throughout the solid, and then derives bulk properties by averaging changes in energy and strain over all component parts of the mosaic. This gives c_σ, α_λ, and isothermal compliances as simple averages of the components; from these other properties may be derived thermodynamically. In contrast, Voigt assumes that any change in *strain* is uniform throughout the solid; this gives c_η, γ_λ and isothermal stiffnesses as simple averages. A popular approximation is that of Hill [Hil52], who takes the mean of the Reuss and Voigt values for any given property.

To illustrate the methods, consider a macroscopically isotropic polycrystal of a single crystalline substance (e.g., [Gib74]). The averages then reduce to averaging equally over all possible orientations of a microcrystal. The Reuss values of c_P, β and χ_T for the polycrystal are then the same as those for a single crystal under hydrostatic pressure; in particular

$$\beta = \alpha_1 + \alpha_2 + \alpha_3 \quad \text{and} \quad \gamma = \sum_{\mu=1}^{6} \chi_\mu^S \gamma_\mu \Big/ \sum_{\mu=1}^{3} \chi_\mu^S \tag{7.1}$$

where the χ_μ are the directional compressibilities defined in Eqs. (2.129) and (2.130). On the other hand, Voigt gives c_V the same as c_η for a single crystal, while

$$\beta = 3 \sum_{\mu=1}^{6} \left(\sum_{\lambda=1}^{3} c_{\lambda\mu}^T \right) \alpha_\mu \Big/ \sum_{\mu=1}^{3} \sum_{\lambda=1}^{3} c_{\lambda\mu}^T \quad \text{and} \quad \gamma = (\gamma_1 + \gamma_2 + \gamma_3)/3 \tag{7.2}$$

Thus whereas Reuss gives the bulk expansion of a polycrystal as a simple average over all directions, Voigt gives more weight to the small expansion in elastically stiff directions. In most polycrystals of a single substance the experimental expansion is closer to the Reuss average. Composites, however, may be tailored to favor the Voigt average. For example, long stiff fibers can greatly reduce the expansion of a less stiff matrix in which they are incorporated, and also reduce the value of the specific heat c_P to a value closer to that of c_V for the pure matrix material.

Similar approximations can be used for models in which the distribution of crystallite orientation is not isotropic, when the polycrystal is said to have 'texture.' The Reuss and Voigt averages are then weighted by the *orientation distribution function* (ODF). Conversely, the theory has been used to estimate the ODF from ultrasonic measurements on the polycrystal.

7.2.2. Use of Aggregate Elasticity

The orientation distribution function takes no account of crystallite shape and the other details of the mosaic structure. But by averaging over the *unknown* geometrical distribution of crystallites it was found possible for a few simple systems to express the thermal expansion of the aggregate in terms of the thermal expansion and elasticity of the separate components *plus* the elasticity of the aggregate. Later Schulgasser [Sch89] showed that these results could be obtained by a delightfully simple physical argument, which demonstrates clearly which kinds of model can be treated in this way. We illustrate Schulgasser's method by using it to derive an expression originally obtained by Levin [Lev67] for the effective coefficient of expansion β_e of a macroscopically isotropic composite with effective compressibility χ_e, composed of two components A and B with isotropic expansion coefficients β_A and β_B and compressibilities χ_A and χ_B.

Schulgasser considers the expansion to proceed in two stages. In the first stage the temperature rise dT is accompanied by a change in pressure dP^* chosen such that *each* component has the same relative expansion $d\ln V$:

$$(d\ln V)_1 = \beta_A dT - \chi_A dP^* = \beta_B dT - \chi_B dP^* \tag{7.3}$$

so that

$$dP^* = \left(\frac{\beta_A - \beta_B}{\chi_A - \chi_B}\right) dT \tag{7.4}$$

and

$$(d\ln V)_1 = \left[\beta_A - \chi_A\left(\frac{\beta_A - \beta_B}{\chi_A - \chi_B}\right)\right] dT = \left(\frac{\beta_B\chi_A - \beta_A\chi_B}{\chi_A - \chi_B}\right) dT \tag{7.5}$$

In this first stage the change in both stress and strain is uniform throughout the composite. In the second (isothermal) stage the applied pressure dP^* is reduced to zero, with a further change in bulk strain for the composite

$$(d\ln V)_2 = \chi_e dP^* = \chi_e\left(\frac{\beta_A - \beta_B}{\chi_A - \chi_B}\right) dT \tag{7.6}$$

The effective thermal expansion coefficient of the composite is thus

$$\beta_e = \left(\frac{\partial \ln V}{\partial T}\right)_P = \left(\frac{\beta_B\chi_A - \beta_A\chi_B}{\chi_A - \chi_B}\right) + \chi_e\left(\frac{\beta_A - \beta_B}{\chi_A - \chi_B}\right) \tag{7.7}$$

This is equivalent to an expression derived earlier showing the departure from additivity [Ros70, Hal76, Mil83]:

$$\beta_e = f_A\beta_A + f_B\beta_B + (\beta_A - \beta_B)\left(\frac{1}{B_e} - \frac{f_A}{B_A} - \frac{f_B}{B_B}\right) \Big/ \left(\frac{1}{B_A} - \frac{1}{B_B}\right) \tag{7.8}$$

where B_A, B_B are the bulk moduli of the components, B_e is the effective bulk modulus of the composite, and f_A, f_B are the volume fractions of each component.

Without further assumptions the method cannot be applied to composites with three or more components, because the condition of uniform expansion in stage 1 could not be satisfied for all three components; nor can it be applied when the components have anisotropic expansion and compressibility. But it can be applied to isotropic polycrystals of a single component with axial symmetry [Mil83, Has84], when dP^* is chosen so that the change in dimensions of each crystallite is isotropic. It can also be applied to the above systems when they have 'texture,' i.e., when they are not macroscopically isotropic [Sch87b]: stage 1 is then the same as before, but in stage 2 the response to the relaxation of pressure is given by the effective directional compressibilities $\chi_{\lambda,e}$ of the aggregate.

Dunn and Leadbetter have combined the method with Voigt–Reuss–Hill averaging of the elasticity to obtain the thermal expansion in terms solely of single crystal properties and the ODF [Dun95].

7.2.3. Other Relations

A number of approximate expressions relating the thermal expansion of different types of composites to the thermoelastic properties of their components are quoted in [Tay98a], including those of Schapery [Sch69] for longitudinal and transverse coefficients of a unidirectional fibrous composite in terms of the expansion coefficients, Young's moduli and Poisson's ratios of matrix and fiber. The review by Hale [Hal76] also lists a number of alternative relations for the axial and transverse values of α for transversely isotropic aligned fiber composites. Some of the upper and lower bounds from these relations are shown in Fig. 7.2, together with experimental values for an aligned glass-fiber composite attributed to Schneider [Sch71].

Note also a formula given by Hartwig and Knaak [Har84] attributed to Schneider [Sch71], for the longitudinal expansion of a uni-directional composite:

$$\alpha_{\parallel} = \alpha_{F\parallel} + \frac{\alpha_M - \alpha_{F\parallel}}{1 + [1.1f/(1-1.1f)](E_{F\parallel}/E_M)} \tag{7.9}$$

where f is relative fiber content per volume, and E_M and $E_{F\parallel}$ are Young's moduli for matrix and fiber respectively.

7.2.4. Theories for Specific Inclusions

Klemens [Kle86] developed a theory for the overall thermal expansion of composites containing either spherical or randomly oriented long cylindrical inclusions in a matrix. The strain field of the inclusions consists of a uniform expansion and short-range strain fields, and these are related by minimizing the elastic free energy. He gives an instructive numerical illustration of the results: for cylindrical and spherical inclusions, the expansion is less than the volume average for low concentrations and more for higher concentrations.

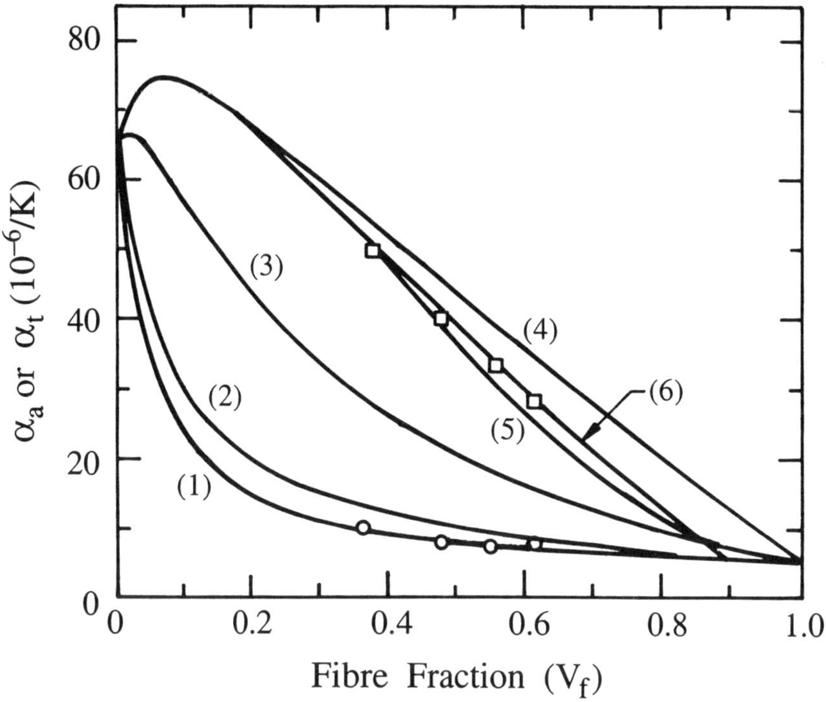

Fig. 7.2. Axial (a) and transverse (t) values of α for glass-fiber composites. Curves 1 and 2 are respectively lower and upper bounds for α_a; 3 and 4 are lower and upper bounds for α_t; 5 and 6 are other approximations. Experimental points from [Sch71] are axial (⊙) and transverse (□) [Hal76].

Klemens [Kle88] generalized this to other shapes of inclusion (i) in matrix (m) and obtained a relation of the same form as the simple mixing rule

$$\overline{\beta} = f\beta_i + (1-f)\beta_m \tag{7.10}$$

but with concentration f being replaced by an effective concentration

$$f' = \frac{fF}{(1+fF)} \tag{7.11}$$

where F is a function of bulk moduli B_i, B_m and shear moduli G_i, G_m of inclusions and matrix as well as f. For low values of f, F is usually > 1 so that $\beta_e > \beta_{\text{simple}}$. For large values of f, $F < 1$, and the simple equation overestimates the effect of inclusions; but departures are usually small.

7.3. EXPERIMENT

7.3.1. Heat Capacity

As we have seen, in the absence of varying local interfacial stresses the specific heat c_P of the composite would be simply the weighted sum of the individual contributions. A composite of 2 components would then have a bulk specific heat

$$c_P = (1-f)c_M + fc_F \tag{7.12}$$

where c_M and c_F are respectively the specific heats c_P per unit volume of matrix and fiber, and f is the volume fraction of fiber (or powder).

But in some materials there are appreciable departures from this approximation. For example, Jäckel [Jac91] measured the specific heat of a glass-fiber epoxy composite and its components from 2 to 80 K and found the calculated values from the above equation were 10 to 20% higher than his measured values.

The relation of experimental values of Θ_0 (and γ_0) to the bulk elastic properties of the composite has been discussed recurrently over many years (e.g., [Bar98, p. 93]. In most measurements the relevant phonon wavelengths even at the lowest temperatures are smaller than the crystallite dimensions, and so the bulk elasticity appears to be strictly irrelevant, although experimentally it has given good agreement for some cubic materials. At lower temperatures (where averaging over individual crystallites becomes invalid) the properties will depend upon the disordered structure of the material, providing intractable theoretical problems similar to those for a glass.

7.3.2. Thermal Expansion

Powder-Filled Composites. Unfilled epoxy resins contract on cooling from room temperature (usually 293 K) to helium temperatures (4 K) by amounts generally in the range 1.0 to 1.4%. The addition of about 40% (by volume) of a filler (silica, alumina powder) can reduce this contraction to about half [Wal94], i.e., to about 0.6% or 60×10^{-4}. Tests by Hartwig [Har77b] with these powders (including zirconium silicate) at concentrations ranging from 10 to 80 vol% of filler showed that $\Delta l/l$ (from 293 to 4 K) decreased roughly linearly with concentration of filler, almost independently of the particular filler. Reduction due to 40 vol% filler reduced the contraction to between 50 and 60% of the unfilled value.

Glass-Fiber Filled Composites. In these, the filling agent is isotropic and is often a glass with a room temperature value of $\alpha \simeq 5\times10^{-6}\text{K}^{-1}$ (and therefore an overall contraction on cooling to 4 K of about 0.05%, similar to Pyrex). Used as filler at 60 vol%, it will reduce the overall contraction from 293 to 4 K of an epoxy resin from over 1% to about 0.2%. Grades of fiberglass suited to various applications include A-glass, S-glass and E-glass, see [Lub82, p. 139]. Of these E-glass is most widely used, particularly for electrical and cryogenic applications. The composition (in wt.%) is about 54% SiO_2, 15% Al_2O_3, 17% CaO, 5% MgO and 8% B_2O_3.

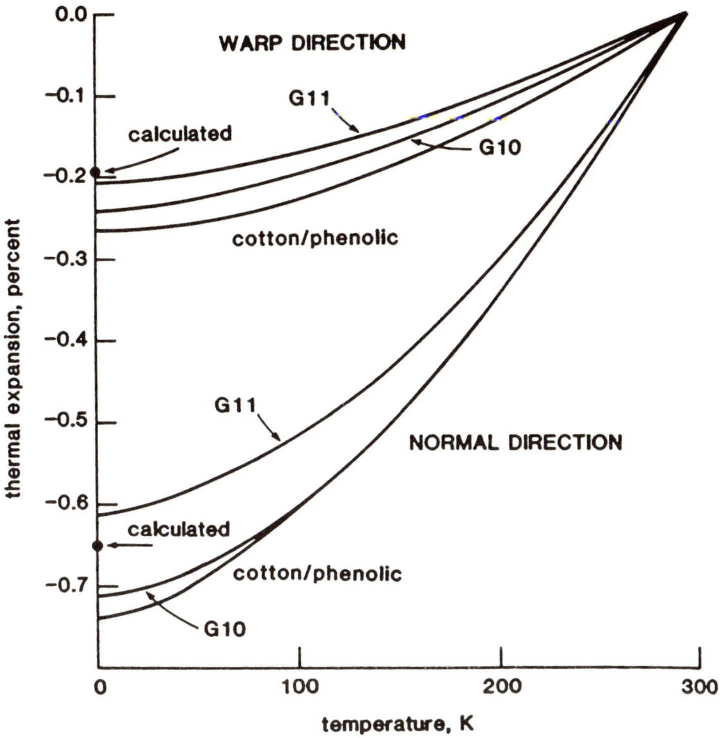

Fig. 7.3. Thermal contraction relative to 293 K of glass-cloth epoxy composites, G10 and G11, and of a cotton–resin composite [Cla81].

Commercial glass-cloth epoxy laminates have become very important as cryogenic structural supports to provide thermal and electrical insulation for large superconducting magnets. This necessitated performance specifications being established by the NBS (now NIST) and the National Electrical Manufacturers Association (NEMA) in the USA. These are based on a woven fabric of silane-finished E-glass in an epoxy resin — heat activated, amine-catalyzed, bisphenolA [Kas80]. These laminates, denoted as G-10CR and G-11CR, have measured contraction ratios (293 to 4 K) of about 0.2% in the warp direction and 0.6–0.7% in the direction normal to the warp of the glass cloth [Kas80].

Clark *et al.* [Cla81] tabulate extensive data for G-10 and G-11 and other cotton–phenolic laminates used in superconducting magnets (Fig. 7.3), including Cu/NbTi wire and NbTi wire interwound with fiberglass cloth. Table C.2 in the Appendix gives values for the contractions of some materials widely used in cryogenics, including G-10.

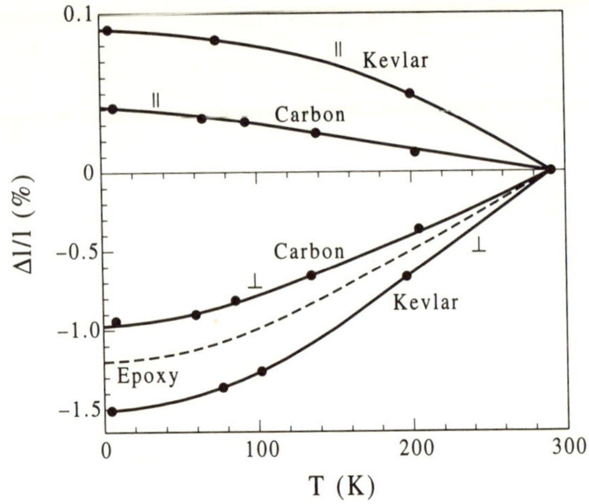

Fig. 7.4. Thermal contractions relative to 293 K observed transverse and parallel to a UD Kevlar fiber in epoxy and to a UD carbon fiber in epoxy. Note scale change above and below the horizontal axis [Har88a].

Carbon-Fiber Composites. Graphite single crystals have hexagonal structure and are highly anisotropic (see Section 5.8.2), with weak Van der Waals forces along the *c*-axis and strong covalent bonding in the *ab*-plane. Fibers are generally small graphitized areas 'cylindrically' aligned [Har84] in the fiber direction so that the fiber axis is roughly equivalent to the *a*-direction and has negative values of α at normal and low temperatures. Thus the expansivity, conductivity and strength of epoxy laminates made from carbon-fiber mats depend critically on fiber alignment and the choice of cross-ply angles. Values of contraction $\Delta l/l$ on cooling to 4 K may be varied from around 1.0% to -0.04% (i.e., an expansion of 4×10^{-4}), depending on choice of alignment [Har88a].

A series of measurements by Yates and collaborators at the University of Salford examined the effect of changing fiber type, fiber weave, orientation, resin type, curing, etc. (e.g., [Yat82]).

Kevlar Composites. Kevlar fibers consist of stretched aramide molecules with strong covalent bonding in the fiber direction only. They are more anisotropic than graphite fiber. The negative coefficient α in the fiber direction is due to transverse vibrations normal to the molecular chain alignment [Har84], as discussed in Section 5.9.

Figure 7.4 shows the differences in $\Delta l/l$ (relative to 293 K) for a Kevlar fiber composite cooled to 4 K. Transverse to the fiber there is a contraction of 1.5%, while in the longitudinal fiber direction there is an *expansion* of about 0.09%.

Chapter 8

Cryocrystals, Clathrates and Curiosities

8.1. CRYOCRYSTALS

8.1.1. Introduction

The term *cryocrystals* is used for the solidified forms of those gases which become solid at low temperatures. The monatomic members of this family (He, Ar, Ne, etc.) have been discussed in Section 5.2 and in the two volumes on rare gas solids edited by Klein and Venables [Kle76, Kle77].

Here we select those molecular gases which are of general cryogenic interest, including hydrogen, nitrogen, oxygen, carbon dioxide, methane. Many of the pioneer investigations of these crystals were done by Professor Manzhelii and his colleagues at the Low Temperature Institute in Kharkov. Some of the crystals, notably H_2, D_2, N_2, CO, CO_2, and N_2O are discussed in detail in *Physics of Cryocrystals* by Manzhelii and Freiman [Man97].

8.1.2. Solid Hydrogen

Both hydrogen and deuterium contain molecules of two kinds, 'ortho' and 'para'; in o-H_2 the spins of the two protons are parallel to each other, and in p-H_2 they are antiparallel (see Section 4.2.3). Due to energy differences between the two states, equilibrium concentrations vary from 100% para-hydrogen at the lowest temperatures to about 20% at room temperature. When cooled and solidified, the transformation to equilibrium is a slow process (in the absence of a catalyst), extending over many days and involving the release of heat of transformation amounting to 670 $J{\cdot}g^{-1}$ for H_2. The crystal structure is hcp with a *c/a* ratio at 0 K of 1.633, very close to the ideal packing ratio, $(8/3)^{1/2}$.

The triple point of equilibrium hydrogen at 13.8033 K is a defining temperature on the International Temperature Scale ITS-90.

The heat capacity of solid para-hydrogen has been measured from near 1 K up to the melting line for a number of capsules sealed at different pressures, corresponding

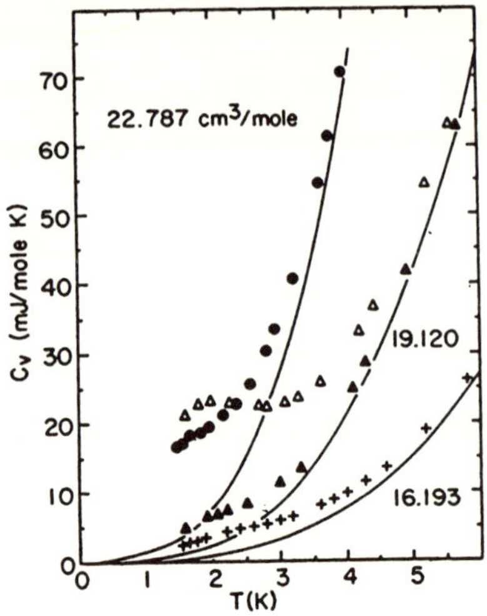

Fig. 8.1. C_V data for para-H_2 containing low levels of ortho-H_2 impurities. Data at 19.120 $cm^3 \cdot mol^{-1}$ are for two different concentrations of ortho-H_2. Solid lines represent lattice contributions [Kra80].

to molar volumes ranging from 22.79 to 16.19 $cm^3 \cdot mol^{-1}$ (see Fig. 8.1 from [Kra80]). The data were analyzed to give the volume and temperature dependence of Θ_D^C, the Grüneisen γ, and the equation of state for p-H_2, as well as the volume dependence of the quadrupole interaction parameter for the residual o-H_2 impurities.

The values of C_V of the nearly pure p-H_2 fit a relation of the form

$$C_V = C_{\text{ortho pairs}} + A_3 T^3 + A_5 T^5 + \ldots$$

The T^3 terms give values of Θ_0 ranging from 127 K (for $V_0 = 22.79$ $cm^3 \cdot mol^{-1}$) to 255 K ($V = 16.19$ $cm^3 \cdot mol^{-1}$). Extrapolation gives values at $P = 0$ ($V = 23.23$ $cm^3 \cdot mol^{-1}$) of $\Theta_0 = 120.6$ K, $\gamma_0 = 2.344$ and $B_0 = 0.176$ GPa.

At higher pressures, Θ_0 and γ_0 are given by [Kra80]

$$\ln \Theta_0 = 8.4528 - 0.614 \ln V - 0.0744V$$

and

$$\gamma_0 = -(d \ln \Theta_0 / d \ln V) = 0.614 + 0.0744V$$

The Kharkov group [Ale89] have made direct measurements of linear expansion of a free-standing rod of para-hydrogen from 0.9 to 5 K. They fitted their results to the equation $\alpha = 1.82 \times 10^{-7} T^3$ K^{-1} and obtained

$$\gamma = 2.0 \pm 0.25$$

They later measured a free standing rod of ortho-deuterium from 1.5 to 5.6 K and fitted to a relation [Sol92] $\alpha = (1.475 \pm 0.04) \times 10^{-7} T^3 + O(T^5) + O(T^{-2})$ K^{-1} for

Table 8.1. Some basic parameters for the N_2-type crystals from [Man97, p. 216, chs. 13,14] and other solid gases. T_t is the triple-point temperature at pressure P_t. $T_{\alpha\beta}$ is the temperature of the α-β phase transition at $P = 0$. Asterisk* denotes value at 51 K [Ste59]

Parameter	N_2	CO	N_2O	CO_2	p-H_2	O_2	CH_4
Θ_0^C (K)	83.6	103.3	141.0	151.8	121	104.5	—
V_0(cm^3/mol)	27.1	27.1	27.0	25.8	23.23	23.5*	31.6
B_0 (GPa)	2.21	2.52	6.70	7.58	0.176	$3._0$*	—
$\gamma_{\sim 10K}$	2.5	2.9	2.1_5	2.1	2.34	—	—
T_t (K)	63.15	68.13	182.35	216.57	13.803	54.3584	90.69
P_t (kPa)	12.5	15.4	87.9	518	7.034	0.1464	11.70
$T_{\alpha\beta}$(K)	35.61	61.55	—	—	—	23.88	—
						43.80($\beta\gamma$)	—

$T > 3.5$ K. The T^3 term in the heat capacity corresponds to $\Theta_0^C = 114$ K. The data below 4 K could be fitted with a lattice term plus a Schottky term with a maximum near 2 K.

X-ray measurements of expansivity of solid p-H_2 show that the anisotropy is not significant between 4 and 9 K but that α_a exceeds α_c for T$\geq$ 10 K [Man97, p. 100].

8.1.3. N_2-Type Solids

The crystals N_2, CO, N_2O and CO_2 are all formed from linear molecules, the molecular units occupying a fcc lattice with the *Pa*3 structure in the orientationally ordered low-pressure phase, see [Man97, Ch. 10]. Another common feature of these solids is that the intermolecular forces are much weaker than the intramolecular ones: the intramolecular vibrational frequencies are an order of magnitude higher than the intermolecular frequencies, corresponding to wave numbers $\sim$ 1000 cm^{-1}. As far as lattice vibrations are concerned, the molecules can be treated as undeformable. The two main types of motion are translational and rotational.

Some of these crystals transform from their low-temperature α-phase (*Pa*3) to a β-phase at higher temperatures. For N_2, the β-phase which is stable above 35.6 K is hexagonal $P6_3/mmc$. For CO, a similar transformation occurs at 61.6 K. There is no clear evidence of other phases in CO_2 and N_2O [Man97]. For residual entropy in CO and N_2O, see Section 2.2.5.

Some important parameters are given in Table 8.1. They include values for the triple point temperature and pressure, T_t and P_t, the α-β transition temperature $T_{\alpha\beta}$, Debye temperature Θ_0^C and bulk modulus B_0. Most values are from the extensive tabulations in [Man97, Chs. 10,12-14] which include references to much experimental data on sound velocities, lattice spacings and heat capacity. Data sources for the other solid gases (H_2, O_2 and CH_4) are discussed below. Triple point data are also given by Jacobsen *et al.* [Jac97]. A few representative values of C_P, C_V, β and γ for α-N_2, β-N_2 and CO_2 in Table 8.2 are from [Man97].

Table 8.2. Thermal data for N_2 and CO_2 at selected temperatures from [Man97, ch. 14]

T (K)	β (10^{-4}/K)	C_P (J/mol·K)	C_V (J/mol·K)	B_S (GPa)	γ
	α-N_2				
10	2.0	4.83	4.81	2.19	2.44
20	7.2	19.9	19.3	2.15	2.13
30	14.1	34.2	31.2	2.00	2.27
35	24.7	44.6	35.9	1.84	2.83
	β-N_2				
40	18.7	37.8	32.8	1.45	2.04
50	20.5	41.5	35.0	1.34	1.84
60	25.3	45.6	35.9	1.10	1.79
	CO_2				
20	0.56	5.14	5.13	7.57	2.13
30	1.42	12.95	12.83	7.55	2.13
40	2.15	19.64	19.29	7.52	2.13
60	3.32	30.05	28.82	7.44	2.14
80	4.00	35.9	33.6	7.34	2.14
100	4.52	39.9	36.4	7.16	2.15
150	5.83	47.7	40.3	6.35	2.17
200	8.70	56.8	41.5	4.92	2.30

8.1.4. Solid Oxygen

The two lower temperature phases of solid oxygen, α and β, are both layered with almost hexagonal packing of the molecules (see [Man97, p. 359]). The transition temperatures under saturated vapor pressure are $T_{\alpha\beta} = 23.88$ K and $T_{\beta\gamma} = 43.80$ K. The triple point is at 54.3584 K, a defining temperature for the ITS-90 Scale. Ancsin [Anc75] measured C_P and the latent heats through the α-β and β-γ transitions because of their importance as temperature-scale fixed points. Burford and Graham [Bur69] measured C_P from 0.8 to 4.2 K and hence Θ_0^C for solid O_2 as well as for N_2, CO and NO. There are also earlier C_P measurements of Giauque and colleagues on these solids (see [Bur69]).

Manzhelii *et al.* [Man66] measured α from 21 to 45 K, noting discontinuities near 24 and 44 K. Stewart [Ste59] measured isothermal compressibilities of O_2 (and methane) over the pressure range 0 to 19 kbar (0 to 1.9 GPa) and temperature range 4 to 120 K, and mapped the phase diagram.

8.1.5. Solid Methane

Methane is another example of weak intermolecular forces so that rotational motion of the molecular units is important even at low temperatures. With its isotopes and isotopic mixtures, which have various symmetries, methane presents an intricate problem and has been the subject of much study, theoretical as well as experimental (see [Par78, Section 9.7], and for thermal expansion only, see [Bar80,

p. 637, p. 670]). As with ortho and para H_2 and D_2, thermodynamic measurements have to take account of slow interconversion between different nuclear spin species. Many different phases are found depending upon isotopic composition as well as on temperature and pressure, and different kinds of anomalies in the heat capacity and thermal expansion due to transitions between them. Thermodynamic properties can be strongly affected by impurities, particularly from Schottky contributions at low temperatures.

CH_4 has fcc structure. Below the melting point, in phase I, the molecules are distributed between twelve equivalent orientations, with large librational and translation amplitudes. At 20 K there is a transition to 'phase II,' which has a remarkable structure: molecules are arranged on the fcc lattice such that on three out of four sites the molecules librate about an ordered orientation, while in the fourth the molecule undergoes hindered rotation. At still lower temperatures ($\sim$ 8 K) there is a broad transition to a state in which all the molecules appear to be rotating, but this behavior is dependent upon isotopic purity and inter-spin conversion.

Early heat capacity measurements on the different isotopic species were performed by Clusius and later greatly extended by A. Sperandio and by J. A. Morrison and colleagues (e.g., [Col63]). Their findings are summarized and discussed in [Par78].

The thermal expansion below 60 K has been measured by the Kharkov group (see [Bar80, p. 679]) and Heberlein and Adams [Heb70]. Values of $\gamma \simeq 1.7$ are obtained for the region from 20 to 60 K. Low temperature values for CD_4 give γ as about 2.0 below 8 K, rising to about 2.8 at 17 K.

8.1.6. Solid Ammonia

Manzhelii and Tolkachev [Bar80, p. 679] measured C_P and α from 24 to 175 K. They fit these to Debye and Einstein terms with $\Theta_0^C \simeq 220$ K and $\gamma_D \simeq 1.35$ from 30 to 50 K. They note that C_V and α increase above 100 K much more rapidly than the Debye model predicts and suggest this contribution is due to hindered rotation of the molecules.

8.2. OTHER ROTATIONALLY DISORDERED CRYSTALS

8.2.1. Introduction

The crystals of general cryogenic interest described in the preceding section are members of a vast group of *orientationally disordered crystals*, in which approximation to spherical or cylindrical shape can permit hindered rotation of a molecule or molecular group. These solids are usually weakly bound, with low melting points and easily deformed, and so used to be called 'plastic crystals' (e.g., [She79]; but this name was obviously inappropriate for a group which contained strongly ionic ammonium salts. Like the cryocrystals, they display a rich variety of phenomena at

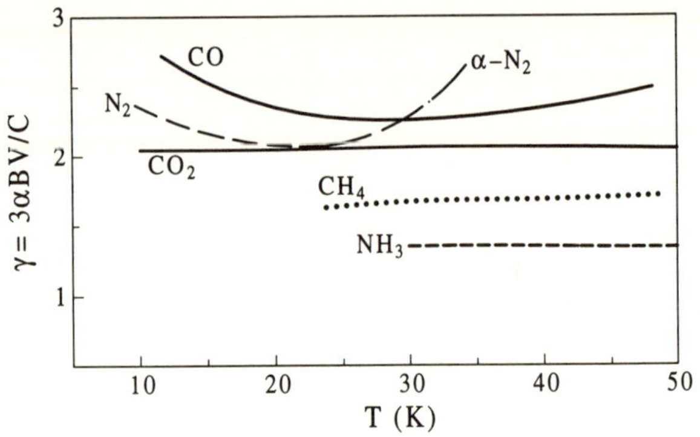

Fig. 8.2. Comparison of $\gamma(T)$ for some cryocrystals (see [Bar80, p. 678] and [Man97]).

cryogenic temperatures, with phase transitions between different kinds of rotational ordering manifest in their thermodynamic properties. The elucidation of this intricate behavior is aided by other measurements, including X-ray and neutron diffraction, NMR, Raman and infra-red spectroscopy, dielectric and acoustic studies, etc. Here we can give only a few brief examples; for references and further discussion see [Par78, She79].

8.2.2. Hydrogen Halides

Solid state transitions have been observed in the three heavier solids, but not in HF (Table 8.3). The explanation for this is to be found in the larger molecular dipole moment of HF, which fixes more firmly the molecular orientation; it keeps its full orientational order up to the melting point, as shown by the large entropy of melting. At low temperatures HCl has an ordered crystal structure with the molecules arranged in zigzag chains, but it loses a large part of its orientational order in a first order transition to an fcc phase at 98.4 K; a large increase of electric permittivity confirms that in this phase molecules can reorientate. In contrast, early heat capacity measurements by Giauque and Wiebe [Gia28a, Gia28b] revealed three separate lambda transitions in HBr (Fig. 8.3). There is an enormous peak in the

Table 8.3. Phase transitions in hydrogen halides; entropies and dipole moments

Halide	HF	HCl	HBr	HI
T_m (K)	190	159	186	222
ΔS_m (J·mol^{-1}·K^{-1})	24.1	12.5	$\sim$ 12.6	—
μ (Debye)	1.91	1.08	0.80	0.42
Transition temperatures (K)	none	98.4	117 113 90	125 69

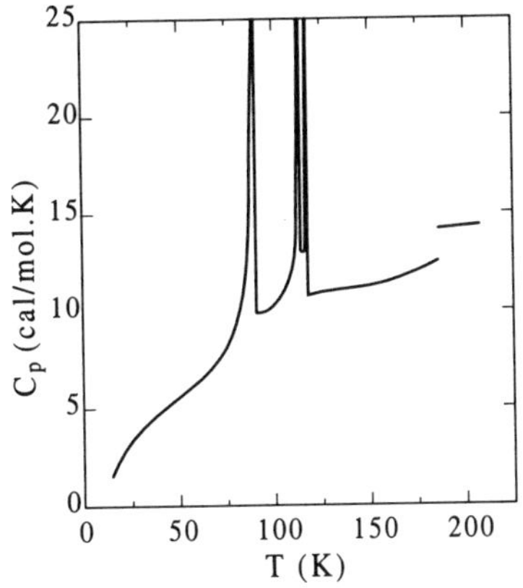

Fig. 8.3. Heat capacity of HBr, showing three solid state phase transitions. Note 1 cal = 4.184 J. From [Gia28b].

permittivity at the 90 K transition, possibly indicating a ferroelectric order–disorder transition. Neutron diffraction on DBr shows that above 90 K each molecule can orient in one of two directions, compared to a unique direction in an ordered zigzag structure below 90 K. HI is rather similar to HBr.

8.2.3. T-Butyl Chloride (TBC) and T-Butyl Bromide (TBB)

There are many rotationally disordered organic solids, of which we take the tertiary butyl halides as an example. Calorimetric data reveals three separate solid state phases for both TBC and TBB, all separated by first order transitions (Table 8.4). The phases are labelled I, II, III, with I adjacent to the melting point. The molecules have the structure $XC(CH_3)_3$, where X is the halogen, with an electric dipole moment along the three-fold symmetry axis. The possibilities for orientational disorder include not only the direction of the dipole but also the orientation of the molecule about the dipolar axis, and perhaps also internal rotation of the methyl groups about their C–C bonds.

X-ray and neutron diffraction show that in both compounds phase I is fcc, with random orientations. For TBC phase II is tetragonal, with the C–Cl bonds fixed along the 4-fold axis and random orientation of the molecules about the bonds; phase III has

Table 8.4. Phase transitions in tertiary butyl halides

	TBC		TBB	
	T_t (K)	ΔS (J·mol^{-1}·K^{-1})	T_t (K)	ΔS (J·mol^{-1}·K^{-1})
III → II	183	10.2	209	27.2
II → I	219	25.8	231	4.6
I → liquid	248	8.0	256	7.5

orthorhombic symmetry. Phases II and III of TBB appear both to be orthorhombic, but with different structures from each other and from phase III of TBC.

The low ΔS on melting shows that both compounds have high orientational disorder in phase I, which for TBB largely persists in stage II, as confirmed by high permittivity in both phases. Phase II for TBC is more ordered, and permittivity shows that the dipoles cannot reorient; but ordering in either compound is not complete until phase III. Further discussion and interpretation of the results is given by Richardson and Taylor [Ric84].

8.2.4. Ammonium Halides

Ammonium salts are hard ionic crystals, but with the possibility of disorder in the orientation of the NH_4^+ ions. In NH_4F there are strong directional hydrogen bonds between the ions, and the crystal is ordered at all temperatures, with the N and F atoms forming a tetrahedral wurtzite structure. The other halides have weaker hydrogen bonds and are rich in phase transitions.

ND_4Br. To illustrate the complexity of behavior in these salts, we take deuterated ammonium bromide, on which there have been extensive measurements. There are four solid state phases, numbered I or α to IV or δ) as the temperature is decreased. I has NaCl structure, with maxima in the smeared deuteron density between the N and Cl sites. The remaining phases have CsCl or distorted CsCl structure, where there are two possible orientations of the tetrahedral deuterons such that they point towards neighboring anions. In phase II (cubic) there is a random distribution of these orientations, in phase III (tetragonal) neighboring ND_4^+ ions are antiparallel in *ab* planes but parallel in the *c* direction, and in phase III (cubic) they are parallel in all directions.

Calorimetric data [Ste68] show a clear first order transition from I to II at 391 K, and lambda peaks at 215.1 K (broad, $\Delta S \simeq R\ln 2$) and 166.7 K (narrow, $\Delta S \simeq 0.15R$), indicating that III to II is a typical order–disorder transition. Surprisingly, X-ray data (see [Hov69] and [Bar98, Fig. 51]) indicate discontinuous jumps in volume at *all* the transitions, and do not reflect in any way the broad peak seen in the heat capacity.

Similar phases are seen in other ammonium halides, although the δ-phase is absent from the iodide and the γ-phase from the chloride. Pressures of the order of kilobars modify behavior considerably.

8.3. CLATHRATES

8.3.1. Introduction

Clathrates, named after the Greek word for cage (*clathros*), are lattice inclusion compounds in which a guest molecule is confined by barriers created by the architecture of the crystalline host. The interaction between guest molecules and host in clathrates is usually weak in comparison with covalent chemical bonding. Many hosts owe their existence and stability to the presence of the guest molecules within the 'cage.'

Technically and cryogenically important are the gas hydrates, because of the significant amounts of natural gas (largely CH_4) which may be locked into clathrate ice (under pressure) in many colder parts of the earth. This has stimulated research on the thermal properties of those ice clathrates which can be made in the laboratory and are stable at atmospheric pressure. These should mirror the properties of natural gas hydrates which are only stable under pressure. Generally the thermal conductivity of the ice clathrates is many times smaller than for normal ice in the range 100 to 200 K, although values of heat capacity (and expansion) are not very different. The thermal conductivity is reduced by rotational disorder.

The book *Disorder in Crystals* [Par78] includes a chapter on clathrates and channel compounds, discussing particularly the evidence for rotational disorder. Before discussing clathrate ices, we describe briefly low temperature data on two organic systems — the quinol clathrates containing rare gas atoms and diatomic molecules, and the clathrates of Dianin's compound.

8.3.2. Gases in β-Quinol

Hydroquinone when crystallized in the β-quinol structure forms a clathrate compound with regularly spaced holes which can trap foreign molecules of suitable size. Although the host β-quinol cannot exist on its own, its properties can be estimated by extrapolation of guest concentration to zero. Measurements of C_P for the trapped molecules show that above 100 K or so they 'rattle around' much as in a gas, but the 'rattling' motion becomes frozen at lower temperatures. The molar contribution to C_P of the monatomic guests Ar and Kr is approximately constant above 100 K, with values of about $2.5R$ (see [Par78, Fig. 11.2]). This lies between the values for a classical particle in a box and that for a three-dimensional classical harmonic oscillator. Below about 50 K, there is a sharp drop in this contribution (Fig. 8.4).

For diatomic guests such as CO, N_2, O_2, there is a hindered rotation of the molecules in the cage in addition to the translational motion. A study of the far-infrared spectra of the monatomic gases showed a single spectral line while the polar diatomic molecules showed an additional line presumed due to hindered rotation. Burgiel *et al.* [Bur65] calculated the specific heat from the infrared data; they found good agreement above 100 K or so, but the calculated rate of change at low temperatures was faster than that observed, due perhaps to coupling effects neglected in their theory. More C_P data below 80 K are needed to elucidate the problem.

8.3.3. Dianin's Compound and Its Clathrates

Dianin's compound (4-p-hydroxyphenyl-2,2,4-trimethylchroman) is a complex organic crystal containing empty cages with relatively narrow waists. It can have the

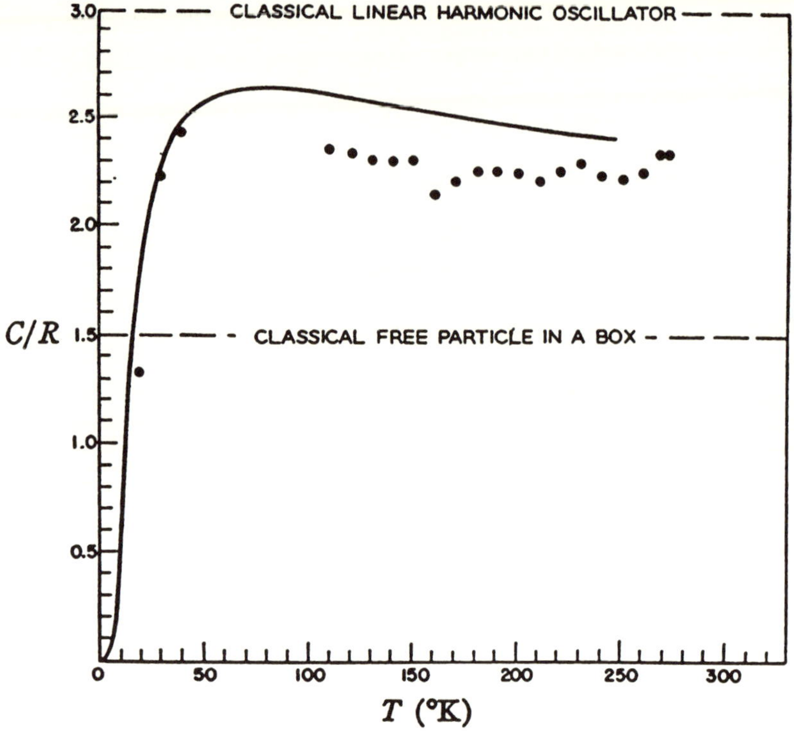

Fig. 8.4. Heat capacity of Ar trapped in β–quinol. Experimental points (see [Par78, p. 724]) are compared with a curve calculated from the infrared spectrum [Bur65].

same crystalline form both when it is pure and when it is a clathrate host, enabling direct comparison to be made of experimental properties. More than fifty different guest species have been identified, but we consider here only the pure compound and its ethanol and carbon tetrachloride adducts. Ethanol is included as a dimer, with one molecule in each half of the cage; in contrast there is a single CCl_4 molecule in each cage, with one Cl as 'foot' in the waist of the cage and the other half of the cage empty.

Heat capacities, lattice parameters and thermal conductivities have been studied down to 50 K by M. A. White and her collaborators (see [Zak91] for references). The effective heat capacity of the guest CH_4 molecules, obtained as C_P(clathrate)$-C_P$(host), was similar to that of the bulk solid, except for the absence of bulk phase transitions which depend upon interactions between neighboring molecules. The heat capacity of guest ethane was closer to the bulk liquid than to the solid.

Estimated Grüneisen functions of the clathrates were similar to that of the pure host (~ 1), except below 100 K where they started to rise sharply with decreasing temperature. Despite uncertainty in both the X-ray and elastic data, this rise appeared to be significant, indicating extra low frequency optic modes with a strong strain dependence.

8.3.4. Clathrate Ice

Specific Heat. Measurements from 100 to 200 K of various ices containing tetrahydrafuran (nominally THF$\cdot 17H_2O$), ethylene oxide (EtO$\cdot 7H_2O$) and a cyclobutane (C$\cdot 16H_2O$) gave values of c_P (per gram) of about 1.0 $J{\cdot}g^{-1}{\cdot}K^{-1}$ at 100 K, increasing to 1.8 $J{\cdot}g^{-1}{\cdot}K^{-1}$ at 260 K; These do not differ by more than 10% from values for normal (hexagonal) ice [Lea82, And83a].

Other heat capacity measurements on THF clathrate hydrate were made from 17 to 261 K by M. A. White and Maclean [Whi85b]. The results were consistent with free or nearly free rotation of the guest molecules above 120 K, but with increasingly hindered rotation below this temperature. Normal hexagonal ice seemed to be a good model for the behavior of the host above 120 K, but not at the lower temperatures.

Thermal Expansion. Measurements of the linear expansion of a rod of tetrahydrafuran ice (THF$\cdot 17H_2O$) from 80 to 265 K gave values which were higher by about $20 \times 10^{-6} K^{-1}$ than for normal ice. The corresponding values of the Grüneisen γ were between 0.8 and 0.9 compared with 0.6 for normal hexagonal ice [Rob84].

8.4. CURIOSITIES

This section includes a number of solids for which there are low temperature data of interest but which do not fit naturally with other major groups.

8.4.1. $Al_{10}V$: An Einstein Solid or 'Particle in a Box'

In Section 5.12 we discussed the effect of introducing a heavy ion impurity into an alkali halide lattice and the possibility of creating a local low-frequency mode which could enhance the heat capacity and expansion at low temperatures, depending on the degree of coupling to the host lattice. Truly localized vibrational modes are unusual because of the strength of the interatomic coupling. With tight coupling, we might expect a 'foreign' ion to move roughly in phase with the neighboring lattice ions: the effect on thermal properties would be less obvious, being distributed over an extended energy and temperature range unlike the localized effect observed in some of the clathrate 'cages' (see Section 8.3).

An example among intermetallic compounds of a localized 'Einstein' mode is $Al_{10}V$, where the specific heat (and electrical resistivity) below ~ 20 K were found to be many times greater than the expected lattice contribution based on a Debye temperature of 390 ± 10 K. C_P data below 5 K (see Fig. 8.5) were fitted well by an

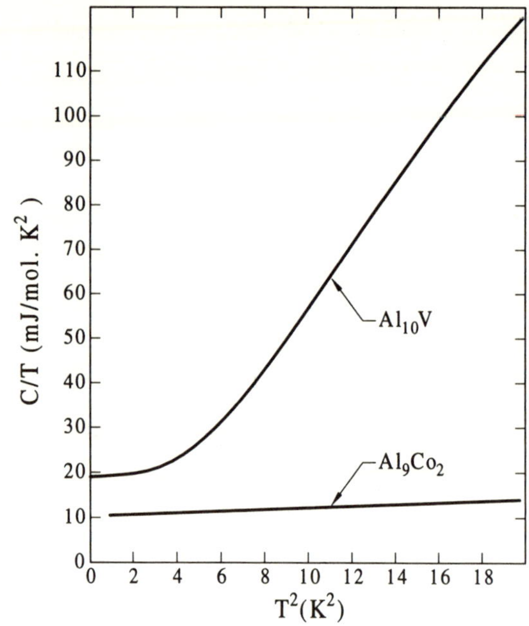

Fig. 8.5. C_P of $Al_{10}V$ (upper curve) compared with Al_9Co_2 plotted as C/T versus T^2 [Cap73, Cap78].

Einstein function with $\Theta_E = 22$ K and a linear term with $\Gamma = 19$ mJ·mol^{-1}·K^{-2} (see [Cap73, Fig. 8.3]). These observations and X-ray evidence indicate that the mode is associated with a 'loose' Al atom occupying a large hole in the $Al_{10}V$ structure, reminiscent of the behavior of the rare gas clathrates (Section 8.3). In the unusual structure of $Al_{10}V$ the cubic unit cell contains 16 formula units (176 atoms) with many Al atoms clustered closely around the V sites. However there are a few extra Al atoms that occupy cage-like sites with 16 Al neighbors at a much larger distance of 3.1 to 3.2 Å [Cap78].

The thermal expansion coefficient of $Al_{10}V$ has a positive 'bump' centered around 10 K which is much larger relatively than that in C_P. A plot of $\ln \alpha T^2$ versus $1/T$ gave a value for the excitation energy associated with this bump which corresponds to an Einstein term of $\Theta_E = 22 \pm 0.3$ K [Leg78]. The magnitude of the bumps in α and in C_P lead to a value of $\gamma_E \simeq 80$. This very large value denotes a large volume dependence of the local energy potential, suggestive of a tunnelling process analogous to that of the off-center Li ions in KCl discussed in Section 5.12.4. Indeed the pseudopotential calculations discussed by Caplin and Nicholson [Cap78] suggest that the Al ions in the Al 'cage' may occupy potential minima offset from the center in [111] directions by about 0.5 Å. This leaves $Al_{10}V$ as an interesting problem with tunnelling as a possible explanation.

8.4.2. α-Cristobalite

We include this cubic form of SiO_2 among the 'Curiosities' because the low temperature values of heat capacity and thermal expansion appear to contradict the behavior of quartz and vitreous silica.

The many forms of silica including the so-called 'stuffed derivatives' such as β-eucryptite (Section 5.7) are an interesting challenge to physicists and crystallographers, as may be seen from reading the volume 29 of Reviews in Mineralogy entitled *Silica* [Hea94]. The low pressure phases include α-quartz (density $\rho = 2.650$ g·cm^{-3}), β-quartz ($\rho = 2.533$), α-cristobalite ($\rho = 2.33$) and tridymite, ($\rho = 2.3$). The α- and β-quartz have structures based on 3, 4, and 6-fold spirals of SiO_4 tetrahedra, some of which can accommodate smaller ions such as Li, e.g., in the LiAlSi family (eucryptite and spodumene).

For α-cristobalite, C_P was measured from 5 to 300 K by Westrum [Wes63] and from 2 to 20 K by Bilir and Phillips [Bil75]. Results showed $\Theta_0^C \approx 475$ K and that Θ^C falls very rapidly with increase in temperature to a minimum of about 270 K near $T \sim \Theta_0/35$. This is clear from the sharp rise in c/T^3 (Fig. 8.5) which is very like that observed for vitreous silica but quite unlike the slower rise seen in quartz or indeed in most other crystals. Bilir and Phillips comment on the similar positioning of the Si atoms in cristobalite to those in silicon and diamond-like crystals, suggesting that dispersive transverse acoustic modes may be important in the determining $c(T)$. However the minimum in Θ^C for the diamond-like crystals is closer to $\Theta_0/15$.

The expansion coefficient, shown as α/T^3 in Fig. 8.6 [Whi76a] behaves in very similar fashion to c/T^3, but there is no sign of negative expansion as occurs in vitreous silica and many diamond-like crystals. Without more details of the frequency spectrum of cristobalite, it is not certain whether TA modes or low-lying optic modes are responsible for the rather unique patterns shown in Fig. 8.6. Inelastic neutron scattering experiments on polycrystalline specimens [Lea69] showed that the lowest energy acoustic branch of the spectrum of cristobalite was very flat with the top frequency being centered around 40 cm^{-1}, slightly lower than that observed for vitreous silica. Leadbetter concluded that there is a broadening of the lowest energy (transverse) acoustic branch in the glassy state due to disorder.

8.4.3. Two Silicates: Mg_2SiO_4 and Zn_2SiO_4

The two silicate minerals forsterite ($2MgO{\cdot}SiO_2$) and willemite ($2ZnO{\cdot}SiO_2$) might seem like close relatives, but their lattice dynamics reflects major differences in their structural arrangement.

Forsterite is one of the olivine family of minerals, $(Mg,Fe)_2SiO_4$, which are important constituents of the earth's upper mantle. It has an orthorhombic Pbnm structure, only slightly distorted from a hexagonal close-packed array of oxygen anions with the orthorhombic b, c axes in the basal plane. The smaller Mg and Si cations sit in octahedral and tetrahedral interstices respectively. The elastic anisotropy is not

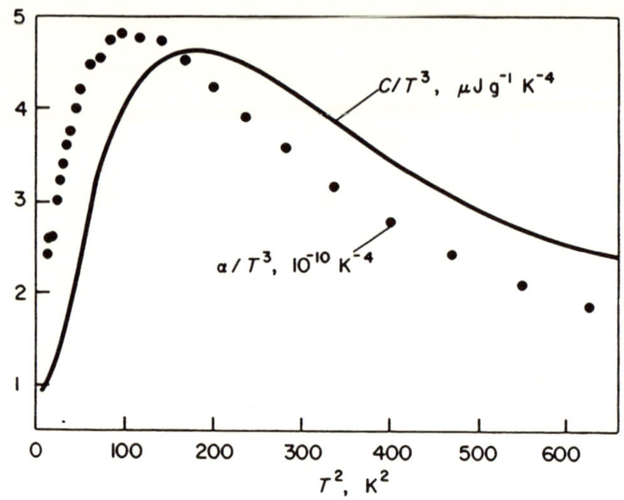

Fig. 8.6. α-cristobalite. A plot of α/T^3 in units of 10^{-10}K^{-4} and c/T^3 in units of μJg^{-1}·K^{-4} [Bil75, Whi76a].

very marked, with linear compressibilities $\chi_b(3.43 \times 10^{-12}) > \chi_c(2.66 \times 10^{-12}) > \chi_a(1.80 \times 10^{-12}\ \text{Pa}^{-1})$ and $B_0 = 130$ GPa [Gra69, Sum77]. Measurements of C_P from 5 to 380 K give $\Theta_0 = 768 \pm 15$ K (cf. Θ_0^{el}=766 K) with $\Theta_{min}^C \simeq 670$ K near $\Theta/15$ [Rob82]. The three principal linear expansion coefficients measured from 2 to 150 K are all positive (see selected values in Table 8.5); they give principal values of the Grüneisen parameters $\gamma_{a,b,c} \simeq 1.2$ at 300 K, and between 1.3 and 1.5 below 30 K [Whi85a].

Willemite. This compound with the phenacite structure can be envisaged in terms of tetrahedral groups with open channels along the c-axis (see [Hea94, p. 84]). The anisotropy and negative values of both $\alpha_\perp$ and $\alpha_\parallel$ at low temperatures are shown by the selected values given in Table 8.5. The volume coefficient remains negative up to about 250 K, reminiscent of the behavior of the tetrahedrally bonded zincblende and wurtzite crystals discussed in Section 5.5.

Low temperature data on C_P are limited to the range 53 to 298 K [Tod51], giving $\Theta^C \simeq 435$ K at 53 K increasing up to 815 K at room temperature. Slack and Huseby [Sla82] give estimates of $\Theta_0 \simeq 700$ K and $B_S = 155$ GPa based on a model of the ZnO_4/SiO_4 arrangement. The values of γ in Table 8.5 are dubious because of the paucity of elastic data, but the trend of $\gamma(T)$ is very like that shown by the zincblende and wurtzite family. The contrast with the more closely packed forsterite (of olivine structure) is obvious and in agreement with the discussion of Section 5.5.

Table 8.5. Selected values of α, C_P and γ for Mg_2SiO_4 ($B_0 = 130$ GPa, $V_0 = 43.5$ cm^3) and Zn_2SiO_4 ($B_S \simeq 155$ GPa, $V \approx 52.5$ cm^3) [Whi85a, Whi88]

T (K)	C_P (J/mol·K)	α_a (10^{-6}/K)	α_b (10^{-6}/K)	α_c (10^{-6}/K)	β (10^{-6}/K)	γ
Mg_2SiO_4						
10	0.030	0.0016	0.0044	0.0027	0.0087	1.4_5
20	0.275	0.013	0.036	0.022	0.071	1.4
30	1.06	0.041	0.131	0.073	0.245	1.3
60	9.8	0.30	1.00	0.56	1.86	1.1
100	32.2	1.39	3.30	2.40	7.09	1.2
150	62	2.95	6.20	4.95	14.1	1.26
300	119	6.6	11.3	10.0	27.9	1.25
Zn_2SiO_4		$\alpha_\perp$		$\alpha_\parallel$		
10	—	−0.03		−0.001	−0.062	—
20	—	−0.21		−0.114	−0.53	—
40	(10.8)	−0.98		−0.58	−2.54	−1.8
60	27.2	−1.55		−0.80	−3.90	−1.15
100	52	−2.00		−0.82	−4.80	−0.75
130	68	−1.80		−0.45	−4.05	−0.48
300	124	0.15		2.00	2.30	0.16

8.4.4. ZrW_2O_8 ... Coupled Polyhedra?

An example of a compound containing fairly rigid polyhedra which shows negative expansivity at normal temperatures is ZrW_2O_8. The WO_4 tetrahedra and ZrO_6 octahedra are linked by common O atoms. Dilatometry and X-ray diffraction measurements above room temperature showed that over the range 320 to 470 K the average value of α is about -10×10^{-6} K^{-1}, apparently decreasing in magnitude at the lower temperature [Mar68]. Recent neutron and X-ray lattice spacing measurements near 4 K, 80 K, 150 K, 200 and 300 K appeared to give a similarly large value of the average α, but the data are insufficiently sensitive to show how α varies below 100 K [Mar96]. Because of motions transverse to the bonds, the open structure can be expected to favor negative expansion, and a model has been proposed in which transverse motions of oxygen atoms along the Zr–O–W linkage cause a coupled rotation of the linked polyhedra, giving cooperative motions of the tetrahedra described as *rigid unit modes* or 'RUMs,' see [Pry96]. Such coupled motions will undoubtedly occur, although modified to some extent by deformation of the tetrahedra. The validity of the model should be clarified by further accurate measurements, especially of dilatometric behavior at low temperatures and elastic moduli. There does appear to be a similarity to the behavior of SiO_4 tetrahedra in β-quartz (to which the model has recently been applied [Wel98]) and the aluminosilicates (Section 5.7.4), where also negative expansion persists to higher temperatures.

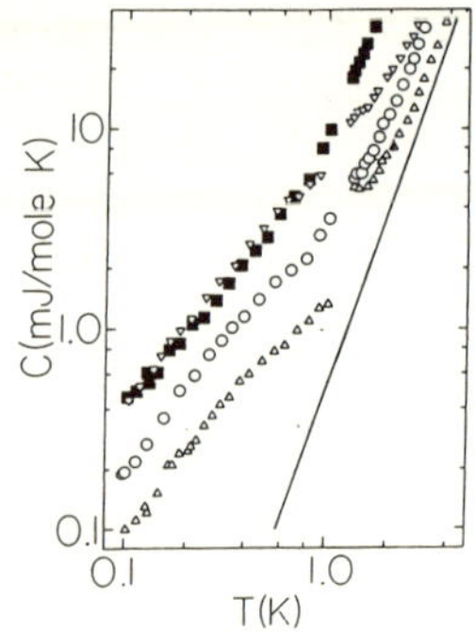

Fig. 8.7. C_P of M β-alumina. M=K (△), M=Na (○), M=Ag (filled square), M=Li (▽), — (Debye background) [McW77, Ant77].

8.4.5. β-Aluminas

β-aluminas containing a few percent of alkali or silver ions have high ionic conductivities, and are of potential use in energy storage systems. There is also a basic interest in the mechanism of conduction/diffusion, whether it is random hopping or a cooperative process. There is thus a need for low temperature measurements.

The heat capacity of β-alumina samples containing about 10 mol.% of M_2O and 90 mol.% Al_2O_3 (M = Ag, Li, Na, K, Rb) has been measured from 1 to 40 K [McW77]. The excess of C_P above a Debye background (based on $\Theta_D = 586$ K) increases progressively from Li (very small excess) through K, Rb, Na to Ag (the largest). For Ag, the excess is about five times the Debye term near 20 K; this behavior is like that of Ag in NaCl (see Section 5.12.3), which was attributed to heavy ion local modes.

At lower temperatures (≤ 3 K) C_P falls more slowly than T^3, suggestive of the tunnelling process evident in disordered solids. The excess in C_P (a possible tunnelling term) was measured below 1 K [Ant77] and expressed (in units of $\text{mJ}\cdot\text{mol}^{-1}\cdot\text{K}^{-1}$) as:

$$6.4T^{1.18}\ (\text{Ag});\ 7.3T^{1.22}\ (\text{Li});\ 3.1T^{1.18}\ (\text{Na});\ 1.3T^{1.12}\ (\text{K})$$

The thermal expansion was measured from about 8 K down to 0.8 K of Na β-alumina, as well as of other disordered solids — silica, epoxy, zirconia, etc. — in a study of the low energy excitations by Ackerman *et al.* [Ack84]. Their crystal was stated to be highly anisotropic, and measurements were only made normal to the c-axis. Results from 1 to 2 K were fitted to the relation

$$\alpha_{\perp} = 0.7 \times 10^{-10}T + 0.07 \times 10^{-10}T^3\ \text{K}^{-1}$$

leading to values of $\gamma = 8$ (for the T term) and $\gamma = 2$ for the T^3 term if B is taken to be 270 GPa. Their values of expansion near 8 K were consistent with unpublished data [And90b] obtained on a polycrystalline rod of Na β-alumina, for

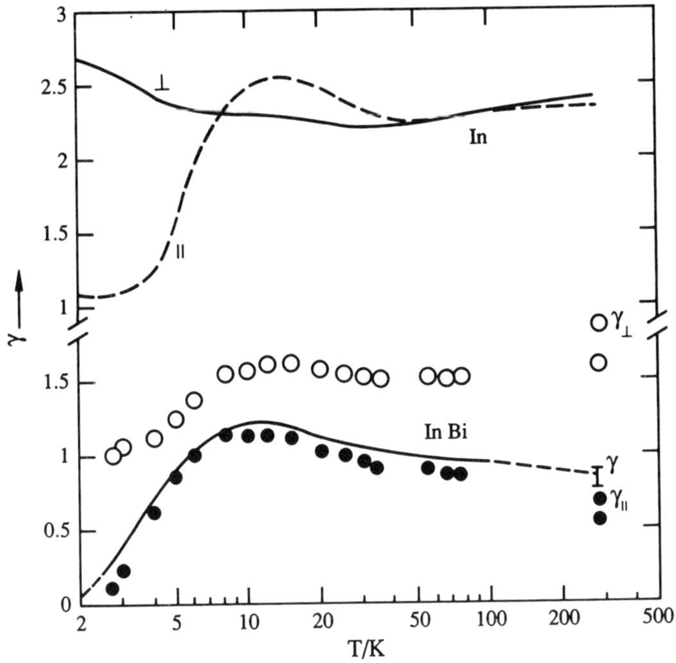

Fig. 8.8. Grüneisen functions for InBi (lower) compared with those for In (upper curves) [Whi90b].

which $\alpha = 2 \times 10^{-11} T^3$ K^{-1} from 5 to 25 K. Some values of α at higher temperatures were 0.37 at 30 K, 1.2 at 75 K and 4.95×10^{-6} K^{-1} at 293 K.

8.4.6. InBi

Indium bismuthide is a 'curiosity' because it appears to break the 'rule' (see Section 5.8) that anisotropic crystals expand more along the softer axes and less in the harder directions. This compound does the opposite. It is a semi-metal crystallizing in the B10 tetragonal structure with layers of In and Bi atoms arranged in planes normal to the c-axis. Each In atom is strongly (tetrahedrally) bonded to two Bi atoms in the plane above and two in the plane below, so that the three planes form a corrugated triple layer, like a sandwich with Bi as the bread and In inside. The sandwiches are piled on each other to form the crystal, and the bonding between neighboring Bi planes is assumed to be weak and responsible for the easy cleavage normal to the tetragonal axis. The elastic moduli measured at room temperature [Akg73] show large anisotropy, with compressibility $\chi_{\|}$ (c-axis) ten times larger than $\chi_{\perp}$ (basal plane). The weakest principal modulus is $c' = 7$ GPa, this being a shear wave propagated in the [110] direction. It softens under pressure, while another shear mode, c_{66}, stiffens under low pressure up to 1.2 GPa and then softens [Fri76].

The heat capacity measured from 1 to 35 K [Mar81] gives a value of $\Theta_0^C = 140$ K and a small electronic term $C_e = 0.8$ mJ·mol^{-1}K^{-1}.

The linear expansion curves $\alpha_\parallel(T)$ and $\alpha_\perp(T)$ look qualitatively similar to those for In (see Fig. 6.11) but are of different magnitude. $\alpha_\parallel$ (soft axis) has a shallow minimum near 4 K and a maximum near 10 K, and then falls monotonically, reaching a value of -85×10^{-6} K^{-1} at room temperature; $\alpha_\perp$ is always positive, reaching 61×10^{-6} K^{-1} at 293 K. The volume coefficient is positive, with γ about 1 below 50 K and about 0.6 at room temperature.

The principal γ values (Fig. 8.8) show that $\gamma_\perp$ exceeds $\gamma_\parallel$ at high temperatures by at least 50%, and at very low temperatures by considerably more. Their rôle vis-a-vis the compliance terms is clear from the relations for axial crystals (see Section 6.3.1):

$$\alpha_\perp = [(s_{11} + s_{12})\gamma_\perp + s_{13}\gamma_\parallel]C/V \tag{8.1}$$

$$\alpha_\parallel = [s_{33}\gamma_\parallel + 2s_{13}\gamma_\perp]C/V \tag{8.2}$$

For InBi, room temperature values of the compliances (in units of 10^{-11}Pa^{-1}) are

$$s_{11} + s_{12} = 3.46, \quad s_{13} = -3.21, \quad s_{33} = 8.83$$

The large negative cross compliance combined with the larger value of $\gamma_\perp$ compared to $\gamma_\parallel$ leads to negative values of $\alpha_\parallel$, despite the large value of s_{33}. The puzzle remains why these principal gammas behave as they do. An explanation would require more detailed knowledge of the phonon spectrum and its strain dependence, and this in turn would require a better understanding of the bonding in this complex layered system.

Chapter 9

Conclusion

This century has seen enormous advance in our knowledge and understanding of the condensed state. The foundations of lattice dynamics were laid by Debye, Born and von Kármán early in the century, and extended later through Brillouin, Peierls and the neutron spectroscopists. They pointed out one of the tasks ahead: to unravel the meaning of the harmonic or approximately harmonic properties of solids, including the heat capacity. And yet notwithstanding the help of high speed computers, the experimental data on crystals and glasses — even the simpler cubic crystals — are not as explicable or predictable as we might like; and despite a century of measurements, precise data is still lacking for many systems of interest. This monograph includes a brief coverage of existing data on various crystal systems, and draws attention to gaps in the understanding of much of the available information at a *quantitative* and sometimes even a qualitative level. The gaps are greater for the anharmonic property of thermal expansion, born from the efforts of Mie and Grüneisen, also early in the century. In particular, the difficulties of inelastic neutron scattering at high pressures have largely prevented the acquisition of data on the volume dependence of dispersion curves (which show the phonon frequency–wave number dependence $\omega(\mathbf{q}, s)$). Apart from some optical zone center frequencies obtainable from spectroscopic measurements, it is usually only at the low frequency limit that we have experimental mode Grüneisen parameters, as obtained from measurements of elastic moduli under pressure.

In the second half of this century powerful computers and improved numerical techniques have enabled detailed studies to be made of the behavior of models of quite complex crystals and even of disordered phases; and *ab initio* techniques for solving many electron systems have reached the point of making reasonably reliable predictions for harmonic interatomic forces in some simple systems. Further improvements in the basic theory underlying the approximations and in the numerical algorithms used may soon extend applications to more complex or larger systems. But we still have little knowledge, empirical or fundamental, of anharmonic forces in solids, apart from those derived from pair potentials. Fermi surface measurements have revealed much about electron energy–wave number relations, but much less about the strain dependence of these properties and their effect on shear moduli and anisotropic effects in the thermal expansion.

For metals the electron gas makes a direct contribution to heat capacity and thermal expansion which can be predominant at low temperatures. Itinerant and collective electron effects are important in superconductors, heavy fermion metals, and magnetic systems such as chromium, rare earths, and invar alloys, resulting in a variety of behavior patterns to challenge the theorist for some time to come. As the late Berndt Matthias was fond of saying, no theorist has yet *predicted* the ordering temperature of a superconductor (or a ferromagnet for that matter). Some properties, both electronic and vibrational, appear to be dependent on such a subtle balance of effects that prediction will only be possible when we can make detailed calculations with much more realistic models and precise *ab initio* approximations than at present available. Low temperature thermodynamic measurements probe the low energy excitations, providing a stringent test of many theories that appear satisfactory at higher temperatures.

We leave this monograph with you not only as an attempt to convey what is known about heat capacity and thermal expansion at low temperatures, but to present some of the outstanding problems for materials ranging from rare gases to high temperature superconductors. As novel materials ‘come on to the market’ they rightly attract attention and present new problems. The familiar things like sodium chloride or graphite are then perhaps ignored for a time; but we should not forget that they are there in the back drawer, still able to provide worthwhile problems and sometimes useful tools for the scientist and the engineer.

Appendix A

Axes and Unit Cells in Crystals

DESCRIPTION OF CRYSTAL STRUCTURE

The notation for describing crystal structure and symmetry is given in Vol. A of *International Tables for X-ray Crystallography* [Hah83]. The crystallographic unit cell is chosen to have the full symmetry of the crystal, and so it is not always the smallest repeating unit. There are conventions for the selection and labelling of the lattice vectors $\mathbf{a}, \mathbf{b}, \mathbf{c}$. They always form a right-handed triad. The description of the structure is completed by specifying the *basis*, comprising the positions of atoms within the unit cell. For many structures these positions are determined or partly determined by symmetry.

THE VOLUME OF THE UNIT CELL

A cell with edges of length a, b, c and angles between the edges $\hat{\alpha}, \hat{\beta}, \hat{\gamma}$ has volume

$$\mathrm{v}_{cell} = abc[1 - \cos^2 \hat{\alpha} - \cos^2 \hat{\beta} - \cos^2 \hat{\gamma} + 2\cos \hat{\alpha} \cos \hat{\beta} \cos \hat{\gamma}]^{\frac{1}{2}} \qquad \text{(A.1)}$$

This reduces to simpler forms for crystals of higher symmetry than triclinic, and in particular to abc for cubic, tetragonal and orthorhombic crystals. From Eq. (A.1) the volumetric coefficient of expansion β can be derived in terms of the cell parameters $a, b, c, \hat{\alpha}, \hat{\beta}, \hat{\gamma}$ and their temperature coefficients (e.g., [Bar98, Section 1.2.4]).

RECTANGULAR CARTESIAN AXES

Tensor properties of crystals such as thermal expansion and elasticity are usually given referred to rectangular Cartesian coordinates. For cubic, tetragonal and orthorhombic crystals the Cartesian axes are taken in the directions of $\mathbf{a}, \mathbf{b}, \mathbf{c}$. For other symmetries the axes recommended by the IRE Piezoelectric Crystals Committee [IRE49, Nye85] are as follows: Oz is taken along **c**; Ox is taken in the **ac** plane perpendicular to Oz, such that it makes an acute angle with **a**; finally, Oy is taken in a direction normal to the **ac** plane such that $Oxyz$ forms a right-handed triad. Using this convention, expressions for the α_λ in terms of the cell parameters and their temperature derivatives are given for all crystal symmetries in [Bar98, Section 1.2.5.4].

Other conventions are sometimes used. For example, a standard crystallographic text [Rol65] takes Oy in the bc plane and Ox perpendicular to this plane; for hexagonal, trigonal and triclinic crystals this would give different Ox and Oy axes from those given by the convention given above. It is not always easy to find out which conventions have been used in published work, and uncertainty has sometimes led to confusion and error (e.g., for the trigonal crystal quartz see [Bar76b, Appendix]).

Appendix B

Manipulating Thermodynamic Expressions

Thermodynamic quantities can be derived as partial derivatives of the internal energy $U(S, V)$ and of the related functions $F(T, V)$, $H(S, P)$ and $G(T, P)$; or more generally as derivatives of similar functions of other or additional variables, e.g., with strain and stress replacing volume and pressure [Cal60]. Relations between such quantities are then obtained as mathematical identities. There are two widely used techniques for doing this.

CHANGING THE ORDER OF DIFFERENTIATION

The order of differentiation does not matter in a multiple derivative, provided that the same set of independent variables is used in each successive partial differentiation. Thus we obtain the well-known Maxwell relations from second derivatives of the energy functions U, F, G, H; for example

$$\left(\frac{\partial S}{\partial V}\right)_T = -\frac{\partial^2 F}{\partial V \partial T} = -\frac{\partial^2 F}{\partial T \partial V} = \left(\frac{\partial P}{\partial T}\right)_V \tag{B.1}$$

as in Eq. (2.11). Similarly we can obtain relations such as Eqs. (2.18)–(2.19); e.g.,

$$\begin{aligned}\left(\frac{\partial C_P}{\partial P}\right)_T &= -T\frac{\partial^3 G}{\partial P \partial T^2} = -T\left[\frac{\partial^2}{\partial T^2}\left(\frac{\partial G}{\partial P}\right)_T\right]_P \\ &= -T\left[\frac{\partial}{\partial T}\left(\frac{\partial V}{\partial T}\right)_P\right]_P = -T\left[\frac{\partial(\beta V)}{\partial T}\right]_P\end{aligned} \tag{B.2}$$

Sometimes preliminary manipulation is needed, as in the proof of the identity in

Eq. (2.21):

$$\delta_T = \left(\frac{\partial \ln \chi_T}{\partial \ln V}\right)_P = \frac{1}{\chi_T}\left(\frac{\partial \chi_T}{\partial T}\right)_P \left(\frac{\partial T}{\partial \ln V}\right)_P$$
$$= -\frac{1}{\beta\chi_T}\frac{\partial^2 \ln V}{\partial T \partial P} = \frac{1}{\beta}\left(\frac{\partial P}{\partial \ln V}\right)_T \left(\frac{\partial \beta}{\partial P}\right)_T = \left(\frac{\partial \ln \beta}{\partial \ln V}\right)_T \tag{B.3}$$

TRANSFORMATIONS BETWEEN DIFFERENT SETS OF INDEPENDENT VARIABLES

If z is a function of a set of independent variables u_i, which are themselves functions of another such set x_j, then

$$\left(\frac{\partial z}{\partial x_k}\right)_{x'} = \sum_i \left(\frac{\partial z}{\partial u_i}\right)_{u'} \left(\frac{\partial u_i}{\partial x_k}\right)_{x'} \tag{B.4}$$

where for example the subscript x' denotes that all the x_j except for x_k are kept constant during differentiation. In thermodynamic applications the sets u_i and x_j often have a variable in common. For example, taking the entropy S successively as a function of T and V and then of T and P, we get

$$C_P = T\left(\frac{\partial S}{\partial T}\right)_P = T\left[\left(\frac{\partial S}{\partial T}\right)_V + \left(\frac{\partial S}{\partial V}\right)_T \left(\frac{\partial V}{\partial T}\right)_P\right] \tag{B.5}$$

$$= C_V + T\,(\beta/\chi_T)\,\beta V = C_V + \beta^2 VT/\chi_T \tag{B.6}$$

in agreement with Eq. (2.10).

A further very useful result is obtained when the set x_j not only has a common member with the set u_i but also contains the variable z being differentiated, so that the left hand side of Eq. (B.4) is identically zero. In a system with only two independent variables, with x, z as the set x_j and x, y as the set u_i, this gives

$$\left(\frac{\partial z}{\partial x}\right)_z = \left(\frac{\partial z}{\partial x}\right)_y + \left(\frac{\partial z}{\partial y}\right)_x \left(\frac{\partial y}{\partial x}\right)_z = 0 \tag{B.7}$$

thus giving

$$\left(\frac{\partial z}{\partial x}\right)_y = -\left(\frac{\partial z}{\partial y}\right)_x \left(\frac{\partial y}{\partial x}\right)_z \tag{B.8}$$

Taking $z = P$, $x = T$ and $y = \ln V$ leads directly to the expression for $(\partial P/\partial T)_V$ in Eq. (2.11). When z and y are replaced by sets of n independent variables the right hand side of this equation involves a sum of products, as for example in the first of Eqs. (2.136):

$$\alpha_\lambda = \left(\frac{\partial \eta_\lambda}{\partial T}\right)_{\sigma,\omega} = -\sum_\mu \left(\frac{\partial \eta_\lambda}{\partial \sigma_\mu}\right)_{\sigma',T,\omega} \left(\frac{\partial \sigma_\mu}{\partial T}\right)_{\eta,\omega} \tag{B.9}$$

$$= \sum_\mu s^T_{\lambda\mu} \gamma_\mu (C_\eta / V) \tag{B.10}$$

where in this appendix (but not in the main text) sums over repeated suffices are written explicitly.

Appendix C

Tables

DATA SOURCES AND MATERIALS LISTED IN TABLES C.1, C.2, C.3

Sources of data for these tables can be found throughout the text. Here we summarize the major compilations which have been used. These also contain original sources and data for a wider range of materials.

1. The CINDAS Series on *Thermophysical Properties of Matter* include Vol. 4 on Specific Heat of Metallic Elements and Alloys [Tou70a], Vol. 5 on Specific Heat of Nonmetallic Solids [Tou70b], Vol. 12 on Thermal Expansion of Metallic Elements and Alloys [Tou75], Vol. 13 on Thermal Expansion of Nonmetallic Solids [Tou77].

2. The NBS Monographs by Corruccini and Gniewek — No. 21 on *Specific Heats and Enthalpies of Technical Solids at Low Temperatures* [Cor60] and No. 29 on *Thermal Expansion of Technical Solids at Low Temperatures* [Cor61].

3. *The American Institute of Physics Handbook* includes evaluated data on heat capacity [Fur72] and thermal expansion [Kir72].

4. Two tabulations on metallic alloys [Cla68] and on 'selected materials' (metals and polymers) [Cla83].

5. The CODATA evaluation of thermophysical properties of key reference materials [Whi97] (see also [Ho98, Ch. 11]).

6. Books on cryogenic techniques, for example [Pob96, Whi79], and on polymers [Har94].

Materials in the tables include the following with approximate alloy compositions expressed in wt%:

- Al2024–T86 (Al + 4.1 Cu, 1.4 Mg, 0.5 Mn, 0.2 Fe, 0.1 Si,0.1 Zn)

- Al5083 (Al + 4.75 Mg, 0.6 Mn, 0.2 Fe, 0.1 Cr, balance Al)
- Brass(65/35) is a yellow brass [Cor61]
- Brass(70/30) data from [Cla83]
- Hastelloy C (Ni + 15.9 Mo, 15.4 Cr, 5 Fe, 3.4 W, 0.8 Si, $\sim$ 0.3 Mn, Co, V)
- Inconel 718 (Ni + 3 Mo, 18 Cr, 17 Fe, 5 Nb+Ta, 0.8 Ti, $\sim$ 0.3 Al, Mn, Si)
- Constantan (60 Cu, 40 Ni)
- Cu/NbTi is multifilamentary NbTi in Cu matrix with volume ratio of 1.8:1
- SnPb(50/50) is tin-lead solder [Zie64]
- S.S.304/316–stainless steels (Fe + 18–20 Cr, 8–10 Ni)
- Ti-6Al-4V is a common Ti-rich alloy
- ZrO_2 (stab) is ZrO_2 + 9 mol% Y_2O_3 (Section 5.6) [Col85a]
- Stycast 2850FT is a powder-filled epoxy (see Section 5.9, [Swe97])
- Polymer G10 is a glass–epoxy laminate (see Section 7.3).

Table C.1. Specific heats c_P ($J \cdot g^{-1} \cdot K^{-1}$) of technical solids

Material	25 K	50 K	75 K	100 K	150 K	200 K	250 K	293 K
METALS [Fur72, Whi97]								
Al	0.0175	0.142	0.322	0.481	0.683	0.797	0.859	0.897
Cu	0.015	0.097	0.187	0.252	0.322	0.355	0.373	0.383
Fe	0.0080	0.051	0.136	0.216	0.324	0.384	0.421	0.444
Nb	0.020	0.085	0.147	0.188	0.230	0.248	0.260	0.262
Ni	0.0098	0.069	0.156	0.232	0.329	0.383	0.416	0.435
Si	0.0085	0.078	0.170	0.260	0.425	0.557	0.649	0.694
Ti(pc)	0.0137	0.098	0.210	0.300	0.408	0.466	0.500	0.518
W	0.0041	0.032	0.064	0.087	0.112	0.123	0.129	0.133
ALLOYS [Tou70a, Whi79]								
Al 2024	—	—	—	0.46	0.65	0.73	0.80	0.84
CuZn(65/35)	0.022	0.118	0.21	0.27	0.33	0.36	0.37	0.377
Constantan	0.013	0.08	0.17	0.24	0.32	0.36	0.38	0.41
Inconel 718	—	0.07	0.16	0.27	0.36	0.40	0.42	0.43
Nb-38Ti	0.03	0.11	0.24	—	—	—	—	—
SnPb(50/50)	0.062	0.116	0.140	0.152	0.163	0.170	0.174	0.178
S.S.304/3168	0.019	0.092	0.19	0.28	0.35	0.42	0.45	0.47
Ti-6Al-4V	—	—	0.21	—	0.40	0.49	0.52	0.55
NON-METALS [Tou70b, Whi97]								
sapphire	0.0014	0.0148	0.0558	0.126	0.313	0.501	0.658	0.763
MgO	0.0019	0.0207	0.085	0.195	0.449	0.661	0.814	0.916
Pyrex	0.043	—	—	0.28	0.406	0.533	0.64	0.72
Silica	0.038	0.111	0.188	0.268	0.420	0.546	0.650	0.728
ZrO_2	0.009	0.041	0.095	0.15	0.26	0.35	0.41	0.45
POLYMERS [Har94, Whi79]								
Epoxy	0.13	0.27	0.39	0.48	—	1.0	—	1.3
Nylon 6	—	—	0.47	—	0.81	1.01	1.2	1.5
Stycast	0.032	0.088	0.15	0.22	—	—	—	—
Teflon	0.10	0.21	0.29	0.39	0.56	0.72	0.87	1.0
G10 (GFRP)	—	0.3	0.4	0.5	—	1.0	—	1.5

Table C.2. Linear thermal contraction $-\Delta l/l_{293}$ (10^{-4}) relative to 293 K, and α_{293} (10^{-6} K^{-1}), for selected technical solids

Material	$-\Delta l/l$ 4 K	$-\Delta l/l$ 40 K	$-\Delta l/l$ 80 K	$-\Delta l/l$ 100 K	$-\Delta l/l$ 150 K	$-\Delta l/l$ 200 K	$-\Delta l/l$ 250 K	α 293 K
METALS [Kir72, Whi79]								
Al	41.4	41.2	39.0	36.9	29.4	20.1	9.6	22.9
Cu	32.6	32.3	30.2	28.3	22.1	14.9	7.1	16.65
Fe	20.4	20.2	19.5	18.5	14.9	10.2	4.5	11.8
Nb	14.3	14.1	12.9	12.1	9.3	6.3	3.0	7.1
Ni	22.9	22.7	21.7	20.6	16.5	11.2	5.4	12.8
Si	2.2	2.2	2.3	2.4	2.4	1.9	1.0	2.56
Ti(pc)	15.1	15.0	14.2	13.4	10.7	7.3	3.5	8.6
W	8.8	8.7	8.1	7.6	5.9	4.0	1.9	4.42
ALLOYS [Cla68, Cla83]								
Al 2024	39.6	39.4	37.2	35.1	27.8	19.0	9.0	21.2
Al 5083	41.5	41.3	39.0	36.8	29.4	20.1	9.7	22.8
brass(65/35)	38.4	38.0	35.0	32.6	25.3	16.9	8.0	19.0
brass(70/30)	36.9	36.6	33.7	31.3	24.5	16.3	7.5	18
$Fe_{64}Ni_{36}$	4.5	4.8	4.8	4.5	3.0	2.0	0.9	~ 1
Hastelloy C	21.8	21.6	20.4	19.3	15.4	10.5	5.2	12.8
Inconel 718	23.8	23.6	22.3	21.1	16.7	11.4	5.3	12.9
Nb-45Ti	18.8	18.4	16.7	15.6	11.7	7.8	3.8	$\simeq 10$
Cu/NbTi	26.5	26.2	24.5	23.1	17.8	11.7	5.4	$\simeq 12$
S.S.304/316	29.7	29.6	27.8	26.0	20.3	13.8	6.6	15.8
Ti-6Al-4V	17.3	17.1	16.2	15.4	11.8	7.8	3.6	8
NON-METALS [Kir72, Whi97]								
sapphire($\parallel$)	7.15	7.15	7.05	6.9	6.1	4.5	2.3	5.80
sapphire($\perp$)	6.05	6.05	5.95	5.85	5.2	3.9	2.0	5.06
MgO	13.9	13.9	13.7	13.3	11.4	8.3	4.2	10.3
Pyrex	5.6	5.7	5.4	5.0	3.95	2.7	1.4	3.0
Silica	−0.05	−0.03	0.01	0.13	0.30	0.31	0.18	0.45
ZrO_2(stab)	13.1	13.0	12.4	11.8	9.8	6.8	3.3	8.0
POLYMERS [Cla81, Cla83]								
Araldite	106	102	93.5	88	71	50.5	25.5	~ 60
Nylon	139	135	125	117	95	67	34	~ 80
Stycast	44	43	40	38	32	22.5	11	30
Teflon	214	206	193	185	160	125	75	~ 200
G10($\perp$)	70.5	69	64	60	49	35	17	$\simeq 40$
G10($\parallel$)	24	23.5	21	20	15.5	11	5.2	$\simeq 12$

Table C.3. Reference Materials. Values of C_P ($J{\cdot}mol^{-1}{\cdot}K^{-1}$) from CODATA [Whi97]; thermal expansion values of α^* ($10^{-6}K^{-1}$) and $\Delta l/l_{293}$ (10^{-6}) from CODATA [Whi97] excepting silica [Hah72, Oka95b]; values of α^* for alumina (sapphire) are the average $\alpha^*_{av} = (2\alpha^*_{\perp} + \alpha^*_{\parallel})/3 = \alpha^*_{55^\circ}$

T (K)	C_P Cu	C_P Al_2O_3	α^* Cu	$\Delta l/l_{293}$ Cu	α^* Si	α^*_{av} Al_2O_3	α^* SiO_2
10	0.0554	0.0087	0.030	−3257	0.0005	—	−0.18
15	0.184	0.0305	0.103	−3257	0.0012	—	−0.44
20	0.462	0.073	0.263	−3256	−0.003	0.004	−0.63
25	0.957	0.145	0.556	−3254	−0.019	0.009	−0.74
30	1.688	0.263	1.00	−3250	−0.053	0.016	−0.80
35	2.628	0.443	1.58	−3244	−0.103	0.028	−0.83
40	3.725	0.698	2.27	−3234	−0.164	0.044	−0.84
45	4.92	1.046	3.05	−3221	-0.22_5	0.068	−0.84
50	6.16	1.506	3.84	−3204	−0.29	0.095	−0.83
60	8.62	2.793	5.46	−3157	−0.40	0.18	−0.80
70	10.89	4.594	6.98	−3095	−0.46	0.29	−0.75
80	12.87	6.902	8.33	−3018	−0.47	0.44	−0.70
90	14.56	9.677	9.49	−2929	−0.43	0.61	−0.62
100	16.00	12.85	10.49	−2829	−0.34	0.81	−0.54
110	17.22	16.34	11.36	−2719	−0.22	1.03	−0.46
120	18.26	20.07	12.05	−2602	−0.06	1.28	−0.38
130	19.13	23.96	12.70	−2478	0.11	1.53	−0.31
140	19.86	27.94	13.19	−2349	0.31	1.80	−0.24
150	20.49	32.00	13.65	−2215	0.49	2.07	−0.17
160	21.03	36.04	14.03	−2077	0.69	2.34	−0.10
180	21.90	43.91	14.67	−1789	1.06	2.90	0.02
200	22.58	51.33	15.19	−1491	1.39	3.42	0.13
220	23.10	58.15	15.62	−1183	1.70	3.89	0.23
240	23.53	64.32	15.96	−867	1.98	4.32	0.32
250	23.74	67.17	16.11	−706	2.10	4.52	0.35
260	23.93	69.86	16.25	−545	2.22	4.69	0.39
280	24.22	74.89	16.50	−218	2.44	5.03	0.45
293.15	24.36	77.95	16.65	0	2.56	5.30	0.48
300	24.44	79.46	16.70	114	2.62	5.40	0.49

Table C.4. Einstein functions. C_V, $(U - U_0)/T$, and S (J·g-at^{-1}·K^{-1}). $\Theta_E = \hbar\omega_E/kT$

Θ_E/T	C_V	$(U-U_0)/T$	S	Θ_E/T	C_V	$(U-U_0)/T$	S
0.0	24.943	24.943	∞	8.2	0.4609	0.0562	0.0630
0.2	24.861	22.532	65.130	8.4	0.3959	0.0471	0.0527
0.4	24.614	20.287	47.965	8.6	0.3398	0.0395	0.0441
0.6	24.209	18.204	38.056	8.8	0.2912	0.0331	0.0369
0.8	23.655	16.282	31.164	9.0	0.2494	0.0277	0.0308
1.0	22.965	14.517	25.958	9.2	0.2134	0.0232	0.0257
1.2	22.154	12.901	21.840	9.4	0.1824	0.0194	0.0215
1.4	21.240	11.430	18.493	9.6	0.1557	0.0162	0.0179
1.6	20.240	10.096	15.721	9.8	0.1329	0.0136	0.0149
1.8	19.174	8.891	13.398	10.0	0.1133	0.0113	0.0125
2.0	18.061	7.808	11.435	10.2	0.0965	0.0095	0.0104
2.2	16.918	6.838	9.767	10.4	0.0821	0.0079	0.0087
2.4	15.764	5.973	8.345	10.6	0.0698	0.0066	0.0072
2.6	14.614	5.203	7.128	10.8	0.0594	0.0055	0.0060
2.8	13.482	4.522	6.087	11.0	0.0504	0.0046	0.0050
3.0	12.379	3.921	5.195	11.2	0.0428	0.0038	0.0042
3.2	11.315	3.392	4.430	11.4	0.0363	0.0032	0.0035
3.4	10.299	2.928	3.775	11.6	0.0308	0.0027	0.0029
3.6	9.336	2.523	3.214	11.8	0.0261	0.0022	0.0024
3.8	8.431	2.169	2.733	12.0	0.0221	0.0018	0.0020
4.0	7.585	1.862	2.323	12.2	0.0187	0.0015	0.0017
4.2	6.801	1.595	1.972	12.4	0.0158	0.0013	0.0014
4.4	6.077	1.364	1.672	12.6	0.0134	0.0011	0.0011
4.6	5.414	1.165	1.417	12.8	0.0113	0.0009	0.0010
4.8	4.808	0.994	1.200	13.0	0.0095	0.0007	0.0008
5.0	4.259	0.846	1.015	13.2	0.0080	0.0006	0.0007
5.2	3.762	0.720	0.857	13.4	0.0068	0.0005	0.0005
5.4	3.315	0.611	0.724	13.6	0.0057	0.0004	0.0005
5.6	2.914	0.518	0.611	13.8	0.0048	0.0003	0.0004
5.8	2.556	0.439	0.515	14.0	0.0041	0.0003	0.0003
6.0	2.237	0.372	0.434	14.2	0.0034	0.0002	0.0003
6.2	1.954	0.314	0.365	14.4	0.0029	0.0002	0.0002
6.4	1.703	0.266	0.307	14.6	0.0024	0.0002	0.0002
6.6	1.482	0.224	0.258	14.8	0.0020	0.0001	0.0001
6.8	1.287	0.189	0.217	15.0	0.0017	0.0001	0.0001
7.0	1.117	0.159	0.182	15.5	0.0011	0.0001	0.0001
7.2	0.967	0.134	0.153	16.0	0.0007	0.0000	0.0000
7.4	0.836	0.113	0.128	16.5	0.0005	0.0000	0.0000
7.6	0.722	0.095	0.107	17.0	0.0003	0.0000	0.0000
7.8	0.622	0.080	0.090	17.5	0.0002	0.0000	0.0000
8.0	0.536	0.067	0.075	18.0	0.0001	0.0000	0.0000

Table C.5. Debye functions. C_V, $(U - U_0)/T$, and S (J·g-at^{-1}·K^{-1})

Θ_D/T	C_V	$(U-U_0)/T$	S	Θ_D/T	C_V	$(U-U_0)/T$	S
0.0	24.943	24.943	∞	8.2	3.2363	0.8512	1.1418
0.2	24.893	23.123	73.430	8.4	3.0404	0.7954	1.0662
0.4	24.745	21.401	56.214	8.6	2.8582	0.7442	0.9968
0.6	24.500	19.778	46.224	8.8	2.6887	0.6969	0.9331
0.8	24.163	18.253	39.219	9.0	2.5309	0.6535	0.8745
1.0	23.739	16.822	33.871	9.2	2.3842	0.6134	0.8204
1.2	23.236	15.485	29.586	9.4	2.2476	0.5764	0.7707
1.4	22.660	14.238	26.047	9.6	2.1203	0.5423	0.7247
1.6	22.021	13.077	23.062	9.8	2.0017	0.5106	0.6822
1.8	21.327	12.000	20.508	10.0	1.8912	0.4814	0.6429
2.0	20.588	11.003	18.299	10.2	1.7882	0.4542	0.6065
2.2	19.814	10.082	16.372	10.4	1.6920	0.4291	0.5727
2.4	19.012	9.231	14.682	10.6	1.6023	0.4055	0.5413
2.6	18.192	8.450	13.193	10.8	1.5184	0.3836	0.5122
2.8	17.363	7.730	11.875	11.0	1.4400	0.3634	0.4850
3.0	16.531	7.073	10.705	11.2	1.3667	0.3444	0.4597
3.2	15.704	6.470	9.665	11.4	1.2980	0.3270	0.4362
3.4	14.887	5.916	8.737	11.6	1.2339	0.3103	0.4141
3.6	14.086	5.413	7.909	11.8	1.1735	0.2951	0.3936
3.8	13.305	4.953	7.169	12.0	1.1170	0.2806	0.3743
4.0	12.548	4.533	6.505	12.2	1.0639	0.2672	0.3563
4.2	11.817	4.148	5.911	12.4	1.0141	0.2547	0.3394
4.4	11.115	3.801	5.378	12.6	0.9672	0.2427	0.3236
4.6	10.444	3.485	4.898	12.8	0.9232	0.2314	0.3087
4.8	9.803	3.195	4.468	13.0	0.8817	0.2209	0.2947
5.0	9.195	2.933	4.080	13.2	0.8426	0.2110	0.2815
5.2	8.619	2.694	3.730	13.4	0.8058	0.2018	0.2691
5.4	8.074	2.477	3.416	13.6	0.7710	0.1931	0.2575
5.6	7.561	2.279	3.131	13.8	0.7382	0.1848	0.2464
5.8	7.078	2.098	2.874	14.0	0.7072	0.1770	0.2360
6.0	6.625	1.936	2.642	14.2	0.6779	0.1696	0.2262
6.2	6.200	1.786	2.432	14.4	0.6502	0.1627	0.2169
6.4	5.803	1.650	2.241	14.6	0.6329	0.1561	0.2082
6.6	5.431	1.526	2.069	14.8	0.5990	0.1499	0.1998
6.8	5.084	1.413	1.912	15.0	0.5755	0.1439	0.1920
7.0	4.761	1.310	1.769	15.5	0.5217	0.1305	0.1740
7.2	4.459	1.215	1.639	16.0	0.4744	0.1186	0.1582
7.4	4.178	1.129	1.521	16.5	0.4326	0.1082	0.1442
7.6	3.917	1.050	1.413	17.0	0.3956	0.0989	0.1319
7.8	3.673	0.9780	1.314	17.5	0.3626	0.0907	0.1209
8.0	3.447	0.9119	1.224	18.0	0.3333	0.0833	0.1111

Appendix D

Commonly Used Symbols

α ... coefficient of linear expansion, $\partial \ln l / \partial T$
α^* ... coefficient of linear expansion, $(1/l_0)\partial l / \partial T$
$\alpha_{a,b,c}$ ... linear coefficients in directions of $\mathbf{a}, \mathbf{b}, \mathbf{c}$
$\mathbf{a}, \mathbf{b}, \mathbf{c}$ or $\mathbf{a}_1, \mathbf{a}_2, \mathbf{a}_3$ ... lattice vectors
β ... coefficient of volume expansion
B_S, B_T ... adiabatic and isothermal bulk moduli
$c_{\lambda\mu}$ ... elastic stiffness moduli
$\tilde{c}_{\lambda\mu}$ ... thermodynamic elastic stiffness moduli
C_V, C_P ... molar heat capacities at constant volume, constant pressure
c_V, c_P ... specific heats per unit mass or unit volume
C_e ... electronic heat capacity
C_m ... magnetic heat capacity
C_{sch} ... Schottky heat capacity
C_{vib} ... lattice vibrational heat capacity
χ_S, χ_T ... adiabatic and isothermal compressibilities
χ_λ, χ_a, etc. ... directional compressibilities
δ_S, δ_T ... Anderson Grüneisen functions
$d\Omega$ ... element of solid angle
$\eta_{\alpha\beta}$ or η_λ ... strain component
E_S, E_T ... adiabatic and isothermal Young's modulus
E_Z ... zero-point vibrational energy
F ... Helmholtz free energy
F_{th} ... thermal Helmholtz free energy
$\gamma(V,T)$ ... thermodynamic Grüneisen function
γ_{vib} ... lattice vibrational Grüneisen function
γ_λ ... anisotropic Grüneisen functions
$\gamma_\perp, \gamma_\parallel$ ... principal anisotropic Grüneisen functions for axial solids
γ_e ... electronic Grüneisen parameter
$\gamma_j, \gamma_{\mathbf{q}s}$ ... mode Grüneisen parameters ($-d \ln \omega_j / d \ln V$, etc.)

γ_0 ... value of γ_{vib} as $T \to 0$
γ_0^{th} and γ_0^{el} ... γ_0 derived respectively from thermal and elastic data
Γ_e ... coefficient of electronic heat capacity
G ... Gibbs free energy
H ... Enthalpy
H, $\mathbf{H}$ ... magnetic fields
k ... Boltzmann Constant
λ ... electron–phonon enhancement factor
l_T ... length at temperature T
ν_j ... frequency of the vibrational mode j
ω_j or $\omega_{\mathbf{q}s}$ ... angular frequency ($2\pi\nu$) of vibrational mode j or $\mathbf{q}s$
N ... number of atoms in material
N_A ... Avogadro's number
Φ_L ... static lattice energy of solid
$\phi(r)$ ... interatomic pair potential
$\mathbf{q}$ ... phonon wave vector
q ... phonon wave number
q ... 'second Grüneisen function' $(\partial \ln \gamma / \partial \ln V)_T$
$R = N_A k$ ... molar gas constant
$\sigma_{\alpha\beta}$ or σ_λ ... stress component
σ ... Poisson's ratio
$s^S_{\lambda\mu}, s^T_{\lambda\mu}$ ... elastic compliances
S ... entropy
Θ, Θ_D ... Debye characteristic temperatures
$\Theta^C(T)$ ... Debye temperature derived from heat capacity
$\Theta^S(T)$ ... Debye temperature derived from entropy
Θ_0 or Θ_0^C ... calorimetric Debye temperature as $T \to 0$
Θ_0^{th} ... Θ_0 derived from thermal data
Θ_0^{el} ... Θ_0 derived from elastic data
Θ_E ... a characteristic temperature for Einstein (optic) mode
Θ_{RT} ... a Debye temperature from room temperature thermal data
Θ_∞^C ... harmonic high temperature limit of $\Theta^C(T)$
T ... thermodynamic temperature (in Kelvin)
T_F ... Fermi temperature
T_N ... transition temperature for antiferromagnetism (Néel point)
T_c ... transition temperature for superconductivity or ferromagnetism (Curie point)
U ... internal energy
U_{th} ... thermal internal free energy
V, V_m ... volume or molar volume
v_a ... volume per atom
v ... velocity of acoustic wave
$< \cdots >$ ... average over vibrational spectrum
$\langle \ldots \rangle$ ... Boltzmann thermal average

References

[Abr65] Abramowitz, M., and Stegun, I. A., *Handbook of Mathematical Functions*, Dover, New York (1965).

[Ack82] Ackerman, D. A., and Anderson, A. C., *Rev. Sci. Instrum.* **53**, 1657 (1982).

[Ack84] Ackerman, D. A., Anderson, A. C., Cotts, E. J., Dobbs, J. N., MacDonald, W. M., and Walker, F. J., *Phys. Rev. B* **29**, 966 (1984).

[Aco96] Acocella, D., Horton, G. K., and Cowley, E. R., *Phys. Lett. A* **212**, 161 (1996).

[Ahl66] Ahlers, G., *Rev. Sci. Instrum.* **37**, 477 (1966).

[Ahl67] Ahlers, G., *Phys. Lett.* **24A**, 152 (1967).

[Ahl70] Ahlers, G., *Phys. Rev. A* **2**, 1505 (1970).

[Akg73] Akgöz, Y. C., Farley, J. M., and Saunders, G. A., *J. Phys. Chem. Solids* **34**, 141 (1973).

[Alc93] Alcock, C. B., Chase, M.W., and Itkin, V. P., *J. Phys. Chem. Ref. Data* **22**, 1 (1993).

[Alc94] Alcock, C. B., Chase, M.W., and Itkin, V. P., *J. Phys. Chem. Ref. Data* **23**, 385 (1994).

[Ale62] Alers, G. A., and Waldorf, D. L., *IBM J. Res. Dev.* **6**, 89 (1962).

[Ale65] Alers, G. A., in *Physical Acoustics* **IIIB**, ed. W. P. Mason, Academic Press, New York (1965) Ch. 1.

[Ale66] Alers, G. A., in *Physical Acoustics* **IVA**, ed. W. P. Mason, Academic Press, New York (1966) Ch. 7.

[Ale89] Aleksandrovskii, A. N., Soldatov, A. V., Manzhelii, V. G., and Palei, V. V., *Sov. J. Low Temp. Phys.* **15**, 492 (1989).

[All87] Allen, M. P., and Tildesley, D. J., *Computer Simulation of Liquids*, Clarendon Press, Oxford (1987).

[All97] Allan, N. L., and Mackrodt, W. C., in *Computer Modelling in Inorganic Chemistry*, ed. C. R. A. Catlow, Academic Press, San Diego (1997) Ch. 10.

[Anc75] Ancsin, J., in *Temperature Measurement, 1975*, ed. B. F. Billing and T. J. Quinn, Conf. Series No. 26, Inst. of Physics, London (1975) p. 57.

[And64] Andres, K., *Phys. Kondens. Materie* **2**, 294 (1964).

[And66a] Anderson, O. L., *Phys. Rev.* **144**, 553 (1966).

[And66b] Anderson, A. C., Roach, W. R., Sarwinski, R. E., and Wheatley, J. C., *Phys. Rev. Lett.* **16**, 263 (1966); also **17**, 374 (1966). .

[And72] Anderson, P. W., Halperin, B. I., and Varma, C. M., *Phil. Mag.* **25**, 1 (1972).

[And82] Anderson, M. R., Harryman, M. B. M., Steinmann, D. K., and White, J. W., *Polymer* **23**, 569 (1982).

[And83a] Andersson, P., and Ross, R. G., *J. Phys. C: Solid State Phys.* **16**, 1423 (1983).

[And83b] Anderson, M. S., and Swenson, C. A., *Phys. Rev. B* **28**, 5395 (1983).

[And85] Anderson, M. S., and Swenson, C. A., *Phys. Rev. B* **31**, 668 (1985).

[And88] Anderson, A. C., in [Cez88], Ch. 2.

[And90a] Anderson, M. S., Swenson, C. A., and Peterson, D. T., *Phys. Rev. B* **41**, 3329 (1990).

[And90b] Andrikidis, C., and White, G. K., CSIRO, Sydney, private communication (1990).

[And95a] Anderson, O. L., *Equations of State of Solids for Geophysics and Ceramic Science*, Oxford University Press, New York (1995).

[And95b] Anderson, O. L., and Isaac, D. G., in *Mineral Physics and Crystallography — A Handbook of Physical Constants*, ed. T. J. Ahrens, American Geophysical Union (1995) p. 64.

[Ang71] Angus, S., and Armstrong, B., (eds.), *International Thermodynamic Tables of the Fluid State, Argon, 1971* (IUPAP), Butterworths, London (1971).

[Ant77] Anthony, P. J., and Anderson, A. C., *Phys. Rev. B* **16**, 3827 (1977).

[Ara77] Arai, N., Sorai, M., Suga, H., and Seki, S., *Bull. Chem. Soc. Jpn.* **50**, 1702 (1977).

[Arg67] Argyle, B. E., Miyata, N., and Schultz, T. D., *Phys. Rev.* **160**, 413 (1967).

[Ash76] Ashcroft, N. W., and Mermin, N. D., *Solid State Physics*, Holt, Rinehart and Winston, New York (1976).

[Bac72] Bachmann, R., DiSalvo, F. J., Jr., Geballe, T. H., Greene, R. L., Howard, R. E., King, C. N., Kirsch, H. C., Lee, K. N., Schwall, R. E., Thomas, H.-U., and Zubeck, R. B., *Rev. Sci. Instrum.* **43**, 205 (1972).

[Bac95] Bach, H., (ed.), *Low Thermal Expansion Glass Ceramics*, Springer, Berlin (1995).

[Bai69] Bailey, A. C., Waterhouse, N., and Yates, B., *J. Phys. C: Solid State Phys.* **2**, 769 (1969).

[Bai70] Bailey, A. C., and Yates, B., *J. Appl. Phys.* **41**, 5088 (1970).

[Ban92] Banerjee, S., Prins, M.W. J., Rajeev, K. P., and Raychaudhuri, A. K., *Pramāna J. Phys.* **39**, 391 (1992).

[Bar57a] Barron, T. H. K., *Ann. Phys. (N. Y.)* **1**, 77 (1957).

[Bar57b] Barron, T. H. K., Berg, W. T., and Morrison, J. A., *Proc. R. Soc. Lond. A* **242**, 478 (1957).

[Bar57c] Barron, T. H. K., and Morrison, J. A., *Canad. J. Phys.* **35**, 799 (1957).

[Bar64] Barron, T. H. K., Leadbetter, A. J., and Morrison, J. A., *Proc. R. Soc. Lond. A* **279**, 62, (CORRIGENDA **289**, 440) (1964).

[Bar66] Barnes, L. J., and Hake, R. R., *Ann. Acad. Sci. Fenn.* **A VI-210**, 78 (1966).

[Bar67] Barron, T. H. K., and Munn, R. W., *Phil. Mag.* **15**, 85 (1967).

[Bar71] Barron, T. H. K., Gibbons, T. G., and Munn, R. W., *J. Phys. C: Solid State Phys.* **4**, 2805 (1971).

[Bar72] Barsch, G. R., and Achar, B. N. N., in [Gra72], p. 211.

[Bar74a] Barron, T. H. K., and Gibbons, T. G., *J. Phys. C: Solid State Phys.* **7**, 3260, 3269 and 3287 (1974).

[Bar74b] Barron, T. H. K., and Klein, M. L., in [Hor74], Ch. 7.

[Bar76a] Barker, J. A., in [Kle76], Ch. 4.

[Bar76b] Barron, T. H. K., Huang, C. C., and Pasternak, A., *J. Phys. C: Solid State Phys.* **9**, 3925 (1976).

[Bar77a] Barron, T. H. K., Birch, J. A., and White, G. K., *J. Phys. C: Solid State Phys.* **10**, 1617 (1977).

[Bar77b] Barron, T. H. K., *Acta Cryst.* **A33**, 602 (1977).

[Bar79] Barron, T. H. K., *J. Phys. C: Solid State Phys.* **12**, L155 (1979).

[Bar80] Barron, T. H. K., Collins, J. G., and White, G. K., *Adv. Phys.* **29**, 609 (1980).

[Bar82] Barron, T. H. K., Collins, J. F., Smith, T. W., and White, G. K., *J. Phys. C: Solid State Phys.* **15**, 4311 (1982).

[Bar85] Barton, M. A., and Stacey, F. D., *Phys. Earth Planet. Inter.* **39**, 167 (1985).

[Bar87] Barron, T. H. K., and Pasternak, A., *J. Phys. C: Solid State Phys.* **20**, 215 (1987).

[Bar88] Barron, R. M., Barron, T. H. K., Mummery, P. M., and Sharkey, M., *Canad. J. Chem.* **66**, 718 (1988).

[Bar89] Barron, T. H. K., and Rogers, K. J., *Molecular Simulation* **4**, 27 (1989).

[Bar98] Barron, T. H. K., in [Ho98], Ch. 1.

[Bas68] Bassett, W. A., Takahashi, T., Mao, H.-K., and Weaver, J. S., *J. Appl. Phys.* **39**, 319 (1968).

[Bas95] Bass, J. D., in *Mineral Physics and Crystallography — A Handbook of Physical Constants*, ed. T. J. Ahrens, American Geophysical Union (1995) p. 45.

[Bat67] Bateman, T. B., *J. Acoust. Soc. Am.* **41**, 1011 (1967).

[Bau73] Baughman, R. H., *J. Chem. Phys.* **58**, 2976 (1973).

[Bea61] Beaumont, R. H., Chihara, H., and Morrison, J. A., *Proc. Phys. Soc.* **78**, 1462 (1961).

[Bec69] Bechmann, R., in *Landolt–Börnstein* **III/2**, Springer, Berlin (1969) p. 102.

[Ber57] Berg, W. T., and Morrison, J. A., *Proc. R. Soc. Lond. A* **242**, 467 (1957).

[Ber70] Bertman, B., Heberlein, D. C., Sandiford, D. J., Shen, L., and Wagner, R. R., *Cryogenics* **10**, 326 (1970).

[Ber76a] Berthold, J. W., Jacobs, S. F., and Norton, M. A., *Appl. Opt.* **15**, 1898 (1976).

[Ber76b] Berman, R., *Thermal Conduction in Solids*, Clarendon Press, Oxford (1976, 1978).

[Bes78] Besley, L. M., and Kemp, R. C., *Cryogenics* **18**, 497 (1978).

[Bet89] Betts, D. S., *An Introduction to Millikelvin Technology*, Cambridge University Press, Cambridge (1989).

[Bey92] Beyermann, W. P., Hundley, M. F., Thompson, J. D., Diederich, F. N., and Grüner, G., *Phys. Rev. Lett.* **68**, 2046 (1992).

[Bil75] Bilir, N., and Phillips, W. A., *Phil. Mag.* **32**, 113 (1975).

[Bil79] Bilz, H., and Kress, W., *Phonon Dispersion Relations in Insulators*, Springer, Berlin (1979).

[Bir96] Birmingham, J. T., Richards, P. L., and Meyer, H., *J. Low Temp. Phys.* **103**, 183 (1996).

[Bla42] Blackman, M., *Proc. R. Soc. Lond. A* **181**, 58 (1942).

[Bla55] Blackman, M., *Handb. Phys.* **7**, (part I), 325 (1955).

[Bla57] Blackman, M., *Proc. Phys. Soc. Lond. B* **70**, 827 (1957).

[Bla75] Blaschko, O., Ernst, G., Quittner, G., Kress, W., and Lechner, R. E., *Phys. Rev. B* **11**, 3960 (1975).

[Blo70] Bloom, D. W., Lowndes, D. H., Jr., and Finegold, L., *Rev. Sci. Instrum.* **41**, 690 (1970).

[Boe77] Boehler, R., Getting, I. C., and Kennedy, G. C., *J. Phys. Chem. Solids* **38**, 233 (1977).

[Bol63] Bolef, D. I., and de Klerk, J., *IEEE Trans. Ultrasonics Eng.* **UE-10**, 27 (1963).

[Bor54] Born, M., and Huang, K., *Dynamical Theory of Crystal Lattices*, Clarendon Press, Oxford (1954).

[Boy83] Boyer, J. D., Lasjaunias, J. C., Fisher, R. A., and Phillips, N. E., *J. Non-Cryst. Solids* **55**, 413 (1983).

[Bro75] Browder, J. S., *J. Phys. Chem. Solids* **36**, 193 (1975).

[Bro89] Brown, J. M., Slutsky, L. J., Nelson, K. A., and Cheng, L.-T., *J. Geophys. Res.* **94**, 9485 (1989).

[Bro90] Brommer, P. E., and Franse, J. J. M., in *Ferromagnetic Materials*, Vol. 5, North-Holland, Amsterdam (1990) p. 323.

[Bru64] Brugger, K., *Phys. Rev.* **133**, A1611 (1964).

[Bru65] Brugger, K., *Phys. Rev.* **137**, A1826 (1965).

[Bru67] Brugger, K., and Fritz, T. C., *Phys. Rev.* **157**, 524 (1967).

[Bru98] Bruno, J. A. O., Allan, N. L., Barron, T. H. K., and Turner, A. D., *Phys. Rev. B* **58**, 8416 (1998).

[Bur65] Burgiel, J. C., Meyer, H., and Richards, P. L., *J. Chem. Phys.* **43**, 4291 (1965).

[Bur69] Burford, J. C., and Graham, G. M., *Canad. J. Phys.* **47**, 23 (1969).

[Bur96] Burkhart, G. J., and Meingast, C., *Phys. Rev. B* **54**, 6865 (1996).

[Cal60] Callen, H. B., *Thermodynamics*, John Wiley and Sons, New York (1960).

[Cap73] Caplin, A. D., Grüner, G., and Dunlop, J. B., *Phys. Rev. Lett.* **30**, 1138 (1973).

[Cap78] Caplin, A. D., and Nicholson, L. K., *J. Phys. F: Met. Phys.* **8**, 51 (1978).

[Car64a] Carr, R. H., and Swenson, C. A., *Cryogenics* **4**, 76 (1964).

[Car64b] Carr, R. H., McCammon, R. D., and White, G. K., *Proc. R. Soc. Lond. A* **280**, 72 (1964).

[Car81] Cartz, L., and Jorgensen, J. D., *J. Appl. Phys.* **52**, 236 (1981).

[Car85] Car, R., and Parrinello, M., *Phys. Rev. Lett.* **55**, 2471 (1985).

[Cas72] Case, C. R., McLean, K. O., Swenson, C. A., and White, G. K., in [Gra72], p. 183.

[Cas84] Castanet, R., *High Temp.-High Press.* **16**, 449 (1984).

[Cez88] Cezairliýan, A., *Specific Heat of Solids*, Vol. I–2 in the CINDAS Data Series on Material Properties, ed. C. Y. Ho, Hemisphere Publishing Corporation, New York (1988) Ch. 9.

[Cha87] Chawla, K. K., in *Composite Materials, Science and Engineering*, ed. B. Ilschner and N. Grant, Springer, New York (1987).

[Cha89] Chae, H. B., and Bretz, M., *J. Low Temp. Phys.* **76**, 199 (1989).

[Cha96] Changyong, S., Jisuo, W., and Lihua, L., *Chinese J. Low Temp. Phys.* **18**, 438 (1996).

[Che82] Cheung, T. D., Chan, W. K., and Kouvel, J. S., *Rev. Sci. Instrum.* **53**, 880 (1982).

[Chi97] Chikazumi, S., *Physics of Ferromagnetism*, Clarendon Press, Oxford (1997).

[Cho75] Choy, C. L., Huq, M., and Moody, D. E., *Phys. Lett.* **54A**, 375 (1975).

[Cho81] Choy, C. L., Chen, F. C., and Young, K., *J. Polym. Sci. Polym. Phys. Ed.* **19**, 335 (1981).

[Chr91] Christensen, R. M., *Mechanics of Composite Materials*, Reprinted by Krieger Publishing Co., Malabar, Florida (1991).

[Cla68] Clark, A. F., *Cryogenics* **8**, 282 (1968).

[Cla78] Claytor, T. N., and Sladek, R. J., *Phys. Rev. B* **18**, 5842 (1978).

[Cla79] Clark, A. F., Reed, R. P., and Hartwig, G., (eds.), *Nonmetallic Materials and Composites at Low Temperatures*, Plenum Press, New York (1979).

[Cla81] Clark, A. F., Fujii, G., and Ranney, M. A., *IEEE Trans. Magn.* **MAG-17**, 2316 (1981).

[Cla83] Clark, A. F., in *Materials at Low Temperatures*, ed. R. P. Reed and A. F. Clark, Am. Soc. Metals, Metals Park, Ohio (1983) Ch. 3.

[Clu49] Clusius, K., Goldmann, J., and Perlick, A., *Zeits. f. Naturf.* **4a**, 424 (1949).

[Coc66] Cochran, J. F., Shiffman, C. A., and Neighbor, J. E., *Rev. Sci. Instrum.* **37**, 499 (1966).

[Coc73] Cochran, W., *The Dynamics of Atoms in Crystals*, E. Arnold, London (1973).

[Coh87] Cohen, E. P., and Giacomo, P., *Physica A* **146**, 1 (also *International Union of Physics and Applied Physics*, Doc. I. U. P. A. P.-25) (1987).

[Col63] Colwell, J. H., Gill, E. K., and Morrison, J. A., *J. Chem. Phys.* **39**, 635 (1963); **42**, 3144 (1965).

[Col73] Collins, J. G., White, G. K., and Swenson, C. A., *J. Low Temp. Phys.* **10**, 69 (1973).

[Col83a] Collocott, S. J., *Aust. J. Phys.* **36**, 573 (1983).

[Col83b] Collocott, S. J., and Collins, J. G., *J. Phys. C: Solid State Phys.* **16**, 6179 (1983).

[Col84a] Collocott, S. J., Collins, J. G., and White, G. K., *Proc. LT17* (Elsevier) Paper FL9, 1375 (1984).

[Col84b] Collocott, S. J., Collins, J. G., Cook, J. S., and White, G. K., unpublished private communication (1984).

[Col85a] Collins, J. G., Collocott, S. J., and White, G. K., *Aust. J. Phys.* **38**, 617 (1985).

[Col85b] Collocott, S. J., *J. Non-Cryst. Solids* **70**, 291 (1985).

[Col86a] Collings, E. W., *Applied Superconductivity, Metallurgy and Physics of Titanium Alloys, Vol. 1, Fundamentals*, Plenum Press, New York (1986).

[Col86b] Collocott, S. J., and White, G. K., *Cryogenics* **26**, 402 (1986).

[Col87a] Collins, J. G., Collocott, S. J., Tainsh, R. J., Andrikidis, C. A., and White, G. K., *Aust. J. Phys.* **40**, 65 (1987).

[Col87b] Collocott, S. J., *J. Phys. C: Solid State Phys.* **20**, 2995 (1987).

[Col87c] Collocott, S. J., White, G. K., Dou, S. X., and Williams, R. K., *Phys. Rev. B* **36**, 5684 (1987).

[Col87d] Collocott, S. J., and Rayne, J. A., *J. Appl. Phys.* **61**, 4404 (1987).

[Col88] Collocott, S. J., Hill, R. W., and Stewart, A. M., *J. Phys. F: Met. Phys.* **18**, L223 (1988).

[Col91] Collocott, S. J., and White, G. K., *Cryogenics* **31**, 102 (1991).

[Cor60] Corruccini, R. J., and Gniewek, J. J., *Specific Heats and Enthalpies of Technical Solids at Low Temperatures*, NBS Monograph 21, U. S. Govt. Printing Office, Washington, D. C. (1960).

[Cor61] Corruccini, R. J., and Gniewek, J. J., *Thermal Expansion of Technical Solids at Low Temperatures*, NBS Monograph 29, U. S. Govt. Printing Office, Washington, D. C. (1961).

[Cor76] Corsan, J. M., and Mitchem, N. I., in *Proc. Sixth Intern. Cryog. Eng. Conf.* IPC Science and Technology Press, UK (1976) p. 342.

[Cou78] Cousins, C. S. G., *J. Phys. C: Solid State Phys.* **11**, 4867, 4881 (1978).

[Cow90] Cowley, E. R., Gross, J., Gong, Z., and Horton, G. K., *Phys. Rev. B* **42**, 3135 (1990).

[Cow95] Cowley, E. R., and Horton, G. K., in [Hor95], Ch. 2.

[Cus87] Cusack, N. E., *The Physics of Structurally Disordered Matter: An Introduction*, Adam Hilger, Bristol (1987).

[Dad81] Dadobaev, G., and Slutsker, A. I., *Sov. Phys. Solid State* **23**, 1131 (1981).

[Dan62a] Daniels, W. B., *Phys. Rev. Lett.* **8**, 3 (1962).

[Dan62b] Dantl, G., *Zeits. f. Physik* **166**, 115 (1962).

[Dav59] Davies, R. O., and Parke, S., *Phil. Mag.* **4**, 341 (1959).

[Deb12] Debye, P., *Ann. d. Phys.* **39**, 789 (1912).

[deB49] de Boer, J., *Rep. Prog. Phys.* **12**, 305 (1949).

[deJ74] de Jongh, L. J., and Miedema, A. R., *Adv. Phys.* **23**, 1 (1974).

[deL54] de Launay, J., *J. Chem. Phys.* **22**, 1676 (1954); **24**, 1071 (1956); **30**, 91 (1959).

[deL56] de Launay, J., *Solid State Phys.* **2**, 219 (1956).

[DeP86] DePuydt, J. M., and Dahlberg, E. D., *Rev. Sci. Instrum.* **57**, 483 (1986).

[DeS53] DeSorbo, W., and Tyler, W. W., *J. Chem. Phys.* **21**, 1660 (1953).

[DeS58] DeSorbo, W., and Nichols, G. E., *J. Phys. Chem. Solids* **6**, 352 (1958).

[deV89] de Visser, A., Franse, J. J. M., and Flouquet, J., *Physica B* **161**, 324 (1989).

[deV90] de Visser, A., Franse, J. J. M., Lacerda, A., Haen, P., and Flouquet, J., *Physica B* **163**, 49 (1990).

[Dic78] Dickens, M. H., and Hutchings, M. T., *J. Phys. C: Solid State Phys.* **11**, 461 (1978).

[Dob86a] Dobbs, J. N., and Anderson, A. C., *Phys. Rev. B* **33**, 4172 (1986).

[Dob86b] Dobbs, J. N., Foote, M. C., and Anderson, A. C., *Phys. Rev. B* **33**, 4178 (1986).

[Dob94] Dobbs, E. R., *Solid Helium Three*, Clarendon Press, Oxford (1994).

[Dol66] Dolling, G., and Cowley, R. A., *Proc. Phys. Soc.* **88**, 463 (1966).

[Dol74] Dolling, G., in [Hor74], Vol. 1, Ch. 10.

[Dom52] Domb, C., and Salter, L. S., *Phil. Mag.* **43**, 1083 (1952).

[Dom60] Domb, C., *Adv. Phys.* **9**, 149 (1960).

[Dom93] Dominec, J., *Supercond. Sci. Technol.* **6**, 153 (1993).

[Dom96] Domb, C., *The Critical Point*, Taylor and Francis, London (1996).

[Dre94] Dresselhaus, M. S., Dresselhaus, G., and Saito, R., in *Physical Properties of High Temperature Superconductors* **4**, ed. D. M. Ginsberg, World Scientific, Singapore (1994) p. 13.

[Dro88] Drotning, W. D., *Int. J. Thermophys.* **9**, 849 (1988).

[duCh66] du Chatenier, F. J., and Miedema, A. R., *Physica* **32**, 403 (1966).

[Dug53] Dugdale, J. S., and MacDonald, D. K. C., *Phys. Rev.* **89**, 832 (1953).

[Dug54] Dugdale, J. S., Morrison, J. A., and Patterson, D., *Proc. R. Soc. Lond. A* **224**, 228 (1954).

[Dug64] Dugdale, J. S., and Franck, J. P., *Phil. Trans. R. Soc. Lond. A* **257**, 1 (1964).

[Dug96] Dugdale, J. S., *Entropy and its Physical Meaning*, Taylor and Francis, London (1996).

[Dul819] Dulong, P. L., and Petit, A. T., *Ann. chim. et phys.* [2] **10**, 395 (1819).

[Dun95] Dunn, M. L., and Leadbetter, H., *J. Appl. Phys.* **78**, 1583 (1995).

[Dut88] Dutzi, J., Pattalwar, S. M., Dixit, R. N., and Shete, S. Y., *Pramāna J. Phys.* **31**, 253 (1988).

[Dym80] Dymond, J. H., and Smith, E. B., *The Virial Coefficients of Pure Gases and Mixtures: A Critical Compilation*, Clarendon Press, Oxford (1980).

[Ear81] Early, S. R., Hellman, F., Marshall, J., and Geballe, T. H., *Physica B* **107**, 327 (1981).

[Edw65] Edwards, D. O., Brewer, D. F., Seligman, P., Skertic, M., and Yaqub, M., *Phys. Rev. Lett.* **15**, 773 (1965).

[Ein07] Einstein, A., *Ann. d. Phys.* **22**, 180 (1907).

[Eki83] Ekin, J. W., in *Materials at Low Temperatures*, ed. R. P. Reed and A. F. Clark, Am. Soc. Metals, Metals Park, Ohio (1983) Ch. 13.

[Eri92] Eriksson, O., Wills, J. M., and Wallace, D., *Phys. Rev. B*, **46**, 5221 (1992).

[Esc95] Escher, U., *Cryogenics* **35**, 775 (1995).

[Euc24] Eucken, A., and Karwat, E., *Z. phys. Chem.* **112**, 467 (1924).

[Eve92] Every, A. G., and McCurdy, A. K., *Second and Higher Elastic Constants, Landolt–Börnstein* **III/29a**, Springer, Berlin (1992).

[Fan93] Fang, Z.-H., and Chen, L.-R., *Phys. Status Solidi (b)* **180**, K5 (1993).

[Faw86] Fawcett, E., Kaiser, A. B., and White, G. K., *Phys. Rev. B* **34**, 6248 (1986).

[Faw88] Fawcett, E., *Rev. Mod. Phys.* **60**, 209 (1988).

[Faw94] Fawcett, E., Alberts, H. L., Gaikin, V. Yu., Noakes, D. R., and Yakhmi, J. V., *Rev. Mod. Phys.* **66**, 25 (1994).

[Fey72] Feynman, R. P., *Statistical Mechanics: A Set of Lectures*, Addison-Wesley, Redwood City, Cal. (1972)

[Fil75] Filler, R. L., Lindenfeld, P., and Deutscher, G., *Rev. Sci. Instrum.* **46**, 439 (1975).

[Fisc88] Fischer, H. E., Watson, S. K., and Cahill, D. G., *Comments Condens. Matter Phys.* **14**, 65 (1988).

[Fisc93] Fischer, G. J., Wang, Z., and Karato, S., *Phys. Chem. Minerals* **20**, 97 (1993).

[Fish88] Fisher, R. A., Gordon, J. E., and Phillips, N. E., *J. Supercond.* **1**, 231 (1988).

[Fisk88] Fisk, Z., Hess, D. W., Pethick, C. J., Pines, D., Smith, J. L., Thompson, J. D., and Willis, J. O., *Science* **239**, No. 4835, 33 (1988).

[Fle79] Fletcher, G. C., and Yahaya, M., *J. Phys. F: Met. Phys.* **9**, 1529 (1979).

[Flu59] Flubacher, P., Leadbetter, A. J., and Morrison, J. A., *Phil. Mag.* **4**, 273 (1959).

[Fom97] Fominaya, F., Fournier, T., Gandit, P., and Chaussy, J., *Rev. Sci. Instrum.* **68**, 4191 (1997).

[For80] Forgan, E. M., and Nedjat, S., *Rev. Sci. Instrum.* **51**, 411 (1980).

[For97] Ford, P. J., and Saunders, G. A., *Contemp. Phys.* **38**, 63 (1997).

[Fra98] Fracchia, R. M., Barrera, G. D., Allan, N. L., Barron, T. H. K., and Mackrodt, W. C., *J. Phys. Chem. Solids* **59**, 435 (1998).

[Fre96] Frenkel, D., and Smit, B., *Understanding Molecular Simulation: From Algorithms to Applications*, Academic Press, San Diego, (1996).

[Fri76] Fritz, I. J., *Sol. State Comm.* **20**, 299 (1976).

[Fuj74] Fujii, Y., Lurie, N. A., Pynn, R., and Shirane, G., *Phys. Rev. B* **10**, 3647 (1974).

[Fur68] Furukawa, G. T., Saba, W. G., and Reilly, M. L., NSRDS–NBS 18, U. S. Govt. Printing Office, Washington, D. C. (1968).

[Fur72] Furukawa, G. T., Douglas, T. B., and Pearlman, N., in *The American Inst. of Physics Handbook*, 3rd edition, McGraw-Hill, New York (1972) Ch. 4e.

[Gau71] Gauster, W. B., *Phys. Rev. B* **4**, 1286 (1971).

[Gau81] Gaur, U., and Wunderlich, B., *J. Phys. Chem. Ref. Data* **10**, 119 (1981); ibid. **10**, 1001 (1981); ibid **10**, 1051 (1981); ibid **11**, 313 (1982); ibid **12**, 29 (1983); ibid **12**, 65 (1983).

[Gia28a] Giauque, W. F., and Wiebe, R., *J. Am. Chem. Soc.* **50**, 101 (1928).

[Gia28b] Giauque, W. F., and Wiebe, R., *J. Am. Chem. Soc.* **50**, 2193 (1928).

[Gib59] Gibbons, D. F., *Phys. Rev.* **115**, 1194 (1959).

[Gib73] Gibbons, T. G., *Phys. Rev. B* **7**, 1410 (1973).

[Gib74] Gibbons, T. G., *J. Chem. Phys.* **60**, 1094 (1974).

[Gil56] Gilvarry, J. J., *Phys. Rev.* **102**, 331 (1956).

[Gly71] Glyde, H. R., and Klein, M. L., *Crit. Rev. Solid State Sci.* **2**, 181 (1971).

[Gly76] Glyde, H. R., in [Kle76], Ch. 7.

[Gme81a] Gmelin, E., and Ripka, K., *Cryogenics* **21**, 117 (1981).

[Gme81b] Gmelin, E., and Rödhammer, P., *J. Phys. E: Sci. Instrum.* **14**, 223 (1981).

[Gme87] Gmelin, E., *Thermochimica Acta* **110**, 183 (1987).

[Gme89] Gmelin, E., in *Studies of High Temperature Superconductors* **2**, ed. A. V. Narlikar, Nova Science, New York (1989) p. 95.

[Gol50] Goldstein, H., *Classical Mechanics*, Addison-Wesley, Cambridge, Mass. (1950).

[Gol92] Goldberg, R. N., and Weir, R. D., *Pure & Appl. Chem.* **64**, 1545 (1992).

[Gon88] Gong, Z., Horton, G. K., and Cowley, E. R., *Phys. Rev. B* **38** 10820 (1988).

[Gop66] Gopal, E. S. R., *Specific Heats at Low Temperatures*, Plenum Press, New York (1966).

[Gop74] Gopal, E. S. R., *Statistical Mechanics and Properties of Matter*, Ellis Horwood, Chichester (1974).

[Gra69] Graham, E. K., and Barsch, G. R., *J. Geophys. Res.* **74**, 5949 (1969).

[Gra72] Graham, M. G., and Hagy, H. E., (eds.), *Thermal Expansion - 1971* (A. I. P. Conf. Proc. No. 3), Am. Inst. Physics, New York (1972).

[Gra84] Gray, C. G., and Gubbins, K. E., *Theory of Molecular Fluids. Vol. 1: Fundamentals*, Clarendon Press, Oxford (1984).

[Gre83] Greywall, D. S., *Phys. Rev. B* **27**, 2747 (1983).

[Gri77] Griessen, R., Lee, M. J. G., and Stanley, D. J., *Phys. Rev. B* **16**, 4385 (1977).

[Gri81] Grimvall, G., *The Electron–Phonon Interaction in Metals*, North-Holland, Amsterdam (1981).

[Gro78] Grosse, R., Krause, P., Meissner, M., and Tausend, A., *J. Phys. C: Solid State Phys.* **11**, 45 (1978).

[Gru12] Grüneisen, E., *Ann. Phys.* **39**, 257 (1912).

[Gsc64] Gschneidner, K. A., Jr., *Solid State Phys.* **16**, 275 (1964).

[Gug94] Gugenberger, F., Meingast, C., Roth, G., Grube, K., Breit, V., Weber, T., Wühl, H., Uchida, S., and Nakamura, Y., *Phys. Rev. B* **49**, 13137 (1994).

[Gui897] Guillaume, C. E., *C. R. Acad. Sci.* **125**, 235 (1897).

[Gui91] Guillermet, A. F., and Grimvall, G., *Phys. Rev. B* **44**, 4332 (1991).

[Gut91] Gutsmiedl, P., Probst, Chr., and Andres, K., *Cryogenics* **31**, 54 (1991).

[Gyo97] Gyorffy, B. L., in [Spr97a], Ch. 4.

[Hah70] Hahn, T. A., *J. Appl. Phys.* **41**, 5096 (1970).

[Hah72] Hahn, T. A., and Kirby, R. K., in [Gra72], p. 13.

[Hah83] Hahn, T., (ed.), *International Tables for Crystallography*, Vol. A, Reidel, Dordrecht (1983).

[Hai72] Haida, O., Matsuo, T., Suga, H., and Seki, S., *Proc. Jpn. Acad.* **48**, 489 (1972).

[Hak68] Hake, R. R., *Phys. Rev.* **166**, 471 (1968).

[Hak69] Hake, R. R., *J. Appl. Phys.* **40**, 5148 (1969).

[Hal76] Hale, D. K., *J. Mater. Sci.* **11**, 2105 (1976).

[Ham96] Hama, J., and Suito, K., *J. Phys.: Condens. Matter* **8**, 67 (1996).

[Han74] Hanson, R. C., Helliwell, K., and Schwab, C., *Phys. Rev. B* **9**, 2649 (1974).

[Han86] Hansen, J. P., and McDonald, I. R., *Theory of Simple Liquids*, 2nd edition, Academic Press, London (1986).

[Har77a] Harper, J. M. E., Greene, R. L., Grant, P. M., and Street, G. B., *Phys. Rev. B* **15**, 539 (1977).

[Har77b] Hartwig, G., *Adv. Cryogen. Eng.* **22**, 283 (1977).

[Har79] Hardy, J. R., and Karo, A. M., *The Lattice Dynamics and Statics of Alkali Halide Crystals*, Plenum Press, New York (1979).

[Har82] Hartwig, G., and Evans, D., (eds.), *Nonmetallic Materials and Composites at Low Temperatures* **2**, Plenum Press, New York (1982).

[Har84] Hartwig, G., and Knaak, S., *Cryogenics* **24**, 639 (1984).

[Har88a] Hartwig, G., *Cryogenics* **28**, 255 (1988).

[Har88b] Hartwig, G., and Evans, D., (eds.), *Nonmetallic Materials and Composites at Low Temperatures* **4**, Plenum Press, New York (1988).

[Har94] Hartwig, G., *Polymer Properties at Room and Cryogenic Temperatures*, Plenum Press, New York (1994).

[Har98] Hartwig, G., and Evans, D., (eds.), *Cryogenics* **38**, *Proc. Conference on Non-Metallic Materials and Composites at Low Temperatures* **8** (1998).

[Has84] Hashin, Z., *J. Mech. Phys. Solids* **32**, 149 (1984).

[Hea66] Hearmon, R. F. S., in *Landolt–Börnstein* **III/1**, Springer, Berlin (1966) p. 1.

[Hea69] Hearmon, R. F. S., in *Landolt–Börnstein* **III/2**, Springer, Berlin (1969) p. 1.

[Hea79] Hearmon, R. F. S., in *Landolt–Börnstein* **III/11**, Springer, Berlin (1979) pp. 1, 245.

[Hea84] Hearmon, R. F. S., in *Landolt–Börnstein* **III/18**, Springer, Berlin (1984) pp. 1, 155.

[Hea94] Heaney, P. J., Prewitt, C. T., and Gibbs, G. V., (eds.), *Silica: Physical Behavior, Geochemistry and Materials Applications*, Reviews in Mineralogy, Vol. 29, Mineralogical Society of America, Washington (1994).

[Heb70] Heberlein, D. C., and Adams, E. D., *J. Low Temp. Phys.* **3**, 115 (1970).

[Hei66] Heiniger, F., Bucher, E., and Müller, J., *Phys. kondens. Mater.* **5**, 243 (1966).

[Hel81] Hellwege, K.-H., and Olsen, J. L. (eds.), *Metals: Phonon States. Electron States and Fermi Surfaces*, *Landolt–Börnstein* **III/13a,b,c**, Springer, Berlin (1981, 1983, 1984).

[Hem84] Hemminger, W., and Höhne, G., *Calorimetry - Fundamentals and Practice*, Verlag Chemie, Weinheim, Germany (1984).

[Hey71] Heydemann, P., in *Physical Acoustics* **VIII**, ed. W. P. Mason and R. N. Thurston, Academic Press, New York (1971) Ch. 4.

[Hil52] Hill, R., *Proc. Phys. Soc.* **A65**, 349 (1952).

[Hil60] Hill, T. L., *An Introduction to Statistical Thermodynamics*, Addison-Wesley, Reading, Mass. (1960).

[Hil68] Hill, R. W., Martin, D. L., and Osborne, D. W., in *Experimental Thermodynamics*, ed. J. P. McCullough and D. W. Scott, Butterworths, London (1968) Vol. 1, Ch. 7.

[Hir70] Hirao, K., Soga, N., and Kunugi, M., *J. Am. Ceram. Soc.* **62**, 570 (1970).

[Ho98] Ho, C. Y., (ed.), *Thermal Expansion of Solids* (Volume Coordinator R. E. Taylor), Vol. I–4 in the CINDAS Data Series on Material Properties, ASM International, Metals Park, Ohio (1998).

[Hol72] Holste, J. C., Cetas, T. C., and Swenson, C. A., *Rev. Sci. Instrum.* **43**, 670 (1972).

[Hol84] Holden, A. J., Heine, V., and Samson, J. H., *J. Phys. F: Met. Phys.* **14**, 1005 (1984).

[Hor74] Horton, G. K., and Maradudin, A. A., (eds.), *Dynamical Properties of Solids*, Vol. 1, North-Holland, Amsterdam (1974).

[Hor75] Horton, G. K., and Maradudin, A. A., (eds.), *Dynamical Properties of Solids*, Vol. 2, North-Holland, Amsterdam (1975).

[Hor95] Horton, G. K., and Maradudin, A. A., (eds.), *Dynamical Properties of Solids*, Vol. 7, North-Holland, Amsterdam (1995).

[Hor96] Horton, G. K., Acocella, D., and Cowley, E. R., *Physica A* **232**, 625 (1996).

[Hor78] Hortal, M., *Phys. Chem. Glasses* **19**, 5 (1978).

[Hov69] Hovi, V., Paavola, K., and Nurmi, E., *Ann. Acad. Sci. Fennicae* (Finland), Series **AVI**, No. 328 (1969).

[Hul73] Hultgren, R., Desai, P. D., Hawkins, D. T., Gleiser, M., Kelley, K. K., and Wagman, D. D., *Selected Values of the Thermodynamic Properties of the Elements*, Am. Soc. Metals, Metals Park, Ohio (1973).

[Hun71] Huntzicker, J. J., and Westrum, E. F., Jr., *J. Chem. Thermodyn.* **3**, 61 (1971).

[Hur74] Hurley, M., and Gerstein, B. C., *J. Chem. Thermodynamics* **6**, 787 (1974).

[Hwa97] Hwang, J. S., Lin, K. J., and Tien, C., *Rev. Sci. Instrum.* **68**, 94 (1997).

[Iba69] Ibach, H., *Phys. Status Solidi* **33**, 257 (1969).

[IRE49] IRE Piezoelectric Crystals Committee, *Proc. IRE* **37**, 1378 (1949).

[Jac81] Jackson, I., Niesler, H., and Weidner, D. J., *J. Geophys. Res.* **86**, 3736 (1981).

[Jac82] Jackson, I., and Niesler, H., *Advances in Earth and Planetary Sciences* **12**, 93 (1982).

[Jac84] Jacobs, S. F., Shough, D., and Connors, C., *Appl. Opt.* **23**, 4237 (1984).

[Jac91] Jäckel, M., *Cryogenics* **35**, 713 (1991).

[Jac97] Jacobsen, R. T., Penoncello, S. G., and Lemmon, E. W., *Thermodynamic Properties of Cryogenic Fluids*, Plenum Press, New York (1997).

[Jam65] James, B. W., and Yates, B., *Cryogenics* **5**, 68 (1965).

[Jef50] Jeffries, H., and Jeffries, B. S., *Methods of Mathematical Physics*, 2nd edition, Cambridge University Press, Cambridge (1950).

[Joh73] Johnson, D. L., Cluxton, D. H., and Young, R. A., *Rev. Sci. Instrum.* **44**, 16 (1973).

[Jon61] Jones, R. V., *J. Sci. Instrum.* **38**, 37 (1961).

[Jon98] Jones, R. O., (Jülich), private communication (1998).

[Jor90] Jorgensen, J. D., Pei, S., Lightfoot, P., Hinks, D. G., Veal, B. W., Dabrowski, B., Paulikas, A. P., Kleb, R., and Brown, I. D., *Physica C* **171**, 93 (1990).

[Jun90a] Junod, A., in *Physical Properties of High Temperature Superconductors* **2**, ed. D. M. Ginsberg, World Scientific, Singapore (1990) p. 13.

[Jun90b] Junod, A., Eckert, A. D., Graf, T., Kaldis, E., Karpinski, J., Rusiecki, R., Sanchez, D., Triscone, G., and Muller, J., *Physica C* **168**, 47 (1990).

[Jun96] Junod, A., in *Studies of High Temperature Superconductors* **19**, ed. A. V. Narlikar, Nova Science Publishers, N. Y. (1996) p. 1.

[Kai85] Kaiser, A. B., and Haines, E. M., *J. Phys. F: Met. Phys.* **15**, 1765 (1985).

[Kan95] Kantorovich, L. N., *Phys. Rev. B* **51**, 3520, 3535 (1995).

[Kas80] Kasen, M. B., MacDonald, G. R., Beekman, D. H., and Schramm, R. E., *Adv. Cryogen. Eng. Materials* **26**, 235 (1980).

[Kel69] Keller, W. E., *Helium-3 and Helium-4*, Plenum Press, New York (1969).

[Ken90] Kenny, T. W., and Richards, P. L., *Rev. Sci. Instrum.* **61**, 822 (1990).

[Kie96] Kierspel, H., Winkelmann, H., Auweiler, T., Schlabitz, W., Büchner, B., Duijn, V. H. M., Hien, N. T., Menovsky, A. A., and Franse, J. J. M., *Physica C* **262**, 177 (1996).

[Kin90] King-Smith, R. D., and Needs, R. J., *J. Phys.: Condens. Matter* **2**, 3431 (1990).

[Kir67] Kirby, R. K., *J. Res. Nat. Bur. Stand.* **71A**, 363 (1967).

[Kir72] Kirby, R. K., Hahn, T. A., and Rothrock, B. D., in *The American Inst. of Physics Handbook*, 3rd edition, McGraw-Hill, New York (1972) Ch. 4f.

[Kit76] Kittel, C., *Introduction to Solid State Physics*, 5th edition, John Wiley & Sons, New York (1976).

[Kla97] Klaasse, J. C. P., *Rev. Sci. Instrum.* **68**, 89 (1997).

[Kle76] Klein, M. L., and Venables, J. A., (eds.), *Rare Gas Solids*, Vol. I, Academic Press, London (1976).

[Kle77] Klein, M. L., and Venables, J. A., (eds.), *Rare Gas Solids*, Vol. II, Academic Press, London (1977).

[Kle86] Klemens, P. G., *Int. J. Thermophys.* **7**, 197 (1986).

[Kle87] Klein, M.W., *Phys. Rev. B* **35**, 1397 (1987).

[Kle88] Klemens, P. G., *Int. J. Thermophys.* **9**, 171 (1988).

[Kna75] Knapp, G. S., Bader, S. D., Culbert, H. V., Fradin, F. Y., and Klippert, T. E., *Phys. Rev. B* **11**, 4331 (1975).

[Kob79] Kobayashi, M., *J. Chem. Phys.* **70**, 509 (1979).

[Kol77] Kollie, T. G., *Phys. Rev. B* **16**, 4872 (1977).

[Kor77] Korpiun, P., and Lüscher, E., in [Kle77], Ch. 12.

[Kos69] Kos, J. F., and Lamarche, J. L. G., *Canad. J. Phys.* **47**, 2509 (1969).

[Kra80] Krause, J. K., and Swenson, C. A., *Phys. Rev. B* **21**, 2533 (1980).

[Kra84] Kraftmakher, Ya. A., in *Compendium of Thermophysical Property Methods*, ed. K. D. Maglić, A. Cezairliÿan and V. E. Peletsky, Vol. 1, Plenum Press, New York (1984) Ch. 15.

[Kre83] Kress, W., in *Phonon States of Alloys. Electron States and Fermi Surfaces of Strained Elements, Landolt Börnstein* **III/13b**, Springer, Berlin (1983) p. 396.

[Kri79] Krishnan, R. S., Srinivasan, R., and Devanarayanan, S., *Thermal Expansion of Crystals*, Pergamon Press, Oxford (1979).

[Kro77] Kroeger, F. R., and Swenson, C. A., *J. Appl. Phys.* **48**, 853 (1977).

[Kru72] Krüger, J., *Phys. Chem. Glasses* **13**, 9 (1972).

[Kub62] Kubaschewski, O., and Hultgren, R., in *Experimental Thermochemistry*, ed. H. A. Skinner, Vol II, Interscience, London (1962).

[Kur72] Kurkjian, C. R., Krause, J. T., McSkimin, H. J., Andreatch, P., and Bateman, T. B., in *Amorphous Materials*, ed. R. W. Douglas and B. Ellis, Wiley, New York (1972) p. 463.

[Kwo90] Kwok, R. S., and Brown, S. E., *Rev. Sci. Instrum.* **61**, 809 (1990).

[Lac94a] Lacks, D. J., and Rutledge, G. C., *Macromolecules* **27**, 7197 (1994).

[Lac94b] Lacks, D. J., and Rutledge, G. C., *J. Phys. Chem.* **98**, 1222 (1994).

[Lan81] Lanchester, P. C., and Baker, D. P., *J. Phys. E: Sci. Instrum.* **14**, 805 (1981).

[Las69] Lasjaunias, J.-C., *C. R. Acad. Sci. Paris B* **269**, 763 (1969).

[Law75] Lawless, W. N., *Cryogenics* **16**, 273 (1975).

[Law80] Lawless, W. N., *Ferroelectrics* **24**, 327 (1980).

[Lea65] Leadbetter, A. J., *Proc. R. Soc. Lond. A* **287**, 403 (1965).

[Lea68] Leadbetter, A. J., *Phys. Chem. Glasses* **9**, 1 (1968).

[Lea69] Leadbetter, A. J., *J. Chem. Phys.* **51**, 779 (1969).

[Lea71] Leadbetter, A. J., and Wycherley, K. E., *Phys. Chem. Glasses* **12**, 41 (1971).

[Lea73] Leadbetter, A. J., and Jeapes, A. P., *J. Phys. C: Solid State Phys.* **6**, 1546 (1973).

[Lea77] Leadbetter, A. J., Jeapes, A. P., Waterfield, C. G., and Wycherley, K. E., *Chem. Phys. Lett.* **52**, 469 (1977).

[Lea82] Leaist, D. C., Murray, J. J., Post, M. L., and Davidson, D. W., *J. Phys. Chem.* **86**, 4175 (1982).

[Leg78] Legg, G. J., and Lanchester, P. C., *J. Phys. F: Met. Phys.* **8**, 2125 (1978).

[LeP80] Le Page, Y., Calvert, L. D., and Gabe, E. J., *J. Phys. Chem. Solids* **41**, 721 (1980).

[Lev67] Levin, V. M., *Mekhanika Tverdogo Tela* **2**, 88 (1967).

[Lew67] Lewis, J. T., Lehoczky, A., and Briscoe, C. V., *Phys. Rev.* **161**, 877 (1967).

[Li96] Li, B., Jackson, I., Gasparik, T., and Liebermann, R. C., *Phys. Earth Plan. Inter.* **98**, 79 (1996).

[Lic98] Lichtenstein, A. I., Jones, R. O., Xu, H., and Heaney, P. J., *Phys. Rev. B* **58**, 6219 (1998).

[Liu91] Liu, S., Horton, G. K., and Cowley, E. R., *Phys. Rev. B* **44**, 11714 (1991).

[Liu95] Liu, M., Finlayson, T. R., and Smith, T. F., *Phys. Rev. B* **52**, 530 (1995).

[Lon54] London, F., *Superfluids, Vol. II*, Wiley, New York (1954).

[Lor67] Lord, A. E., *J. Phys. Chem. Solids* **28**, 517 (1967).

[Lor94] Loram, J. W., Mirza, K. A., Cooper, J. R., Liang, W. Y., and Wade, J. M., *J. Supercond.* **7**, 243 (1994).

[Lou74] Lounasmaa, O. V., *Experimental Principles and Methods Below 1 K*, Academic Press, London (1974).

[Lub82] Lubin, G., (ed.), *Handbook of Composites*, Van Nostrand–Reinhold, New York (1982).

[Luo85] Luong, N. H., Franse, J. J. M., and Hien, T. D., *J. Phys. F: Met. Phys.* **15**, 1751 (1985).

[Lyo79] Lyon, K. G., Salinger, G. L., and Swenson, C. A., *Phys. Rev. B* **19**, 4231 (1979).

[Mac85] MacDonald, W. M., Anderson, A. C., and Schroeder, J., *Phys. Rev. B* **31**, 1090 (1985).

[Mad82] Madelung, O., (ed.), *Physics of Group IV Elements and III-V Compounds, Landolt–Börnstein* **III/17a**, Springer, Berlin (1982).

[Mad88] Madsen, J., and Trefny, J., *J. Phys. E: Sci. Instrum.* **21**, 363 (1988).

[Mad96] Madden, P. A., and Wilson, M., *Chem. Soc. Reviews* **25**, 339 (1996).

[Man59] Manchester, F. D., *Canad. J. Phys.* **37**, 989 (1959).

[Man66] Manzhelii, V. G., Tolkachev, A. M., and Voitovich, E. I., *Phys. Status Solidi* **13**, 351 (1966).

[Man97] Manzhelii, V. G., and Freiman, Y. A., (eds.), *Physics of Cryocrystals*, English version edited by M. L. Klein and A. A. Maradudin, Am. Inst. Physics, New York (1997).

[Mar60] Martin, D. L., *Canad. J. Phys.* **38**, 17 (1960).

[Mar67] Martin, D. L., *Rev. Sci. Instrum.* **38**, 1738 (1967).

[Mar68] Martinek, C., and Hummel, F. A., *J. Am. Ceram. Soc.* **51**, 227 (1968).

[Mar71] Maradudin, A. A., Montroll, E. W., Weiss, G. H., and Ipatova, I. P., *Theory of Lattice Dynamics in the Harmonic Approximation*, Suppl. 3 of *Solid State Physics*, 2nd edition, Academic Press, New York (1971).

[Mar73] Martin, D. L., Bradley, L. L. T., Cazemier, W. J., and Snowdon, R. L., *Rev. Sci. Instrum.* **44**, 675 (1973).

[Mar75] Martin, D. L., *Rev. Sci. Instrum.* **46**, 657 (1975).

[Mar79] Martin, D. L., *J. Phys. E: Sci. Instrum.* **12**, 478 (1979).

[Mar81] Martin, D. L., *Canad. J. Phys.* **59**, 567 (1981).

[Mar87a] Marsh, K. N., (ed.), *Recommended Reference Materials for the Realization of Physicochemical Properties*, Blackwells Scientific, Oxford (1987) Section 9.

[Mar87b] Martin, D. L., *Canad. J. Phys.* **65**, 1104 (1987).

[Mar88] Martin, D. L., in [Cez88], Ch. 3.

[Mar96] Mary, T. A., Evans, J. S. O., Vogt, T., and Sleight, A. W., *Science* **272**, 90 (1996).

[Mar97] Marone, M. J., and Payne, J. E., *Rev. Sci. Instrum.* **68**, 4516 (1997).

[May92] Maynard, J. D., McKenna, M. J., Migliori, A., and Visscher, W. M., in *Physical Acoustics*, **XX**, ed. M. Levy, Academic Press, San Diego (1992) Ch. 8.

[McG95] McGurn, A. R., in [Hor95], Ch. 1.

[McL72] McLean, K. O., Swenson, C. A., and Case, C. R., *J. Low Temp. Phys.* **7**, 77 (1972).

[McM94] McMenamin, C. S., Brewer, D. F., Hargreaves, T. E., Hussey, N. E., Moreno, C., Thomson, A. L., and Bird, J. P., *Physica B* **194**, 21 (1994).

[McM96] McMenamin, C. S., Al-Khawaja, S., Brewer, D. F., Chastin, S., Ekosipedidis, G., and Thomson, A. L., *Czech. J. Phys.* **46**, Suppl. S5 2781 (1996).

[McS53] McSkimin, H. J., *J. Appl. Phys.* **24**, 988 (1953).

[McS64] McSkimin, H. J., in *Physical Acoustics* **IA**, ed. W. P. Mason, Academic Press, New York (1964) Ch. 4.

[McW77] McWhan, D. B., Varma, C. M., Hsu, F. L. S., and Remeika, J. P., *Phys. Rev. B* **15**, 553 (1977).

[Mei78] Meissner, M., Tausend, A., and Wobig, D., *Phys. Status Solidi (a)* **49**, 59 (1978).

[Mei91] Meingast, C., Kraut, O., Wolf, T., Wühl, H., Erb, A., and Müller-Vogt, G., *Phys. Rev. Lett.* **67**, 1634 (1991).

[Mei93] Meingast, C., Karpinski, J., Jilek, E., and Kaldis, E., *Physica C* **209**, 591 (1993).

[Mei96] Meingast, C., Junod, A., and Walker, E., *Physica C* **272**, 106 (1996).

[Mer73] Merchant, H. D., Srivastava, K. K., and Pandey, H. D., *Crit. Rev. Solid State Sci.* **3**, 451 (1973); also Rapp, J. E., and Merchant, H. D., *J. Appl. Phys.* **44**, 3919 (1973).

[Met53] Metropolis, N., Rosenbluth, A. W., Rosenbluth, M. N., Teller, A. H., and Teller, E., *J. Chem. Phys.* **21**, 1087 (1953).

[Mig93] Migliori, A., Sarrao, J. L., Visscher, W. M., Bell, T. M., Ming Lei, Fisk, Z., and Leisure, R. G., *Physica* **B183**, 1 (1993).

[Mig97] Migliori, A., and Sarrao, J. L., *Resonant Ultrasound Spectroscopy*, Wiley, New York (1997).

[Mii88] Miiller, A. P., in [Cez88], Ch. 1.

[Mil83] Milton, G. W., private communication (1983).

[Min94] Minakov, A. A., *IEEE Trans. Magn.* **MAG-30**, 1058 (1994).

[Moe57] Moelwyn-Hughes, E. A., *Physical Chemistry*, Pergamon, London (1957).

[Mon67] Montgomery, H., Pells, G. P., and Wray, E. M., *Proc. R. Soc. Lond. A* **301**, 261 (1967).

[Moo62] Moore, W. J., *Physical Chemistry*, 4th edition, Prentice-Hall, Englewood Cliffs, N. J. (1962).

[Mor81a] Morgan, G. J., White, G. K., and Collins, J. G., *Phil. Mag. B* **43**, 1039 (1981).

[Mor81b] Moriya, T., (ed.), *Electron Correlation and Magnetism in Narrow-Band Systems*, Springer, Berlin (1981).

[Mor85] Moriya, T., *Spin Fluctuations in Itinerant Electron Magnetism*, Springer, Berlin (1985).

[Mos77] Moses, D., Ben-Aroya, O., and Lupu, N., *Rev. Sci. Instrum.* **48**, 1098 (1977).

[Mou78] Mountain, D. S., and Webber, J. M. B., *Proc. Soc. Photo-Opt. Inst. Engrs.* **164**, 189 (1978).

[Mun68] Munn, R. W., *Phil. Mag.* **17**, 433 (1968).

[Mun69] Munn, R. W., *Adv. Phys.* **18**, 515 (1969).

[Mun72] Munn, R. W., *J. Phys. C: Solid State Phys.* **5**, 535 (1972).

[Nak94] Nakamura, Y., *Hyperfine Interact.* **83**, 55 (1994).

[Nar70] Narayanamurti, V., and Pohl, R. O., *Rev. Mod. Phys.* **42**, 201 (1970).

[Nel98] Nelson, R. A., *Physics Today*, BG13 (August 1998).

[Nie76] Niebel, K. F., and Venables, J. A., in [Kle76], Ch. 9.

[Nie89] Niesler, H., and Jackson, I., *J. Acoust. Soc. Am.* **86**, 1573 (1989).

[Nik80] Niksch, M., Lüthi, B., and Andres, K., *Phys. Rev. B* **22**, 5774 (1980).

[Nor58] Norwood, M. H., and Briscoe, C. V., *Phys. Rev.* **112**, 45 (1958).

[Nye85] Nye, J. F., *Physical Properties of Crystals*, (reprinted with suppl. material), Clarendon Press, Oxford (1985).

[Oka91] Okaji, M., and Birch, K. P., *Metrologia* **28**, 27 (1991).

[Oka94] Okaji, M., Nara, K., Kato, H., Michishita, K., and Kubo, Y., *Cryogenics* **34**, 163 (1994).

[Oka95a] Okada, T., Reed, R. P., and Nishima, S., (eds.), *Nonmetallic Materials and Composites at Low Temperatures* **7**, Plenum Press, New York (1995).

[Oka95b] Okaji, M., Yamada, N., Nara, K., and Kato, H., *Cryogenics* **35**, 887 (1995).

[Oka97a] Okaji, M., Yamada, N., Kato, H., and Nara, K., *Cryogenics* **37**, 251 (1997).

[Oka97b] Okaji, M., and Yamada, N., *High Temp.-High Press.* **29**, 89 (1997).

[Osb67] Osborne, D. W., Flotow, H. E., and Schreiner, F., *Rev. Sci. Instrum.* **38**, 159 (1967).

[Ott77] Ott, H. R., and Lüthi, B., *Z. Phys. B* **28**, 141 (1977).

[Ott85] Ott, H. R., Chandrasekhar, B. S., and Seeber, B., *Phys. Rev. B* **31**, 2700 (1985).

[Ott87] Ott, H. R., *Prog. Low Temp. Phys.* **11**, 215 (1987).

[Ove65] Overton, W. C., and Schuch, A. F., Los Alamos Report, LA3162 (U. S. Atomic Energy Commission) (1965).

[Pac63] Packard, J. R., and Swenson, C. A., *J. Phys. Chem. Solids*, **24**, 1405 (1963).

[Pap90] Papadakis, E. P., in *Physical Acoustics* **XIX**, ed. R. N. Thurston and A. D. Pierce, Academic Press, San Diego (1990) Ch. 2.

[Par51] Partington, J. R., *An Advanced Treatise on Physical Chemistry. Vol. Two: The Properties of Liquids*, Longmans, Green & Co., London (1951).

[Par78] Parsonage, N. G., and Staveley, L. A. K., *Disorder in Crystals*, Clarendon Press, Oxford (1978) Ch. 11.

[Pas98] Pasler, V., Schweiss, P., Meingast, C., Obst, B., Wühl, H., Rykov, A. I., and Tajima, S., *Phys. Rev. Lett.* **81**, 1094 (1998).

[Pat55] Patterson, D., *Canad. J. Chem.* **33**, 1079 (1955).

[Pay64] Payne, R. T., *Phys. Rev. Lett.* **13**, 53 (1964).

[Pea67] Pearson, W. B., *Handbook of Lattice Spacings of Metals and Alloys*, Pergamon, Oxford, Vols. I and II (1958, 1967).

[Pec97] Pecharsky, V. K., Moorman, J. O., and Gschneidner, K. A., Jr., *Rev. Sci. Instrum.* **68**, 4196 (1997).

[Pet95] Pettifor, D. G., *Bonding and Structure of Molecules and Solids*, Clarendon Press, Oxford (1995).

[Phi64] Phillips, N. E., *Phys. Rev.* **134**, A385 (1964).

[Phi71] Phillips, N. E., *Crit. Rev. Solid State Sci.* **2**, 467 (1971).

[Phi72] Phillips, W. A., *J. Low Temp. Phys.* **7**, 351 (1972).

[Phi73] Phillips, J. C., *Bonds and Bands in Semiconductors*, Academic Press, New York (1973).

[Phi81] Phillips, W. A., (ed.), *Amorphous Solids, Low Temperature Properties*, Springer, Berlin (1981).

[Phi92] Phillips, N. E., Fisher, R. A., and Gordon, J. E., *Prog. Low Temp. Phys.* **13**, 267 (1992).

[Pil73] Pillars, W. W., and Peacor, D. R., *Am. Mineral.* **58**, 681 (1973).

[Pip64] Pippard, A. B., *Elements of Classical Thermodynamics*, Cambridge University Press, Cambridge (1964).

[Pob96] Pobell, F., *Matter and Methods at Low Temperatures*, 2nd edition, Springer, Berlin (1996).

[Pom50] Pomeranchuk, I., *Zh. éksp. teor. Fiz* **20**, 919 (1950).

[Pop75] Popov, V. P., and Pervakov, V. A., *Phys. Status Solidi (b)* **71**, K95 (1975).

[Pop76] Popov, V. P., Pervakov, V. A., and Kuzmenko, I. N., *Phys. Metals Metallogr.* **41**, 201 (1976).

[Pot83] Pott, R., and Schefzyk, R., *J. Phys. E: Sci. Instrum.* **16**, 444 (1983).

[Pre90] Preston-Thomas, H., *Metrologia* **27**, pp. 1, 107 (1990).

[Pry96] Pryde, A. K. A., Hammonds, K. D., Dove, M. T., Heine, V., Gale, J. D., and Warren, M. C., *J. Phys.: Condens. Matter* **8**, 10973 (1996).

[Qui90] Quinn, T. J., *Temperature*, 2nd edition, Academic Press, London (1990).

[Ram92] Ramirez, A. P., Rosseinsky, M. J., Murphy, D. W., and Haddon, R. C., *Phys. Rev. Lett.* **69**, 1687 (1992).

[Ray56] Rayne, J. A., and Kemp, W. R. G., *Phil. Mag.* **1**, 918 (1956).

[Ray59] Rayne, J. A., *Phys. Rev.* **115**, 63 (1959).

[Ray61] Rayne, J. A., and Chandrasekhar, B. S., *Phys. Rev.* **122**, 1714 (1961).

[Ray81] Rayne, J. A., Collins, J. G., and White, G. K., *J. Appl. Phys.* **52**, 1977 (1981).

[Reg86] Regelsberger, M., Wernhardt, R., and Rosenberg, M., *J. Phys. E: Sci. Instrum.* **19**, 525 (1986).

[Reu29] Reuss, A., *Z. Angew. Math. Mech.* **9**, 49 (1929).

[Ric67] Rice, O. K., *Statistical Mechanics, Thermodynamics and Kinetics*, W. H. Freeman, San Francisco (1967).

[Ric84] Richardson, R. M., and Taylor, P., *Molecular Physics* **52**, 525 (1984).

[Ric88] Richardson, R. C., and Smith, E. N., (eds.), *Experimental Techniques in Condensed Matter Physics at Low Temperatures*, Addison-Wesley Inc. (1988).

[Rob72] Roberts, B. W., in *The American Inst. of Physics Handbook*, 3rd edition, McGraw-Hill, New York (1972) Ch. 9g.

[Rob76] Robie, R. A., Hemingway, B. S., and Wilson, W. H., *J. Res. U. S. Geol. Survey* **4**, 631 (1976).

[Rob81] Roberts, R. B., *J. Phys. E: Sci. Instrum.* **14**, 1386 (1981).

[Rob82] Robie, R. A., Hemingway, B. S., and Takei, H., *Am. Mineral.* **67**, 470 (1982).

[Rob84] Roberts, R. B., Andrikidis, C., Tainsh, R. J., and White, G. K., in *Proc. ICEC 10*, ed. H. Collan, P. Berglund and M. Krusius, Butterworths, Guildford, Surrey (1984) p. 499.

[Rol65] Rollett, J. S., *Computing Methods in Crystallography*, Pergamon, Oxford (1965).

[Ros70] Rosen, B. W., and Hashin, Z., *Int. J. Eng. Sci.* **8**, 157 (1970).

[Ros73] Rose-Innes, A. C., *Low Temperature Laboratory Techniques*, English Universities Press, London (1973).

[Row69] Rowlinson, J. S., *Liquids and Liquid Mixtures*, 2nd edition, Butterworths, London (1969).

[Rub97] Rubin, L. G., *Cryogenics* **37**, 341 (1997).

[Rus49] Rushbrooke, G. S., *Introduction to Statistical Mechanics*, Clarendon Press, Oxford (1949).

[Sak66] Sakurada, I., Ito, T., and Nakumae, K., *J. Polym. Sci. C* **15** 75 (1966).

[Sal65] Salter, L. S., *Adv. Phys.* **14**, 1 (1965).

[Sam71] Samara, G. A., *Ferroelectrics* **2**, 177 (1971).

[Sam73] Samara, G. A., and Morosin, B., *Phys. Rev. B* **8**, 1256 (1973).

[San95] Sannabhadti, R. S., Ramakrishnan, S., Ghosh, K., Patade, A., and Chandra, G., *Pramāna - J. Phys.* **45**, 81 (1995).

[Sch69] Schapery, J., *Composite Mater.* **2**, 311 (1969).

[Sch71] Schneider, W., *Kunstoffe* **61**, 273 (1971).

[Sch74] Schutz, R. J., *Rev. Sci. Instrum.* **45**, 548 (1974).

[Sch75] Schwall, R. E., Howard, R. E., and Stewart, G. R., *Rev. Sci. Instrum.* **46**, 1054 (1975).

[Sch81] Schober, H., and Dederichs, P. H., in *Phonon States of Elements. Electron States and Fermi Surfaces of Alloys*, *Landolt–Börnstein* **III/13a**, Springer, Berlin (1981) p. 1.

[Sch87a] Schmiedeshoff, G. M., Fortune, N. A., Brooks, J. S., and Stewart, G. R., *Rev. Sci. Instrum.* **58**, 1743 (1987).

[Sch87b] Schulgasser, K., *J. Mech. Phys. Solids* **35**, 35 (1987).

[Sch89] Schulgasser, K., *J. Mater. Sci. Lett.* **8**, 228 (1989).

[Sch92] Schilling, J. S., and Klotz, S., in *Physical Properties of High Temperature Superconductors* **3**, ed. D. M. Ginsberg, World Scientific, Singapore (1992) p. 59.

[Sco78] Scott, T. E., in *Handbook on the Physics and Chemistry of Rare Earths* **1**, ed. K. A. Gschneidner and L. Eyring, North-Holland, Amsterdam (1978) p. 591.

[Scu65] Scurlock, R. G., and Wray, E. M., *J. Sci. Instrum.* **42**, 421 (1965).

[She72] Sheard, F. W., in [Gra72], p. 155.

[She79] Sherwood, J. N., (ed.), *The Plastically Crystalline State*, John Wiley and Sons, Chichester (1979).

[Shi74] Shimizu, M., *Phys. Lett. A* **50**, 93 (1974).

[Sho60] Shoenberg, D., *Superconductivity*, Cambridge University Press, Cambridge (1960).

[Shu85] Shukla, R. C., and Cowley, E. R., *Phys. Rev. B* **31**, 372 (1985).

[Sim37] Simon, F. E., *Physica* **4**, 1089 (1937).

[Sim56] Simon, F. E., *Year Book of the Physical Society*, London (1956) p. 1.

[Sim60] Simmons, R. O., and Balluffi, R. W., *Phys. Rev.* **117**, 52, **119**, 600 (1960); **125**, 862 (1962); **129**, 1533 (1963).

[Sim63] Simmons, R. O., and Balluffi, R. W., *Phys. Rev.* **129**, 1533 (1963).

[Sim71] Simmons, G., and Wang, H., *Single Crystal Elastic Constants and Calculated Aggregate Properties: a Handbook*, M. I. T. Press, Cambridge, Mass. (1971).

[Sim78] Simpson, M. A., and Smith, T. F., *J. Low Temp. Phys.* **32**, 57 (1978).

[Sim98] Sims, C. E., Barrera, G. D., Allan, N. L., and Mackrodt, W. C., *Phys. Rev. B* **57**, 11164 (1998).

[Sko67] Skove, M. J., and Powell, B. E., *J. Appl. Phys.* **38**, 404 (1967); Powell, B. E., and Skove, M. J., *ibid.* **38**, 404 (1967).

[Sla39a] Slater, J. C., *Introduction to Chemical Physics*, McGraw-Hill, New York (1939).

[Sla39b] Slansky, C. M., and Coulter, L. V., *J. Am. Chem. Soc.* **61**, 564 (1939).

[Sla75] Slack, G. A., and Bartram, S. F., *J. Appl. Phys.* **46**, 89 (1975).

[Sla82] Slack, G. A., and Huseby, I. C., *J. Appl. Phys.* **53**, 6817 (1982).

[Sol92] Soldatov, A. V., Aleksandrovskii, A. N., Manzhelii, V. G., Palei, V. V., and Strzhemechnyi, M. A., *Sov. J. Low Temp. Phys.* **18**, 116 (1992).

[Spa67] Sparks, P. W., and Swenson, C. A., *Phys. Rev.* **163**, 779 (1967).

[Spe96] Spetzler, H., Shen, A., Chen, G., Herrmannsdoerfer, G., Schulze, H., and Weigel, R., *Phys. Earth Plan. Inter.* **98**, 93 (1996).

[Spr97a] Springford, M., (ed.), *Electron, a Centenary Volume*, Cambridge University Press, Cambridge (1997).

[Spr97b] Springford, M., in [Spr97a], Ch. 8.

[Sta71] Stanley, H. E., *Introduction to Phase Transitions and Critical Phenomena*, Clarendon Press, Oxford (1971).

[Ste59] Stewart, J. W., *J. Phys. Chem. Solids* **12**, 122 (1959).

[Ste68] Stephenson, C. C., and Karo, A. M., *J. Chem. Phys.* **48**, 104 (1968).

[Ste73] Stephens, R. B., *Phys. Rev. B* **8**, 2896 (1973).

[Ste81] Steinemann, S. G., and Fisher, E. S., in *Treatise on Materials Science and Technology*, Vol. 21, ed. F. Y. Fredin, Academic Press, New York (1981) p. 223.

[Ste83] Stewart, G. R., *Rev. Sci. Instrum.* **54**, 1 (1983).

[Ste89] Stewart, A. M., and Collocott, S. J., *J. Phys.: Condens. Matter* **1**, 677 (1989).

[Sto64] Stout, J. W., and Hadley, W. B., *J. Chem. Phys.* **40**, 55 (1964).

[Sto68] Stout, J. W., in *Experimental Thermodynamics*, ed. J. P. McCullough and D. W. Scott, Vol. 1, Butterworths, London (1968) Ch. 6.

[Stu71] Stull, D. R., and Prophet, H., (eds.), *JANAF Thermochemical Tables*, NSRDS–NBS 37, U. S. Govt. Printing Office, Washington (1971); see also JANAF Supplements, ed. M.W. Chase *et al.*, in *J. Phys. Chem. Ref. Data* **3**, 311 (1974) and **4**, 1 (1975).

[Sul68] Sullivan, P. F., and Seidel, G., *Phys. Rev.* **173**, 679 (1968).

[Sum77] Sumino, Y., Nishizawa, O., Goto, T., Ohno, I., and Ozima, M., *J. Phys. Earth* **25**, 377 (1977).

[Sum84] Sumino, Y., and Anderson, O. L., in *Handbook of Physical Properties of Rocks*, Vol. III, CRC Press, Boca Raton, Florida (1984) p. 39.

[Sun78] Sundstrom, L. J., in *Handbook on the Physics and Chemistry of Rare Earths*, Vol. 1, ed. K. A. Gschneidner and L. Eyring, North-Holland, Amsterdam (1978) p. 379.

[Sut93] Sutton, A. P., *Electronic Structure of Materials*, Clarendon Press, Oxford (1993).

[Swe68] Swenson, C. A., *J. Phys. Chem. Solids* **29**, 1337 (1968).

[Swe77] Swenson, C. A., in [Kle77], Ch. 13.

[Swe83] Swenson, C. A., *J. Phys. Chem. Ref. Data* **12**, 179 (1983).

[Swe89] Swenson, C. A., McCallum, R. W., and No, K., *Phys. Rev. B* **40**, 8861 (1989).

[Swe96] Swenson, C. A., *Phys. Rev. B* **53**, 3669 (1996).

[Swe97] Swenson, C. A., *Rev. Sci. Instrum.* **68**, 1312 (1997).

[Swe98] Swenson, C. A., in [Ho98], Ch. 8.

[Tab91] Tabor, D., *Gases, Liquids and Solids, and Other States of Matter*, 3rd edition, Cambridge University Press, Cambridge (1991).

[Tas90] Tashiro, K., Ozawa, N., Sugihara, K., and Tsuzuku, T., *J. Phys. Soc. Jpn.* **59**, 4022 (1990).

[Tay75] Taylor, D. W., in [Hor75], Ch. 5.

[Tay97a] Taylor, M. B., Barrera, G. D., Allan, N. L., and Barron, T. H. K., *Phys. Rev. B* **56**, 14380 (1997).

[Tay97b] Taylor, M. B., Barrera, G. D., Allan, N. L., Barron, T. H. K., and Mackrodt, W. C., *Faraday Discuss.* **106**, 377 (1997).

[Tay98a] Taylor, R. E., in [Ho98], Ch. 2.

[Tay98b] Taylor, M. B., Barrera, G. D., Allan, N. L., Barron, T. H. K., and Mackrodt, W. C., *Comput. Phys. Commun.* **109**, 135 (1998).

[Thi13] Thirring, H., *Phys. Z.* **14**, 867 (1913); **15**, 127, 180 (1914).

[Tho58] Thompson, A. M., *IRE Trans. Instrum.* **I-7**, 245 (1958).

[Thu64] Thurston, R. N., and Brugger, K., *Phys. Rev.* **133**, A1604 (1964).

[Thu65a] Thurston, R. N., *J. Acoust. Soc. Am.* **37**, 348 (1965).

[Thu65b] Thurston, R. N., *Proc. IEEE* **53**, 1320 (1965).

[Tim70] Timmesfeld, K. H., and Elliott, R. J., *Phys. Status Solidi* **42**, 859 (1970).

[Tod51] Todd, S. S., *J. Am. Chem. Soc.* **73**, 3277 (1951).

[Tol38] Tolman, R. C., *The Principles of Statistical Mechanics*, Clarendon Press, Oxford (1938).

[Tol75] Tolkachev, A. M., Aleksandrovskii, A. N., and Kuchnev, V. I., *Cryogenics* **15**, 547 (1975).

[Tou70a] Touloukian, Y. S., and Ho, C. Y., (eds.), *Thermophysical Properties of Matter*, Volume 4 — Specific Heat — Metallic Elements and Alloys, Plenum Press, New York (1970).

[Tou70b] Touloukian, Y. S., and Ho, C. Y., (eds.), *Thermophysical Properties of Matter*, Volume 5 — Specific Heat — Nonmetallic Solids, Plenum Press, New York, (1970).

[Tou70c] Touloukian, Y. S., and Ho, C. Y., (eds.), *Thermophysical Properties of Matter*, Volume 6 — Specific Heat — Nonmetallic Liquids and Gases, Plenum Press, New York, (1970).

[Tou75] Touloukian, Y. S., and Ho, C. Y., (eds.), *Thermophysical Properties of Matter*, Volume 12 — Thermal Expansion — Metallic Elements and Alloys, Plenum Press, New York (1975).

[Tou77] Touloukian, Y. S., and Ho, C. Y., (eds.), *Thermophysical Properties of Matter*, Volume 13 — Thermal Expansion — Nonmetallic Solids, Plenum Press, New York (1977).

[Van63] Van der Hoeven, B. J. C., and Keesom, P. H., *Phys. Rev.* **130**, 1318 (1963).

[Van74] Van Degrift, C. T., *Rev. Sci. Instrum.* **45**, 1171 (1974).

[Van86] Van Sciver, S. W., *Helium Cryogenics*, Plenum Press, New York (1986).

[Var96] Varandani, D., Bandyopadhyay, A. K., Yadav, V. S., Gmelin, E., and Narlikar, A. V., *Meas. Sci. Technol.* **7**, 511 (1996).

[Vea63] Veal, B. W., and Rayne, J. A., *Phys. Rev.* **130**, 2156 (1963).

[Vea64] Veal, B. W., and Rayne, J. A., *Phys. Rev.* **135**, A442 (1964).

[Ven75] Venkataraman, G., Feldkamp, L. A., and Sahni, V. C., *Dynamics of Perfect Crystals*, M. I. T. Press, Cambridge (Mass) (1975).

[Vil80] Villar, R., Hortal, M., and Vieira, S., *Rev. Sci. Instrum.* **51**, 27 (1980).

[Vin87a] Vinet, P., Smith, J. R., Ferrante, J., and Rose, J. H., *Phys. Rev. B* **35**, 1945 (1987).

[Vin87b] Vinet, P., Ferrante, J., Rose, J. H., and Smith, J. R., *J. Geophys. Res.* **92**, 9319 (1987).

[Voi889] Voigt, W., *Ann. Phys.* **38**, 573 (1889).

[Vol90] Vollhardt, D., and Wölfle, P., *The Superfluid Phases of Helium 3*, Taylor and Francis, London (1990).

[Wal70] Wallace, D. C., *Solid State Phys.* **25**, 301 (1970).

[Wal72] Wallace, D. C., *Thermodynamics of Crystals*, Wiley, New York (1972).

[Wal75] Wallace, R. F., in [Hor75], Ch. 7.

[Wal91] Wallace, D. C., *Proc. R. Soc. Lond. A* **433**, 631 (1991).

[Wal92] Wallace, D. C., *Phys. Rev. B* **46**, 5242 (1992).

[Wal93] Wall, A., Parker, S. C., and Watson, G. W., *Phys. Chem. Minerals* **20**, 69 (1993).

[Wal94] Walsh, R. P., and Reed, R. P., *Adv. Cryog. Eng.* **40**, 1145 (1994).

[Was89] Wasserman, E. F., *Physica Scripta* **T25**, 209 (1989).

[Was90] Wasserman, E. F., in *Ferromagnetic Materials*, ed. K. H. J. Buschow and E. P. Wohlfarth, Vol. 5, North-Holland, Amsterdam (1990). Ch. 3.

[Wat89] Watson, S. K., Cahill, D. G., and Pohl, R. O., *Phys. Rev. B* **40**, 6381 (1989).

[Web98] Webb, Sharon, Australian National University, private communication (1998).

[Wei75] Weidner, D. J., Swyler, K., and Carleton, H. R., *Geophys. Res. Lett.* **2**, 189 (1975).

[Wei77] Weinstein, B. A., *Solid State Comm.* **24**, 595 (1977).

[Wel98] Welche, P. R. L., Heine, V., and Dove, M. T., *Physics and Chemistry of Minerals* **26**, 63 (1998).

[Wes63] Westrum, E. F., Jr., private communication (1963); see also review [Lea68].

[Wes68] Westrum, E. F., Jr., Furukawa, G. T., and McCullough, J. P., in *Experimental Thermodynamics*, ed. J. P. McCullough and D. W. Scott, Vol. 1, Butterworths, London (1968) Ch. 5.

[Wes88] Westrum, E. F., Jr., in [Cez88], Ch. 4.

[Whi61] White, G. K., *Cryogenics* **1**, 151 (1961).

[Whi62] White, G. K., *Cryogenics* **2**, 292 (1962).

[Whi70] White, G. K., and Pawlowicz, A. T., *J. Low Temp. Phys.* **2**, 631 (1970).

[Whi72a] White, G. K., in [Gra72], p. 59.

[Whi72b] White, G. K., and Collins, J. G., *J. Low Temp. Phys* **7**, 43 (1972).

[Whi73] White, G. K., and Collins, J. G., *Proc. R. Soc. Lond. A* **333**, 237 (1973).

[Whi74] White, G. K., and Sheard, F. W., *J. Low Temp. Phys.* **14**, 445 (1974).

[Whi75] White, T. J., Davis, J. H., and Walter, H. U., *J. Appl. Phys.* **46**, 11 (1975).

[Whi76a] White, G. K., *Cryogenics* **16**, 487 (corrigenda, p. 606) (1976).

[Whi76b] White, G. K., Smith, T. F., and Birch, J. A., *J. Chem. Phys.* **65**, 554 (1976).

[Whi77] White, G. K., Birch, J. A., and Manghnani, M. H., *J. Non-Cryst. Solids* **23**, 99 (1977).

[Whi79] White, G. K., *Experimental Techniques in Low Temperature Physics*, 3rd edition, Clarendon Press, Oxford (1979, 1987).

[Whi80] White, G. K., *J. Phys. C: Solid State Phys.* **13**, 4905 (1980).

[Whi81] White, G. K., *J. Phys. C: Solid State Phys.* **14**, L297 (1981).

[Whi84a] White, G. K., and Choy, C. L., *J. Polym. Sci. Polym. Phys. Ed.* **22**, 835 (1984).

[Whi84b] White, G. K., and Collocott, S. J., *J. Phys. Chem. Ref. Data* **13**, 1251 (1984).

[Whi84c] White, G. K., Collocott, S. J., and Cook, J. S., *Phys. Rev. B* **29**, 4778 (1984).

[Whi85a] White, G. K., Roberts, R. B., and Collins, J. G., *High Temp.-High Press.* **17**, 61 (1985).

[Whi85b] White, M. A., and MacLean, M. T., *J. Phys. Chem.* **89**, 1380 (1985).

[Whi87] White, G. K., and Collocott, S. J., *J. Phys. C: Solid State Phys.* **20**, L813 (1987).

[Whi88] White, G. K., and Roberts, R. B., *Aust. J. Phys.* **41**, 791 (1988).

[Whi89a] White, G. K., *J. Phys.: Condens. Matter* **1**, 6987 (1989).

[Whi89b] White, G. K., *High Temp.-High Press.* **21**, 233 (1989).

[Whi90a] White, G. K., Collocott, S. J., and Collins, J. G., *J. Phys.: Condens. Matter* **2**, 7715 (1990).

[Whi90b] White, G. K., Collins, J. G., and Schirber, J. E., *Aust. J. Phys.* **43**, 93 (1990).

[Whi93a] White, G. K., in *Studies of High Temperature Superconductors* **9**, ed. A. V. Narlikar, Nova Science, N. Y. (1993) p. 131.

[Whi93b] White, G. K., *Thermochimica Acta* **218**, 83 (1993).

[Whi93c] White, G. K., *Aust. J. Phys.* **46**, 707 (1993).

[Whi97] White, G. K., and Minges, M. L., (eds.), *Int. J. Thermophys.* **18**, 1269 (1997) (An Update of CODATA Bulletin Nó. 59, Pergamon Press, 1985).

[Whi98] White, G. K., in [Ho98], Ch. 11.

[Wil67] Wilks, J., *Properties of Liquid and Solid Helium*, Clarendon Press, Oxford (1967).

[Wil83] Wilson, M. N., *Superconducting Magnets*, Clarendon Press, Oxford (1983).

[Wil87] Wilks, J., and Betts, D. S., *An Introduction to Liquid Helium*, 2nd edition, Clarendon Press, Oxford (1987).

[Woh75] Wohlfarth, E. P., *IEEE Trans. Magn.* **MAG-11**, 1638 (1975).

[Wol59] Wolcott, N. M., *J. Chem. Phys.* **31**, 536 (1959).

[Wri76] Wright, A. F., *Fourth Int. Conf. Phys. Non-Crystalline Solids*, ed. G. H. Frischat, Trans. Tech. Publications (1976) p. 598.

[Wri84] Wright, O. B., and Phillips, W. A., *Phil. Mag. B* **50**, 63 (1984).

[Xu91] Xu, C. H., Wang, C. Z., Chan, C. T., and Ho, H. K., *Phys. Rev. B* **43**, 5024 (1991).

[Yam97] Yamada, N., Okaji, M., Nara, K., Kato, H., Michishita, K., and Kubo, Y., *Physica C* **282-287**, 1441 (1997).

[Yat72] Yates, B., *Thermal Expansion*, Plenum Press, New York (1972).

[Yat82] Yates, B., Rogers, K. F., Kingston-Lee, D. M., and Phillips, L. N., *J. Mater. Sci.* **17**, 1880 (1982).

[Yin82] Yin, M. T., and Cohen, M. L., *Phys. Rev. B* **26** 3259 (1982).

[Zak91] Zakrzewski, M., and White, M. A., *J. Phys.: Condens. Matter* **3**, 6703 (1991).

[Zie64] Ziegler, W. T., and Mullins, J. C., *Cryogenics* **4**, 39 (1964).

[Zel71] Zeller, R. C., and Pohl, R. O., *Phys. Rev. B* **4**, 2029 (1971).

[Zhd66] Zhdanova, V. V., Lukina, V. I., and Novikova, S. I., *Phys. Status Solidi* **13**, K19 (1966).

Index